# All Prime Numbers from 1-1000000

(And the Java code used to find them)

Max Mercer

# Table of Contents

# List of Prime Numbers from 1-1000000

(1), 2, 3, 5, 7, 11, 13, 17, 19, 23, 29, 31, 37, 41, 43, 47, 53, 59, 61, 67, 71, 73, 79, 83, 89, 97, 101, 103, 107, 109, 113, 127, 131, 137, 139, 149, 151, 157, 163, 167, 173, 179, 181, 191, 193, 197, 199, 211, 223, 227, 229, 233, 239, 241, 251, 257, 263, 269, 271, 277, 281, 283, 293, 307, 311, 313, 317, 331, 337, 347, 349, 353, 359, 367, 373, 379, 383, 389, 397, 401, 409, 419, 421, 431, 433, 439, 443, 449, 457, 461, 463, 467, 479, 487, 491, 499, 503, 509, 521, 523, 541, 547, 557, 563, 569, 571, 577, 587, 593, 599, 601, 607, 613, 617, 619, 631, 641, 643, 647, 653, 659, 661, 673, 677, 683, 691, 701, 709, 719, 727, 733, 739, 743, 751, 757, 761, 769, 773, 787, 797, 809, 811, 821, 823, 827, 829, 839, 853, 857, 859, 863, 877, 881, 883, 887, 907, 911, 919, 929, 937, 941, 947, 953, 967, 971, 977, 983, 991, 997, 1009, 1013, 1019, 1021, 1031, 1033, 1039, 1049, 1051, 1061, 1063, 1069, 1087, 1091, 1093, 1097, 1103, 1109, 1117, 1123, 1129, 1151, 1153, 1163, 1171, 1181, 1187, 1193, 1201, 1213, 1217, 1223, 1229, 1231, 1237, 1249, 1259, 1277, 1279, 1283, 1289, 1291, 1297, 1301, 1303, 1307, 1319, 1321, 1327, 1361, 1367, 1373, 1381, 1399, 1409, 1423, 1427, 1429, 1433, 1439, 1447, 1451, 1453, 1459, 1471, 1481, 1483, 1487, 1489, 1493, 1499, 1511, 1523, 1531, 1543, 1549, 1553, 1559, 1567, 1571, 1579, 1583, 1597, 1601, 1607, 1609, 1613, 1619, 1621, 1627, 1637, 1657, 1663, 1667, 1669, 1693, 1697, 1699, 1709, 1721, 1723, 1733, 1741, 1747, 1753, 1759, 1777, 1783, 1787, 1789, 1801, 1811, 1823, 1831, 1847, 1861, 1867, 1871, 1873, 1877, 1879, 1889, 1901, 1907, 1913, 1931, 1933, 1949, 1951, 1973, 1979, 1987, 1993, 1997, 1999, 2003, 2011, 2017, 2027, 2029, 2039, 2053, 2063, 2069, 2081, 2083, 2087, 2089, 2099, 2111, 2113, 2129, 2131, 2137, 2141, 2143, 2153, 2161, 2179, 2203, 2207, 2213, 2221, 2237, 2239, 2243, 2251, 2267, 2269, 2273, 2281, 2287, 2293, 2297, 2309, 2311, 2333, 2339, 2341, 2347, 2351, 2357, 2371, 2377, 2381, 2383, 2389, 2393, 2399, 2411, 2417, 2423, 2437, 2441, 2447, 2459, 2467, 2473, 2477, 2503, 2521, 2531, 2539, 2543, 2549, 2551, 2557, 2579, 2591, 2593, 2609, 2617, 2621, 2633, 2647, 2657, 2659, 2663, 2671, 2677, 2683, 2687, 2689, 2693, 2699, 2707, 2711, 2713, 2719, 2729, 2731, 2741, 2749, 2753, 2767, 2777, 2789, 2791, 2797, 2801, 2803, 2819, 2833, 2837, 2843, 2851, 2857, 2861, 2879, 2887, 2897, 2903, 2909, 2917, 2927, 2939, 2953, 2957, 2963, 2969, 2971, 2999, 3001, 3011, 3019, 3023, 3037, 3041, 3049, 3061, 3067, 3079, 3083, 3089, 3109, 3119, 3121, 3137, 3163, 3167, 3169, 3181, 3187, 3191, 3203, 3209, 3217, 3221, 3229, 3251, 3253, 3257, 3259, 3271, 3299, 3301, 3307, 3313, 3319, 3323, 3329, 3331, 3343, 3347, 3359, 3361, 3371, 3373, 3389, 3391, 3407, 3413, 3433, 3449, 3457, 3461, 3463, 3467, 3469, 3491, 3499, 3511, 3517, 3527, 3529, 3533, 3539, 3541, 3547, 3557, 3559, 3571, 3581, 3583, 3593, 3607, 3613, 3617, 3623, 3631, 3637, 3643, 3659, 3671, 3673, 3677, 3691, 3697, 3701, 3709, 3719, 3727, 3733, 3739, 3761, 3767, 3769, 3779, 3793, 3797, 3803, 3821, 3823, 3833, 3847, 3851, 3853, 3863, 3877, 3881, 3889, 3907, 3911, 3917, 3919, 3923, 3929, 3931, 3943, 3947, 3967, 3989, 4001, 4003, 4007, 4013, 4019, 4021, 4027, 4049, 4051, 4057, 4073, 4079, 4091, 4093, 4099, 4111, 4127, 4129, 4133, 4139, 4153, 4157, 4159, 4177, 4201, 4211, 4217, 4219, 4229, 4231, 4241, 4243, 4253, 4259, 4261, 4271, 4273, 4283, 4289, 4297, 4327, 4337, 4339, 4349, 4357, 4363, 4373, 4391, 4397, 4409, 4421, 4423, 4441, 4447, 4451, 4457, 4463, 4481, 4483, 4493, 4507, 4513, 4517, 4519, 4523, 4547, 4549, 4561, 4567, 4583, 4591, 4597, 4603, 4621, 4637, 4639, 4643, 4649, 4651, 4657, 4663, 4673, 4679, 4691, 4703, 4721, 4723, 4729, 4733, 4751, 4759, 4783, 4787, 4789, 4793, 4799, 4801, 4813, 4817, 4831, 4861, 4871, 4877, 4889, 4903, 4909, 4919, 4931, 4933, 4937, 4943, 4951, 4957, 4967, 4969, 4973, 4987, 4993, 4999, 5003, 5009, 5011, 5021, 5023, 5039, 5051, 5059, 5077, 5081, 5087, 5099, 5101, 5107, 5113, 5119, 5147, 5153, 5167, 5171, 5179, 5189, 5197, 5209, 5227, 5231, 5233, 5237, 5261, 5273, 5279, 5281, 5297, 5303, 5309, 5323, 5333, 5347, 5351, 5381, 5387, 5393, 5399, 5407, 5413, 5417, 5419, 5431, 5437, 5441, 5443, 5449, 5471, 5477, 5479, 5483, 5501, 5503, 5507, 5519, 5521, 5527, 5531, 5557, 5563, 5569, 5573, 5581, 5591, 5623, 5639, 5641, 5647, 5651, 5653, 5657, 5659, 5669, 5683, 5689, 5693, 5701, 5711, 5717, 5737, 5741, 5743, 5749, 5779, 5783, 5791, 5801, 5807, 5813, 5821, 5827, 5839, 5843, 5849, 5851, 5857, 5861, 5867, 5869, 5879, 5881, 5897, 5903, 5923, 5927, 5939, 5953, 5981, 5987, 6007, 6011, 6029, 6037, 6043, 6047, 6053, 6067, 6073, 6079, 6089, 6091, 6101, 6113, 6121, 6131, 6133, 6143, 6151, 6163, 6173, 6197, 6199, 6203, 6211, 6217, 6221, 6229, 6247, 6257, 6263, 6269, 6271, 6277, 6287, 6299, 6301, 6311, 6317, 6323, 6329, 6337, 6343, 6353, 6359, 6361, 6367, 6373, 6379, 6389, 6397, 6421, 6427, 6449, 6451, 6469, 6473, 6481, 6491, 6521, 6529, 6547, 6551, 6553, 6563, 6569, 6571, 6577, 6581, 6599, 6607, 6619, 6637, 6653, 6659, 6661, 6673, 6679, 6689, 6691, 6701, 6703, 6709, 6719, 6733, 6737, 6761, 6763, 6779, 6781, 6791, 6793, 6803, 6823, 6827, 6829, 6833, 6841, 6857, 6863, 6869, 6871, 6883, 6899, 6907, 6911, 6917,

6947, 6949, 6959, 6961, 6967, 6971, 6977, 6983, 6991, 6997, 7001, 7013, 7019, 7027, 7039, 7043, 7057, 7069, 7079,
7103, 7109, 7121, 7127, 7129, 7151, 7159, 7177, 7187, 7193, 7207, 7211, 7213, 7219, 7229, 7237, 7243, 7247, 7253,
7283, 7297, 7307, 7309, 7321, 7331, 7333, 7349, 7351, 7369, 7393, 7411, 7417, 7433, 7451, 7457, 7459, 7477, 7481,
7487, 7489, 7499, 7507, 7517, 7523, 7529, 7537, 7541, 7547, 7549, 7559, 7561, 7573, 7577, 7583, 7589, 7591, 7603,
7607, 7621, 7639, 7643, 7649, 7669, 7673, 7681, 7687, 7691, 7699, 7703, 7717, 7723, 7727, 7741, 7753, 7757, 7759,
7789, 7793, 7817, 7823, 7829, 7841, 7853, 7867, 7873, 7877, 7879, 7883, 7901, 7907, 7919, 7927, 7933, 7937, 7949,
7951, 7963, 7993, 8009, 8011, 8017, 8039, 8053, 8059, 8069, 8081, 8087, 8089, 8093, 8101, 8111, 8117, 8123, 8147,
8161, 8167, 8171, 8179, 8191, 8209, 8219, 8221, 8231, 8233, 8237, 8243, 8263, 8269, 8273, 8287, 8291, 8293, 8297,
8311, 8317, 8329, 8353, 8363, 8369, 8377, 8387, 8389, 8419, 8423, 8429, 8431, 8443, 8447, 8461, 8467, 8501, 8513,
8521, 8527, 8537, 8539, 8543, 8563, 8573, 8581, 8597, 8599, 8609, 8623, 8627, 8629, 8641, 8647, 8663, 8669, 8677,
8681, 8689, 8693, 8699, 8707, 8713, 8719, 8731, 8737, 8741, 8747, 8753, 8761, 8779, 8783, 8803, 8807, 8819, 8821,
8831, 8837, 8839, 8849, 8861, 8863, 8867, 8887, 8893, 8923, 8929, 8933, 8941, 8951, 8963, 8969, 8971, 8999, 9001,
9007, 9011, 9013, 9029, 9041, 9043, 9049, 9059, 9067, 9091, 9103, 9109, 9127, 9133, 9137, 9151, 9157, 9161, 9173,
9181, 9187, 9199, 9203, 9209, 9221, 9227, 9239, 9241, 9257, 9277, 9281, 9283, 9293, 9311, 9319, 9323, 9337, 9341,
9343, 9349, 9371, 9377, 9391, 9397, 9403, 9413, 9419, 9421, 9431, 9433, 9437, 9439, 9461, 9463, 9467, 9473, 9479,
9491, 9497, 9511, 9521, 9533, 9539, 9547, 9551, 9587, 9601, 9613, 9619, 9623, 9629, 9631, 9643, 9649, 9661, 9677,
9679, 9689, 9697, 9719, 9721, 9733, 9739, 9743, 9749, 9767, 9769, 9781, 9787, 9791, 9803, 9811, 9817, 9829, 9833,
9839, 9851, 9857, 9859, 9871, 9883, 9887, 9901, 9907, 9923, 9929, 9931, 9941, 9949, 9967, 9973, 10007, 10009,
10037, 10039, 10061, 10067, 10069, 10079, 10091, 10093, 10099, 10103, 10111, 10133, 10139, 10141, 10151,
10159, 10163, 10169, 10177, 10181, 10193, 10211, 10223, 10243, 10247, 10253, 10259, 10267, 10271, 10273,
10289, 10301, 10303, 10313, 10321, 10331, 10333, 10337, 10343, 10357, 10369, 10391, 10399, 10427, 10429,
10433, 10453, 10457, 10459, 10463, 10477, 10487, 10499, 10501, 10513, 10529, 10531, 10559, 10567, 10589,
10597, 10601, 10607, 10613, 10627, 10631, 10639, 10651, 10657, 10663, 10667, 10687, 10691, 10709, 10711,
10723, 10729, 10733, 10739, 10753, 10771, 10781, 10789, 10799, 10831, 10837, 10847, 10853, 10859, 10861,
10867, 10883, 10889, 10891, 10903, 10909, 10937, 10939, 10949, 10957, 10973, 10979, 10987, 10993, 11003,
11027, 11047, 11057, 11059, 11069, 11071, 11083, 11087, 11093, 11113, 11117, 11119, 11131, 11149, 11159,
11161, 11171, 11173, 11177, 11197, 11213, 11239, 11243, 11251, 11257, 11261, 11273, 11279, 11287, 11299,
11311, 11317, 11321, 11329, 11351, 11353, 11369, 11383, 11393, 11399, 11411, 11423, 11437, 11443, 11447,
11467, 11471, 11483, 11489, 11491, 11497, 11503, 11519, 11527, 11549, 11551, 11579, 11587, 11593, 11597,
11617, 11621, 11633, 11657, 11677, 11681, 11689, 11699, 11701, 11717, 11719, 11731, 11743, 11777, 11779,
11783, 11789, 11801, 11807, 11813, 11821, 11827, 11831, 11833, 11839, 11863, 11867, 11887, 11897, 11903,
11909, 11923, 11927, 11933, 11939, 11941, 11953, 11959, 11969, 11971, 11981, 11987, 12007, 12011, 12037,
12041, 12043, 12049, 12071, 12073, 12097, 12101, 12107, 12109, 12113, 12119, 12143, 12149, 12157, 12161,
12163, 12197, 12203, 12211, 12227, 12239, 12241, 12251, 12253, 12263, 12269, 12277, 12281, 12289, 12301,
12323, 12329, 12343, 12347, 12373, 12377, 12379, 12391, 12401, 12409, 12413, 12421, 12433, 12437, 12451,
12457, 12473, 12479, 12487, 12491, 12497, 12503, 12511, 12517, 12527, 12539, 12541, 12547, 12553, 12569,
12577, 12583, 12589, 12601, 12611, 12613, 12619, 12637, 12641, 12647, 12653, 12659, 12671, 12689, 12697,
12703, 12713, 12721, 12739, 12743, 12757, 12763, 12781, 12791, 12799, 12809, 12821, 12823, 12829, 12841,
12853, 12889, 12893, 12899, 12907, 12911, 12917, 12919, 12923, 12941, 12953, 12959, 12967, 12973, 12979,
12983, 13001, 13003, 13007, 13009, 13033, 13037, 13043, 13049, 13063, 13093, 13099, 13103, 13109, 13121,
13127, 13147, 13151, 13159, 13163, 13171, 13177, 13183, 13187, 13217, 13219, 13229, 13241, 13249, 13259,
13267, 13291, 13297, 13309, 13313, 13327, 13331, 13337, 13339, 13367, 13381, 13397, 13399, 13411, 13417,
13421, 13441, 13451, 13457, 13463, 13469, 13477, 13487, 13499, 13513, 13523, 13537, 13553, 13567, 13577,
13591, 13597, 13613, 13619, 13627, 13633, 13649, 13669, 13679, 13681, 13687, 13691, 13693, 13697, 13709,
13711, 13721, 13723, 13729, 13751, 13757, 13759, 13763, 13781, 13789, 13799, 13807, 13829, 13831, 13841,
13859, 13873, 13877, 13879, 13883, 13901, 13903, 13907, 13913, 13921, 13931, 13933, 13963, 13967, 13997,
13999, 14009, 14011, 14029, 14033, 14051, 14057, 14071, 14081, 14083, 14087, 14107, 14143, 14149, 14153,
14159, 14173, 14177, 14197, 14207, 14221, 14243, 14249, 14251, 14281, 14293, 14303, 14321, 14323, 14327,
14341, 14347, 14369, 14387, 14389, 14401, 14407, 14411, 14419, 14423, 14431, 14437, 14447, 14449, 14461,

14479, 14489, 14503, 14519, 14533, 14537, 14543, 14549, 14551, 14557, 14561, 14563, 14591, 14593, 14621,
14627, 14629, 14633, 14639, 14653, 14657, 14669, 14683, 14699, 14713, 14717, 14723, 14731, 14737, 14741,
14747, 14753, 14759, 14767, 14771, 14779, 14783, 14797, 14813, 14821, 14827, 14831, 14843, 14851, 14867,
14869, 14879, 14887, 14891, 14897, 14923, 14929, 14939, 14947, 14951, 14957, 14969, 14983, 15013, 15017,
15031, 15053, 15061, 15073, 15077, 15083, 15091, 15101, 15107, 15121, 15131, 15137, 15139, 15149, 15161,
15173, 15187, 15193, 15199, 15217, 15227, 15233, 15241, 15259, 15263, 15269, 15271, 15277, 15287, 15289,
15299, 15307, 15313, 15319, 15329, 15331, 15349, 15359, 15361, 15373, 15377, 15383, 15391, 15401, 15413,
15427, 15439, 15443, 15451, 15461, 15467, 15473, 15493, 15497, 15511, 15527, 15541, 15551, 15559, 15569,
15581, 15583, 15601, 15607, 15619, 15629, 15641, 15643, 15647, 15649, 15661, 15667, 15671, 15679, 15683,
15727, 15731, 15733, 15737, 15739, 15749, 15761, 15767, 15773, 15787, 15791, 15797, 15803, 15809, 15817,
15823, 15859, 15877, 15881, 15887, 15889, 15901, 15907, 15913, 15919, 15923, 15937, 15959, 15971, 15973,
15991, 16001, 16007, 16033, 16057, 16061, 16063, 16067, 16069, 16073, 16087, 16091, 16097, 16103, 16111,
16127, 16139, 16141, 16183, 16187, 16189, 16193, 16217, 16223, 16229, 16231, 16249, 16253, 16267, 16273,
16301, 16319, 16333, 16339, 16349, 16361, 16363, 16369, 16381, 16411, 16417, 16421, 16427, 16433, 16447,
16451, 16453, 16477, 16481, 16487, 16493, 16519, 16529, 16547, 16553, 16561, 16567, 16573, 16603, 16607,
16619, 16631, 16633, 16649, 16651, 16657, 16661, 16673, 16691, 16693, 16699, 16703, 16729, 16741, 16747,
16759, 16763, 16787, 16811, 16823, 16829, 16831, 16843, 16871, 16879, 16883, 16889, 16901, 16903, 16921,
16927, 16931, 16937, 16943, 16963, 16979, 16981, 16987, 16993, 17011, 17021, 17027, 17029, 17033, 17041,
17047, 17053, 17077, 17093, 17099, 17107, 17117, 17123, 17137, 17159, 17167, 17183, 17189, 17191, 17203,
17207, 17209, 17231, 17239, 17257, 17291, 17293, 17299, 17317, 17321, 17327, 17333, 17341, 17351, 17359,
17377, 17383, 17387, 17389, 17393, 17401, 17417, 17419, 17431, 17443, 17449, 17467, 17471, 17477, 17483,
17489, 17491, 17497, 17509, 17519, 17539, 17551, 17569, 17573, 17579, 17581, 17597, 17599, 17609, 17623,
17627, 17657, 17659, 17669, 17681, 17683, 17707, 17713, 17729, 17737, 17747, 17749, 17761, 17783, 17789,
17791, 17807, 17827, 17837, 17839, 17851, 17863, 17881, 17891, 17903, 17909, 17911, 17921, 17923, 17929,
17939, 17957, 17959, 17971, 17977, 17981, 17987, 17989, 18013, 18041, 18043, 18047, 18049, 18059, 18061,
18077, 18089, 18097, 18119, 18121, 18127, 18131, 18133, 18143, 18149, 18169, 18181, 18191, 18199, 18211,
18217, 18223, 18229, 18233, 18251, 18253, 18257, 18269, 18287, 18289, 18301, 18307, 18311, 18313, 18329,
18341, 18353, 18367, 18371, 18379, 18397, 18401, 18413, 18427, 18433, 18439, 18443, 18451, 18457, 18461,
18481, 18493, 18503, 18517, 18521, 18523, 18539, 18541, 18553, 18583, 18587, 18593, 18617, 18637, 18661,
18671, 18679, 18691, 18701, 18713, 18719, 18731, 18743, 18749, 18757, 18773, 18787, 18793, 18797, 18803,
18839, 18859, 18869, 18899, 18911, 18913, 18917, 18919, 18947, 18959, 18973, 18979, 19001, 19009, 19013,
19031, 19037, 19051, 19069, 19073, 19079, 19081, 19087, 19121, 19139, 19141, 19157, 19163, 19181, 19183,
19207, 19211, 19213, 19219, 19231, 19237, 19249, 19259, 19267, 19273, 19289, 19301, 19309, 19319, 19333,
19373, 19379, 19381, 19387, 19391, 19403, 19417, 19421, 19423, 19427, 19429, 19433, 19441, 19447, 19457,
19463, 19469, 19471, 19477, 19483, 19489, 19501, 19507, 19531, 19541, 19543, 19553, 19559, 19571, 19577,
19583, 19597, 19603, 19609, 19661, 19681, 19687, 19697, 19699, 19709, 19717, 19727, 19739, 19751, 19753,
19759, 19763, 19777, 19793, 19801, 19813, 19819, 19841, 19843, 19853, 19861, 19867, 19889, 19891, 19913,
19919, 19927, 19937, 19949, 19961, 19963, 19973, 19979, 19991, 19993, 19997, 20011, 20021, 20023, 20029,
20047, 20051, 20063, 20071, 20089, 20101, 20107, 20113, 20117, 20123, 20129, 20143, 20147, 20149, 20161,
20173, 20177, 20183, 20201, 20219, 20231, 20233, 20249, 20261, 20269, 20287, 20297, 20323, 20327, 20333,
20341, 20347, 20353, 20357, 20359, 20369, 20389, 20393, 20399, 20407, 20411, 20431, 20441, 20443, 20477,
20479, 20483, 20507, 20509, 20521, 20533, 20543, 20549, 20551, 20563, 20593, 20599, 20611, 20627, 20639,
20641, 20663, 20681, 20693, 20707, 20717, 20719, 20731, 20743, 20747, 20749, 20753, 20759, 20771, 20773,
20789, 20807, 20809, 20849, 20857, 20873, 20879, 20887, 20897, 20899, 20903, 20921, 20929, 20939, 20947,
20959, 20963, 20981, 20983, 21001, 21011, 21013, 21017, 21019, 21023, 21031, 21059, 21061, 21067, 21089,
21101, 21107, 21121, 21139, 21143, 21149, 21157, 21163, 21169, 21179, 21187, 21191, 21193, 21211, 21221,
21227, 21247, 21269, 21277, 21283, 21313, 21317, 21319, 21323, 21341, 21347, 21377, 21379, 21383, 21391,
21397, 21401, 21407, 21419, 21433, 21467, 21481, 21487, 21491, 21493, 21499, 21503, 21517, 21521, 21523,
21529, 21557, 21559, 21563, 21569, 21577, 21587, 21589, 21599, 21601, 21611, 21613, 21617, 21647, 21649,

21661, 21673, 21683, 21701, 21713, 21727, 21737, 21739, 21751, 21757, 21767, 21773, 21787, 21799, 21803, 21817, 21821, 21839, 21841, 21851, 21859, 21863, 21871, 21881, 21893, 21911, 21929, 21937, 21943, 21961, 21977, 21991, 21997, 22003, 22013, 22027, 22031, 22037, 22039, 22051, 22063, 22067, 22073, 22079, 22091, 22093, 22109, 22111, 22123, 22129, 22133, 22147, 22153, 22157, 22159, 22171, 22189, 22193, 22229, 22247, 22259, 22271, 22273, 22277, 22279, 22283, 22291, 22303, 22307, 22343, 22349, 22367, 22369, 22381, 22391, 22397, 22409, 22433, 22441, 22447, 22453, 22469, 22481, 22483, 22501, 22511, 22531, 22541, 22543, 22549, 22567, 22571, 22573, 22613, 22619, 22621, 22637, 22639, 22643, 22651, 22669, 22679, 22691, 22697, 22699, 22709, 22717, 22721, 22727, 22739, 22741, 22751, 22769, 22777, 22783, 22787, 22807, 22811, 22817, 22853, 22859, 22861, 22871, 22877, 22901, 22907, 22921, 22937, 22943, 22961, 22963, 22973, 22993, 23003, 23011, 23017, 23021, 23027, 23029, 23039, 23041, 23053, 23057, 23059, 23063, 23071, 23081, 23087, 23099, 23117, 23131, 23143, 23159, 23167, 23173, 23189, 23197, 23201, 23203, 23209, 23227, 23251, 23269, 23279, 23291, 23293, 23297, 23311, 23321, 23327, 23333, 23339, 23357, 23369, 23371, 23399, 23417, 23431, 23447, 23459, 23473, 23497, 23509, 23531, 23537, 23539, 23549, 23557, 23561, 23563, 23567, 23581, 23593, 23599, 23603, 23609, 23623, 23627, 23629, 23633, 23663, 23669, 23671, 23677, 23687, 23689, 23719, 23741, 23743, 23747, 23753, 23761, 23767, 23773, 23789, 23801, 23813, 23819, 23827, 23831, 23833, 23857, 23869, 23873, 23879, 23887, 23893, 23899, 23909, 23911, 23917, 23929, 23957, 23971, 23977, 23981, 23993, 24001, 24007, 24019, 24023, 24029, 24043, 24049, 24061, 24071, 24077, 24083, 24091, 24097, 24103, 24107, 24109, 24113, 24121, 24133, 24137, 24151, 24169, 24179, 24181, 24197, 24203, 24223, 24229, 24239, 24247, 24251, 24281, 24317, 24329, 24337, 24359, 24371, 24373, 24379, 24391, 24407, 24413, 24419, 24421, 24439, 24443, 24469, 24473, 24481, 24499, 24509, 24517, 24527, 24533, 24547, 24551, 24571, 24593, 24611, 24623, 24631, 24659, 24671, 24677, 24683, 24691, 24697, 24709, 24733, 24749, 24763, 24767, 24781, 24793, 24799, 24809, 24821, 24841, 24847, 24851, 24859, 24877, 24889, 24907, 24917, 24919, 24923, 24943, 24953, 24967, 24971, 24977, 24979, 24989, 25013, 25031, 25033, 25037, 25057, 25073, 25087, 25097, 25111, 25117, 25121, 25127, 25147, 25153, 25163, 25169, 25171, 25183, 25189, 25219, 25229, 25237, 25243, 25247, 25253, 25261, 25301, 25303, 25307, 25309, 25321, 25339, 25343, 25349, 25357, 25367, 25373, 25391, 25409, 25411, 25423, 25439, 25447, 25453, 25457, 25463, 25469, 25471, 25523, 25537, 25541, 25561, 25577, 25579, 25583, 25589, 25601, 25603, 25609, 25621, 25633, 25639, 25643, 25657, 25667, 25673, 25679, 25693, 25703, 25717, 25733, 25741, 25747, 25759, 25763, 25771, 25793, 25799, 25801, 25819, 25841, 25847, 25849, 25867, 25873, 25889, 25903, 25913, 25919, 25931, 25933, 25939, 25943, 25951, 25969, 25981, 25997, 25999, 26003, 26017, 26021, 26029, 26041, 26053, 26083, 26099, 26107, 26111, 26113, 26119, 26141, 26153, 26161, 26171, 26177, 26183, 26189, 26203, 26209, 26227, 26237, 26249, 26251, 26261, 26263, 26267, 26293, 26297, 26309, 26317, 26321, 26339, 26347, 26357, 26371, 26387, 26393, 26399, 26407, 26417, 26423, 26431, 26437, 26449, 26459, 26479, 26489, 26497, 26501, 26513, 26539, 26557, 26561, 26573, 26591, 26597, 26627, 26633, 26641, 26647, 26669, 26681, 26683, 26687, 26693, 26699, 26701, 26711, 26713, 26717, 26723, 26729, 26731, 26737, 26759, 26777, 26783, 26801, 26813, 26821, 26833, 26839, 26849, 26861, 26863, 26879, 26881, 26891, 26893, 26903, 26921, 26927, 26947, 26951, 26953, 26959, 26981, 26987, 26993, 27011, 27017, 27031, 27043, 27059, 27061, 27067, 27073, 27077, 27091, 27103, 27107, 27109, 27127, 27143, 27179, 27191, 27197, 27211, 27239, 27241, 27253, 27259, 27271, 27277, 27281, 27283, 27299, 27329, 27337, 27361, 27367, 27397, 27407, 27409, 27427, 27431, 27437, 27449, 27457, 27479, 27481, 27487, 27509, 27527, 27529, 27539, 27541, 27551, 27581, 27583, 27611, 27617, 27631, 27647, 27653, 27673, 27689, 27691, 27697, 27701, 27733, 27737, 27739, 27743, 27749, 27751, 27763, 27767, 27773, 27779, 27791, 27793, 27799, 27803, 27809, 27817, 27823, 27827, 27847, 27851, 27883, 27893, 27901, 27917, 27919, 27941, 27943, 27947, 27953, 27961, 27967, 27983, 27997, 28001, 28019, 28027, 28031, 28051, 28057, 28069, 28081, 28087, 28097, 28099, 28109, 28111, 28123, 28151, 28163, 28181, 28183, 28201, 28211, 28219, 28229, 28277, 28279, 28283, 28289, 28297, 28307, 28309, 28319, 28349, 28351, 28387, 28393, 28403, 28409, 28411, 28429, 28433, 28439, 28447, 28463, 28477, 28493, 28499, 28513, 28517, 28537, 28541, 28547, 28549, 28559, 28571, 28573, 28579, 28591, 28597, 28603, 28607, 28619, 28621, 28627, 28631, 28643, 28649, 28657, 28661, 28663, 28669, 28687, 28697, 28703, 28711, 28723, 28729, 28751, 28753, 28759, 28771, 28789, 28793, 28807, 28813, 28817, 28837, 28843, 28859, 28867, 28871, 28879, 28901, 28909, 28921, 28927, 28933, 28949, 28961, 28979, 29009, 29017, 29021, 29023, 29027, 29033, 29059, 29063, 29077, 29101, 29123, 29129, 29131,

29137, 29147, 29153, 29167, 29173, 29179, 29191, 29201, 29207, 29209, 29221, 29231, 29243, 29251, 29269,
29287, 29297, 29303, 29311, 29327, 29333, 29339, 29347, 29363, 29383, 29387, 29389, 29399, 29401, 29411,
29423, 29429, 29437, 29443, 29453, 29473, 29483, 29501, 29527, 29531, 29537, 29567, 29569, 29573, 29581,
29587, 29599, 29611, 29629, 29633, 29641, 29663, 29669, 29671, 29683, 29717, 29723, 29741, 29753, 29759,
29761, 29789, 29803, 29819, 29833, 29837, 29851, 29863, 29867, 29873, 29879, 29881, 29917, 29921, 29927,
29947, 29959, 29983, 29989, 30011, 30013, 30029, 30047, 30059, 30071, 30089, 30091, 30097, 30103, 30109,
30113, 30119, 30133, 30137, 30139, 30161, 30169, 30181, 30187, 30197, 30203, 30211, 30223, 30241, 30253,
30259, 30269, 30271, 30293, 30307, 30313, 30319, 30323, 30341, 30347, 30367, 30389, 30391, 30403, 30427,
30431, 30449, 30467, 30469, 30491, 30493, 30497, 30509, 30517, 30529, 30539, 30553, 30557, 30559, 30577,
30593, 30631, 30637, 30643, 30649, 30661, 30671, 30677, 30689, 30697, 30703, 30707, 30713, 30727, 30757,
30763, 30773, 30781, 30803, 30809, 30817, 30829, 30839, 30841, 30851, 30853, 30859, 30869, 30871, 30881,
30893, 30911, 30931, 30937, 30941, 30949, 30971, 30977, 30983, 31013, 31019, 31033, 31039, 31051, 31063,
31069, 31079, 31081, 31091, 31121, 31123, 31139, 31147, 31151, 31153, 31159, 31177, 31181, 31183, 31189,
31193, 31219, 31223, 31231, 31237, 31247, 31249, 31253, 31259, 31267, 31271, 31277, 31307, 31319, 31321,
31327, 31333, 31337, 31357, 31379, 31387, 31391, 31393, 31397, 31469, 31477, 31481, 31489, 31511, 31513,
31517, 31531, 31541, 31543, 31547, 31567, 31573, 31583, 31601, 31607, 31627, 31643, 31649, 31657, 31663,
31667, 31687, 31699, 31721, 31723, 31727, 31729, 31741, 31751, 31769, 31771, 31793, 31799, 31817, 31847,
31849, 31859, 31873, 31883, 31891, 31907, 31957, 31963, 31973, 31981, 31991, 32003, 32009, 32027, 32029,
32051, 32057, 32059, 32063, 32069, 32077, 32083, 32089, 32099, 32117, 32119, 32141, 32143, 32159, 32173,
32183, 32189, 32191, 32203, 32213, 32233, 32237, 32251, 32257, 32261, 32297, 32299, 32303, 32309, 32321,
32323, 32327, 32341, 32353, 32359, 32363, 32369, 32371, 32377, 32381, 32401, 32411, 32413, 32423, 32429,
32441, 32443, 32467, 32479, 32491, 32497, 32503, 32507, 32531, 32533, 32537, 32561, 32563, 32569, 32573,
32579, 32587, 32603, 32609, 32611, 32621, 32633, 32647, 32653, 32687, 32693, 32707, 32713, 32717, 32719,
32749, 32771, 32779, 32783, 32789, 32797, 32801, 32803, 32831, 32833, 32839, 32843, 32869, 32887, 32909,
32911, 32917, 32933, 32939, 32941, 32957, 32969, 32971, 32983, 32987, 32993, 32999, 33013, 33023, 33029,
33037, 33049, 33053, 33071, 33073, 33083, 33091, 33107, 33113, 33119, 33149, 33151, 33161, 33179, 33181,
33191, 33199, 33203, 33211, 33223, 33247, 33287, 33289, 33301, 33311, 33317, 33329, 33331, 33343, 33347,
33349, 33353, 33359, 33377, 33391, 33403, 33409, 33413, 33427, 33457, 33461, 33469, 33479, 33487, 33493,
33503, 33521, 33529, 33533, 33547, 33563, 33569, 33577, 33581, 33587, 33589, 33599, 33601, 33613, 33617,
33619, 33623, 33629, 33637, 33641, 33647, 33679, 33703, 33713, 33721, 33739, 33749, 33751, 33757, 33767,
33769, 33773, 33791, 33797, 33809, 33811, 33827, 33829, 33851, 33857, 33863, 33871, 33889, 33893, 33911,
33923, 33931, 33937, 33941, 33961, 33967, 33997, 34019, 34031, 34033, 34039, 34057, 34061, 34123, 34127,
34129, 34141, 34147, 34157, 34159, 34171, 34183, 34211, 34213, 34217, 34231, 34253, 34259, 34261, 34267,
34273, 34283, 34297, 34301, 34303, 34313, 34319, 34327, 34337, 34351, 34361, 34367, 34369, 34381, 34403,
34421, 34429, 34439, 34457, 34469, 34471, 34483, 34487, 34499, 34501, 34511, 34513, 34519, 34537, 34543,
34549, 34583, 34589, 34591, 34603, 34607, 34613, 34631, 34649, 34651, 34667, 34673, 34679, 34687, 34693,
34703, 34721, 34729, 34739, 34747, 34757, 34759, 34763, 34781, 34807, 34819, 34841, 34843, 34847, 34849,
34871, 34877, 34883, 34897, 34913, 34919, 34939, 34949, 34961, 34963, 34981, 35023, 35027, 35051, 35053,
35059, 35069, 35081, 35083, 35089, 35099, 35107, 35111, 35117, 35129, 35141, 35149, 35153, 35159, 35171,
35201, 35221, 35227, 35251, 35257, 35267, 35279, 35281, 35291, 35311, 35317, 35323, 35327, 35339, 35353,
35363, 35381, 35393, 35401, 35407, 35419, 35423, 35437, 35447, 35449, 35461, 35491, 35507, 35509, 35521,
35527, 35531, 35533, 35537, 35543, 35569, 35573, 35591, 35593, 35597, 35603, 35617, 35671, 35677, 35729,
35731, 35747, 35753, 35759, 35771, 35797, 35801, 35803, 35809, 35831, 35837, 35839, 35851, 35863, 35869,
35879, 35897, 35899, 35911, 35923, 35933, 35951, 35963, 35969, 35977, 35983, 35993, 35999, 36007, 36011,
36013, 36017, 36037, 36061, 36067, 36073, 36083, 36097, 36107, 36109, 36131, 36137, 36151, 36161, 36187,
36191, 36209, 36217, 36229, 36241, 36251, 36263, 36269, 36277, 36293, 36299, 36307, 36313, 36319, 36341,
36343, 36353, 36373, 36383, 36389, 36433, 36451, 36457, 36467, 36469, 36473, 36479, 36493, 36497, 36523,
36527, 36529, 36541, 36551, 36559, 36563, 36571, 36583, 36587, 36599, 36607, 36629, 36637, 36643, 36653,
36671, 36677, 36683, 36691, 36697, 36709, 36713, 36721, 36739, 36749, 36761, 36767, 36779, 36781, 36787,

36791, 36793, 36809, 36821, 36833, 36847, 36857, 36871, 36877, 36887, 36899, 36901, 36913, 36919, 36923,
36929, 36931, 36943, 36947, 36973, 36979, 36997, 37003, 37013, 37019, 37021, 37039, 37049, 37057, 37061,
37087, 37097, 37117, 37123, 37139, 37159, 37171, 37181, 37189, 37199, 37201, 37217, 37223, 37243, 37253,
37273, 37277, 37307, 37309, 37313, 37321, 37337, 37339, 37357, 37361, 37363, 37369, 37379, 37397, 37409,
37423, 37441, 37447, 37463, 37483, 37489, 37493, 37501, 37507, 37511, 37517, 37529, 37537, 37547, 37549,
37561, 37567, 37571, 37573, 37579, 37589, 37591, 37607, 37619, 37633, 37643, 37649, 37657, 37663, 37691,
37693, 37699, 37717, 37747, 37781, 37783, 37799, 37811, 37813, 37831, 37847, 37853, 37861, 37871, 37879,
37889, 37897, 37907, 37951, 37957, 37963, 37967, 37987, 37991, 37993, 37997, 38011, 38039, 38047, 38053,
38069, 38083, 38113, 38119, 38149, 38153, 38167, 38177, 38183, 38189, 38197, 38201, 38219, 38231, 38237,
38239, 38261, 38273, 38281, 38287, 38299, 38303, 38317, 38321, 38327, 38329, 38333, 38351, 38371, 38377,
38393, 38431, 38447, 38449, 38453, 38459, 38461, 38501, 38543, 38557, 38561, 38567, 38569, 38593, 38603,
38609, 38611, 38629, 38639, 38651, 38653, 38669, 38671, 38677, 38693, 38699, 38707, 38711, 38713, 38723,
38729, 38737, 38747, 38749, 38767, 38783, 38791, 38803, 38821, 38833, 38839, 38851, 38861, 38867, 38873,
38891, 38903, 38917, 38921, 38923, 38933, 38953, 38959, 38971, 38977, 38993, 39019, 39023, 39041, 39043,
39047, 39079, 39089, 39097, 39103, 39107, 39113, 39119, 39133, 39139, 39157, 39161, 39163, 39181, 39191,
39199, 39209, 39217, 39227, 39229, 39233, 39239, 39241, 39251, 39293, 39301, 39313, 39317, 39323, 39341,
39343, 39359, 39367, 39371, 39373, 39383, 39397, 39409, 39419, 39439, 39443, 39451, 39461, 39499, 39503,
39509, 39511, 39521, 39541, 39551, 39563, 39569, 39581, 39607, 39619, 39623, 39631, 39659, 39667, 39671,
39679, 39703, 39709, 39719, 39727, 39733, 39749, 39761, 39769, 39779, 39791, 39799, 39821, 39827, 39829,
39839, 39841, 39847, 39857, 39863, 39869, 39877, 39883, 39887, 39901, 39929, 39937, 39953, 39971, 39979,
39983, 39989, 40009, 40013, 40031, 40037, 40039, 40063, 40087, 40093, 40099, 40111, 40123, 40127, 40129,
40151, 40153, 40163, 40169, 40177, 40189, 40193, 40213, 40231, 40237, 40241, 40253, 40277, 40283, 40289,
40343, 40351, 40357, 40361, 40387, 40423, 40427, 40429, 40433, 40459, 40471, 40483, 40487, 40493, 40499,
40507, 40519, 40529, 40531, 40543, 40559, 40577, 40583, 40591, 40597, 40609, 40627, 40637, 40639, 40693,
40697, 40699, 40709, 40739, 40751, 40759, 40763, 40771, 40787, 40801, 40813, 40819, 40823, 40829, 40841,
40847, 40849, 40853, 40867, 40879, 40883, 40897, 40903, 40927, 40933, 40939, 40949, 40961, 40973, 40993,
41011, 41017, 41023, 41039, 41047, 41051, 41057, 41077, 41081, 41113, 41117, 41131, 41141, 41143, 41149,
41161, 41177, 41179, 41183, 41189, 41201, 41203, 41213, 41221, 41227, 41231, 41233, 41243, 41257, 41263,
41269, 41281, 41299, 41333, 41341, 41351, 41357, 41381, 41387, 41389, 41399, 41411, 41413, 41443, 41453,
41467, 41479, 41491, 41507, 41513, 41519, 41521, 41539, 41543, 41549, 41579, 41593, 41597, 41603, 41609,
41611, 41617, 41621, 41627, 41641, 41647, 41651, 41659, 41669, 41681, 41687, 41719, 41729, 41737, 41759,
41761, 41771, 41777, 41801, 41809, 41813, 41843, 41849, 41851, 41863, 41879, 41887, 41893, 41897, 41903,
41911, 41927, 41941, 41947, 41953, 41957, 41959, 41969, 41981, 41983, 41999, 42013, 42017, 42019, 42023,
42043, 42061, 42071, 42073, 42083, 42089, 42101, 42131, 42139, 42157, 42169, 42179, 42181, 42187, 42193,
42197, 42209, 42221, 42223, 42227, 42239, 42257, 42281, 42283, 42293, 42299, 42307, 42323, 42331, 42337,
42349, 42359, 42373, 42379, 42391, 42397, 42403, 42407, 42409, 42433, 42437, 42443, 42451, 42457, 42461,
42463, 42467, 42473, 42487, 42491, 42499, 42509, 42533, 42557, 42569, 42571, 42577, 42589, 42611, 42641,
42643, 42649, 42667, 42677, 42683, 42689, 42697, 42701, 42703, 42709, 42719, 42727, 42737, 42743, 42751,
42767, 42773, 42787, 42793, 42797, 42821, 42829, 42839, 42841, 42853, 42859, 42863, 42899, 42901, 42923,
42929, 42937, 42943, 42953, 42961, 42967, 42979, 42989, 43003, 43013, 43019, 43037, 43049, 43051, 43063,
43067, 43093, 43103, 43117, 43133, 43151, 43159, 43177, 43189, 43201, 43207, 43223, 43237, 43261, 43271,
43283, 43291, 43313, 43319, 43321, 43331, 43391, 43397, 43399, 43403, 43411, 43427, 43441, 43451, 43457,
43481, 43487, 43499, 43517, 43541, 43543, 43573, 43577, 43579, 43591, 43597, 43607, 43609, 43613, 43627,
43633, 43649, 43651, 43661, 43669, 43691, 43711, 43717, 43721, 43753, 43759, 43777, 43781, 43783, 43787,
43789, 43793, 43801, 43853, 43867, 43889, 43891, 43913, 43933, 43943, 43951, 43961, 43963, 43969, 43973,
43987, 43991, 43997, 44017, 44021, 44027, 44029, 44041, 44053, 44059, 44071, 44087, 44089, 44101, 44111,
44119, 44123, 44129, 44131, 44159, 44171, 44179, 44189, 44201, 44203, 44207, 44221, 44249, 44257, 44263,
44267, 44269, 44273, 44279, 44281, 44293, 44351, 44357, 44371, 44381, 44383, 44389, 44417, 44449, 44453,
44483, 44491, 44497, 44501, 44507, 44519, 44531, 44533, 44537, 44543, 44549, 44563, 44579, 44587, 44617,

44621, 44623, 44633, 44641, 44647, 44651, 44657, 44683, 44687, 44699, 44701, 44711, 44729, 44741, 44753,
44771, 44773, 44777, 44789, 44797, 44809, 44819, 44839, 44843, 44851, 44867, 44879, 44887, 44893, 44909,
44917, 44927, 44939, 44953, 44959, 44963, 44971, 44983, 44987, 45007, 45013, 45053, 45061, 45077, 45083,
45119, 45121, 45127, 45131, 45137, 45139, 45161, 45179, 45181, 45191, 45197, 45233, 45247, 45259, 45263,
45281, 45289, 45293, 45307, 45317, 45319, 45329, 45337, 45341, 45343, 45361, 45377, 45389, 45403, 45413,
45427, 45433, 45439, 45481, 45491, 45497, 45503, 45523, 45533, 45541, 45553, 45557, 45569, 45587, 45589,
45599, 45613, 45631, 45641, 45659, 45667, 45673, 45677, 45691, 45697, 45707, 45737, 45751, 45757, 45763,
45767, 45779, 45817, 45821, 45823, 45827, 45833, 45841, 45853, 45863, 45869, 45887, 45893, 45943, 45949,
45953, 45959, 45971, 45979, 45989, 46021, 46027, 46049, 46051, 46061, 46073, 46091, 46093, 46099, 46103,
46133, 46141, 46147, 46153, 46171, 46181, 46183, 46187, 46199, 46219, 46229, 46237, 46261, 46271, 46273,
46279, 46301, 46307, 46309, 46327, 46337, 46349, 46351, 46381, 46399, 46411, 46439, 46441, 46447, 46451,
46457, 46471, 46477, 46489, 46499, 46507, 46511, 46523, 46549, 46559, 46567, 46573, 46589, 46591, 46601,
46619, 46633, 46639, 46643, 46649, 46663, 46679, 46681, 46687, 46691, 46703, 46723, 46727, 46747, 46751,
46757, 46769, 46771, 46807, 46811, 46817, 46819, 46829, 46831, 46853, 46861, 46867, 46877, 46889, 46901,
46919, 46933, 46957, 46993, 46997, 47017, 47041, 47051, 47057, 47059, 47087, 47093, 47111, 47119, 47123,
47129, 47137, 47143, 47147, 47149, 47161, 47189, 47207, 47221, 47237, 47251, 47269, 47279, 47287, 47293,
47297, 47303, 47309, 47317, 47339, 47351, 47353, 47363, 47381, 47387, 47389, 47407, 47417, 47419, 47431,
47441, 47459, 47491, 47497, 47501, 47507, 47513, 47521, 47527, 47533, 47543, 47563, 47569, 47581, 47591,
47599, 47609, 47623, 47629, 47639, 47653, 47657, 47659, 47681, 47699, 47701, 47711, 47713, 47717, 47737,
47741, 47743, 47777, 47779, 47791, 47797, 47807, 47809, 47819, 47837, 47843, 47857, 47869, 47881, 47903,
47911, 47917, 47933, 47939, 47947, 47951, 47963, 47969, 47977, 47981, 48017, 48023, 48029, 48049, 48073,
48079, 48091, 48109, 48119, 48121, 48131, 48157, 48163, 48179, 48187, 48193, 48197, 48221, 48239, 48247,
48259, 48271, 48281, 48299, 48311, 48313, 48337, 48341, 48353, 48371, 48383, 48397, 48407, 48409, 48413,
48437, 48449, 48463, 48473, 48479, 48481, 48487, 48491, 48497, 48523, 48527, 48533, 48539, 48541, 48563,
48571, 48589, 48593, 48611, 48619, 48623, 48647, 48649, 48661, 48673, 48677, 48679, 48731, 48733, 48751,
48757, 48761, 48767, 48779, 48781, 48787, 48799, 48809, 48817, 48821, 48823, 48847, 48857, 48859, 48869,
48871, 48883, 48889, 48907, 48947, 48953, 48973, 48989, 48991, 49003, 49009, 49019, 49031, 49033, 49037,
49043, 49057, 49069, 49081, 49103, 49109, 49117, 49121, 49123, 49139, 49157, 49169, 49171, 49177, 49193,
49199, 49201, 49207, 49211, 49223, 49253, 49261, 49277, 49279, 49297, 49307, 49331, 49333, 49339, 49363,
49367, 49369, 49391, 49393, 49409, 49411, 49417, 49429, 49433, 49451, 49459, 49463, 49477, 49481, 49499,
49523, 49529, 49531, 49537, 49547, 49549, 49559, 49597, 49603, 49613, 49627, 49633, 49639, 49663, 49667,
49669, 49681, 49697, 49711, 49727, 49739, 49741, 49747, 49757, 49783, 49787, 49789, 49801, 49807, 49811,
49823, 49831, 49843, 49853, 49871, 49877, 49891, 49919, 49921, 49927, 49937, 49939, 49943, 49957, 49991,
49993, 49999, 50021, 50023, 50033, 50047, 50051, 50053, 50069, 50077, 50087, 50093, 50101, 50111, 50119,
50123, 50129, 50131, 50147, 50153, 50159, 50177, 50207, 50221, 50227, 50231, 50261, 50263, 50273, 50287,
50291, 50311, 50321, 50329, 50333, 50341, 50359, 50363, 50377, 50383, 50387, 50411, 50417, 50423, 50441,
50459, 50461, 50497, 50503, 50513, 50527, 50539, 50543, 50549, 50551, 50581, 50587, 50591, 50593, 50599,
50627, 50647, 50651, 50671, 50683, 50707, 50723, 50741, 50753, 50767, 50773, 50777, 50789, 50821, 50833,
50839, 50849, 50857, 50867, 50873, 50891, 50893, 50909, 50923, 50929, 50951, 50957, 50969, 50971, 50989,
50993, 51001, 51031, 51043, 51047, 51059, 51061, 51071, 51109, 51131, 51133, 51137, 51151, 51157, 51169,
51193, 51197, 51199, 51203, 51217, 51229, 51239, 51241, 51257, 51263, 51283, 51287, 51307, 51329, 51341,
51343, 51347, 51349, 51361, 51383, 51407, 51413, 51419, 51421, 51427, 51431, 51437, 51439, 51449, 51461,
51473, 51479, 51481, 51487, 51503, 51511, 51517, 51521, 51539, 51551, 51563, 51577, 51581, 51593, 51599,
51607, 51613, 51631, 51637, 51647, 51659, 51673, 51679, 51683, 51691, 51713, 51719, 51721, 51749, 51767,
51769, 51787, 51797, 51803, 51817, 51827, 51829, 51839, 51853, 51859, 51869, 51871, 51893, 51899, 51907,
51913, 51929, 51941, 51949, 51971, 51973, 51977, 51991, 52009, 52021, 52027, 52051, 52057, 52067, 52069,
52081, 52103, 52121, 52127, 52147, 52153, 52163, 52177, 52181, 52183, 52189, 52201, 52223, 52237, 52249,
52253, 52259, 52267, 52289, 52291, 52301, 52313, 52321, 52361, 52363, 52369, 52379, 52387, 52391, 52433,
52453, 52457, 52489, 52501, 52511, 52517, 52529, 52541, 52543, 52553, 52561, 52567, 52571, 52579, 52583,

52609, 52627, 52631, 52639, 52667, 52673, 52691, 52697, 52709, 52711, 52721, 52727, 52733, 52747, 52757, 52769, 52783, 52807, 52813, 52817, 52837, 52859, 52861, 52879, 52883, 52889, 52901, 52903, 52919, 52937, 52951, 52957, 52963, 52967, 52973, 52981, 52999, 53003, 53017, 53047, 53051, 53069, 53077, 53087, 53089, 53093, 53101, 53113, 53117, 53129, 53147, 53149, 53161, 53171, 53173, 53189, 53197, 53201, 53231, 53233, 53239, 53267, 53269, 53279, 53281, 53299, 53309, 53323, 53327, 53353, 53359, 53377, 53381, 53401, 53407, 53411, 53419, 53437, 53441, 53453, 53479, 53503, 53507, 53527, 53549, 53551, 53569, 53591, 53593, 53597, 53609, 53611, 53617, 53623, 53629, 53633, 53639, 53653, 53657, 53681, 53693, 53699, 53717, 53719, 53731, 53759, 53773, 53777, 53783, 53791, 53813, 53819, 53831, 53849, 53857, 53861, 53881, 53887, 53891, 53897, 53899, 53917, 53923, 53927, 53939, 53951, 53959, 53987, 53993, 54001, 54011, 54013, 54037, 54049, 54059, 54083, 54091, 54101, 54121, 54133, 54139, 54151, 54163, 54167, 54181, 54193, 54217, 54251, 54269, 54277, 54287, 54293, 54311, 54319, 54323, 54331, 54347, 54361, 54367, 54371, 54377, 54401, 54403, 54409, 54413, 54419, 54421, 54437, 54443, 54449, 54469, 54493, 54497, 54499, 54503, 54517, 54521, 54539, 54541, 54547, 54559, 54563, 54577, 54581, 54583, 54601, 54617, 54623, 54629, 54631, 54647, 54667, 54673, 54679, 54709, 54713, 54721, 54727, 54751, 54767, 54773, 54779, 54787, 54799, 54829, 54833, 54851, 54869, 54877, 54881, 54907, 54917, 54919, 54941, 54949, 54959, 54973, 54979, 54983, 55001, 55009, 55021, 55049, 55051, 55057, 55061, 55073, 55079, 55103, 55109, 55117, 55127, 55147, 55163, 55171, 55201, 55207, 55213, 55217, 55219, 55229, 55243, 55249, 55259, 55291, 55313, 55331, 55333, 55337, 55339, 55343, 55351, 55373, 55381, 55399, 55411, 55439, 55441, 55457, 55469, 55487, 55501, 55511, 55529, 55541, 55547, 55579, 55589, 55603, 55609, 55619, 55621, 55631, 55633, 55639, 55661, 55663, 55667, 55673, 55681, 55691, 55697, 55711, 55717, 55721, 55733, 55763, 55787, 55793, 55799, 55807, 55813, 55817, 55819, 55823, 55829, 55837, 55843, 55849, 55871, 55889, 55897, 55901, 55903, 55921, 55927, 55931, 55933, 55949, 55967, 55987, 55997, 56003, 56009, 56039, 56041, 56053, 56081, 56087, 56093, 56099, 56101, 56113, 56123, 56131, 56149, 56167, 56171, 56179, 56197, 56207, 56209, 56237, 56239, 56249, 56263, 56267, 56269, 56299, 56311, 56333, 56359, 56369, 56377, 56383, 56393, 56401, 56417, 56431, 56437, 56443, 56453, 56467, 56473, 56477, 56479, 56489, 56501, 56503, 56509, 56519, 56527, 56531, 56533, 56543, 56569, 56591, 56597, 56599, 56611, 56629, 56633, 56659, 56663, 56671, 56681, 56687, 56701, 56711, 56713, 56731, 56737, 56747, 56767, 56773, 56779, 56783, 56807, 56809, 56813, 56821, 56827, 56843, 56857, 56873, 56891, 56893, 56897, 56909, 56911, 56921, 56923, 56929, 56941, 56951, 56957, 56963, 56983, 56989, 56993, 56999, 57037, 57041, 57047, 57059, 57073, 57077, 57089, 57097, 57107, 57119, 57131, 57139, 57143, 57149, 57163, 57173, 57179, 57191, 57193, 57203, 57221, 57223, 57241, 57251, 57259, 57269, 57271, 57283, 57287, 57301, 57329, 57331, 57347, 57349, 57367, 57373, 57383, 57389, 57397, 57413, 57427, 57457, 57467, 57487, 57493, 57503, 57527, 57529, 57557, 57559, 57571, 57587, 57593, 57601, 57637, 57641, 57649, 57653, 57667, 57679, 57689, 57697, 57709, 57713, 57719, 57727, 57731, 57737, 57751, 57773, 57781, 57787, 57791, 57793, 57803, 57809, 57829, 57839, 57847, 57853, 57859, 57881, 57899, 57901, 57917, 57923, 57943, 57947, 57973, 57977, 57991, 58013, 58027, 58031, 58043, 58049, 58057, 58061, 58067, 58073, 58099, 58109, 58111, 58129, 58147, 58151, 58153, 58169, 58171, 58189, 58193, 58199, 58207, 58211, 58217, 58229, 58231, 58237, 58243, 58271, 58309, 58313, 58321, 58337, 58363, 58367, 58369, 58379, 58391, 58393, 58403, 58411, 58417, 58427, 58439, 58441, 58451, 58453, 58477, 58481, 58511, 58537, 58543, 58549, 58567, 58573, 58579, 58601, 58603, 58613, 58631, 58657, 58661, 58679, 58687, 58693, 58699, 58711, 58727, 58733, 58741, 58757, 58763, 58771, 58787, 58789, 58831, 58889, 58897, 58901, 58907, 58909, 58913, 58921, 58937, 58943, 58963, 58967, 58979, 58991, 58997, 59009, 59011, 59021, 59023, 59029, 59051, 59053, 59063, 59069, 59077, 59083, 59093, 59107, 59113, 59119, 59123, 59141, 59149, 59159, 59167, 59183, 59197, 59207, 59209, 59219, 59221, 59233, 59239, 59243, 59263, 59273, 59281, 59333, 59341, 59351, 59357, 59359, 59369, 59377, 59387, 59393, 59399, 59407, 59417, 59419, 59441, 59443, 59447, 59453, 59467, 59471, 59473, 59497, 59509, 59513, 59539, 59557, 59561, 59567, 59581, 59611, 59617, 59621, 59627, 59629, 59651, 59659, 59663, 59669, 59671, 59693, 59699, 59707, 59723, 59729, 59743, 59747, 59753, 59771, 59779, 59791, 59797, 59809, 59833, 59863, 59879, 59887, 59921, 59929, 59951, 59957, 59971, 59981, 59999, 60013, 60017, 60029, 60037, 60041, 60077, 60083, 60089, 60091, 60101, 60103, 60107, 60127, 60133, 60139, 60149, 60161, 60167, 60169, 60209, 60217, 60223, 60251, 60257, 60259, 60271, 60289, 60293, 60317, 60331, 60337, 60343, 60353, 60373, 60383, 60397, 60413, 60427, 60443, 60449, 60457, 60493, 60497, 60509, 60521, 60527, 60539, 60589, 60601,

60607, 60611, 60617, 60623, 60631, 60637, 60647, 60649, 60659, 60661, 60679, 60689, 60703, 60719, 60727,
60733, 60737, 60757, 60761, 60763, 60773, 60779, 60793, 60811, 60821, 60859, 60869, 60887, 60889, 60899,
60901, 60913, 60917, 60919, 60923, 60937, 60943, 60953, 60961, 61001, 61007, 61027, 61031, 61043, 61051,
61057, 61091, 61099, 61121, 61129, 61141, 61151, 61153, 61169, 61211, 61223, 61231, 61253, 61261, 61283,
61291, 61297, 61331, 61333, 61339, 61343, 61357, 61363, 61379, 61381, 61403, 61409, 61417, 61441, 61463,
61469, 61471, 61483, 61487, 61493, 61507, 61511, 61519, 61543, 61547, 61553, 61559, 61561, 61583, 61603,
61609, 61613, 61627, 61631, 61637, 61643, 61651, 61657, 61667, 61673, 61681, 61687, 61703, 61717, 61723,
61729, 61751, 61757, 61781, 61813, 61819, 61837, 61843, 61861, 61871, 61879, 61909, 61927, 61933, 61949,
61961, 61967, 61979, 61981, 61987, 61991, 62003, 62011, 62017, 62039, 62047, 62053, 62057, 62071, 62081,
62099, 62119, 62129, 62131, 62137, 62141, 62143, 62171, 62189, 62191, 62201, 62207, 62213, 62219, 62233,
62273, 62297, 62299, 62303, 62311, 62323, 62327, 62347, 62351, 62383, 62401, 62417, 62423, 62459, 62467,
62473, 62477, 62483, 62497, 62501, 62507, 62533, 62539, 62549, 62563, 62581, 62591, 62597, 62603, 62617,
62627, 62633, 62639, 62653, 62659, 62683, 62687, 62701, 62723, 62731, 62743, 62753, 62761, 62773, 62791,
62801, 62819, 62827, 62851, 62861, 62869, 62873, 62897, 62903, 62921, 62927, 62929, 62939, 62969, 62971,
62981, 62983, 62987, 62989, 63029, 63031, 63059, 63067, 63073, 63079, 63097, 63103, 63113, 63127, 63131,
63149, 63179, 63197, 63199, 63211, 63241, 63247, 63277, 63281, 63299, 63311, 63313, 63317, 63331, 63337,
63347, 63353, 63361, 63367, 63377, 63389, 63391, 63397, 63409, 63419, 63421, 63439, 63443, 63463, 63467,
63473, 63487, 63493, 63499, 63521, 63527, 63533, 63541, 63559, 63577, 63587, 63589, 63599, 63601, 63607,
63611, 63617, 63629, 63647, 63649, 63659, 63667, 63671, 63689, 63691, 63697, 63703, 63709, 63719, 63727,
63737, 63743, 63761, 63773, 63781, 63793, 63799, 63803, 63809, 63823, 63839, 63841, 63853, 63857, 63863,
63901, 63907, 63913, 63929, 63949, 63977, 63997, 64007, 64013, 64019, 64033, 64037, 64063, 64067, 64081,
64091, 64109, 64123, 64151, 64153, 64157, 64171, 64187, 64189, 64217, 64223, 64231, 64237, 64271, 64279,
64283, 64301, 64303, 64319, 64327, 64333, 64373, 64381, 64399, 64403, 64433, 64439, 64451, 64453, 64483,
64489, 64499, 64513, 64553, 64567, 64577, 64579, 64591, 64601, 64609, 64613, 64621, 64627, 64633, 64661,
64663, 64667, 64679, 64693, 64709, 64717, 64747, 64763, 64781, 64783, 64793, 64811, 64817, 64849, 64853,
64871, 64877, 64879, 64891, 64901, 64919, 64921, 64927, 64937, 64951, 64969, 64997, 65003, 65011, 65027,
65029, 65033, 65053, 65063, 65071, 65089, 65099, 65101, 65111, 65119, 65123, 65129, 65141, 65147, 65167,
65171, 65173, 65179, 65183, 65203, 65213, 65239, 65257, 65267, 65269, 65287, 65293, 65309, 65323, 65327,
65353, 65357, 65371, 65381, 65393, 65407, 65413, 65419, 65423, 65437, 65447, 65449, 65479, 65497, 65519,
65521, 65537, 65539, 65543, 65551, 65557, 65563, 65579, 65581, 65587, 65599, 65609, 65617, 65629, 65633,
65647, 65651, 65657, 65677, 65687, 65699, 65701, 65707, 65713, 65717, 65719, 65729, 65731, 65761, 65777,
65789, 65809, 65827, 65831, 65837, 65839, 65843, 65851, 65867, 65881, 65899, 65921, 65927, 65929, 65951,
65957, 65963, 65981, 65983, 65993, 66029, 66037, 66041, 66047, 66067, 66071, 66083, 66089, 66103, 66107,
66109, 66137, 66161, 66169, 66173, 66179, 66191, 66221, 66239, 66271, 66293, 66301, 66337, 66343, 66347,
66359, 66361, 66373, 66377, 66383, 66403, 66413, 66431, 66449, 66457, 66463, 66467, 66491, 66499, 66509,
66523, 66529, 66533, 66541, 66553, 66569, 66571, 66587, 66593, 66601, 66617, 66629, 66643, 66653, 66683,
66697, 66701, 66713, 66721, 66733, 66739, 66749, 66751, 66763, 66791, 66797, 66809, 66821, 66841, 66851,
66853, 66863, 66877, 66883, 66889, 66919, 66923, 66931, 66943, 66947, 66949, 66959, 66973, 66977, 67003,
67021, 67033, 67043, 67049, 67057, 67061, 67073, 67079, 67103, 67121, 67129, 67139, 67141, 67153, 67157,
67169, 67181, 67187, 67189, 67211, 67213, 67217, 67219, 67231, 67247, 67261, 67271, 67273, 67289, 67307,
67339, 67343, 67349, 67369, 67391, 67399, 67409, 67411, 67421, 67427, 67429, 67433, 67447, 67453, 67477,
67481, 67489, 67493, 67499, 67511, 67523, 67531, 67537, 67547, 67559, 67567, 67577, 67579, 67589, 67601,
67607, 67619, 67631, 67651, 67679, 67699, 67709, 67723, 67733, 67741, 67751, 67757, 67759, 67763, 67777,
67783, 67789, 67801, 67807, 67819, 67829, 67843, 67853, 67867, 67883, 67891, 67901, 67927, 67931, 67933,
67939, 67943, 67957, 67961, 67967, 67979, 67987, 67993, 68023, 68041, 68053, 68059, 68071, 68087, 68099,
68111, 68113, 68141, 68147, 68161, 68171, 68207, 68209, 68213, 68219, 68227, 68239, 68261, 68279, 68281,
68311, 68329, 68351, 68371, 68389, 68399, 68437, 68443, 68447, 68449, 68473, 68477, 68483, 68489, 68491,
68501, 68507, 68521, 68531, 68539, 68543, 68567, 68581, 68597, 68611, 68633, 68639, 68659, 68669, 68683,
68687, 68699, 68711, 68713, 68729, 68737, 68743, 68749, 68767, 68771, 68777, 68791, 68813, 68819, 68821,

68863, 68879, 68881, 68891, 68897, 68899, 68903, 68909, 68917, 68927, 68947, 68963, 68993, 69001, 69011, 69019, 69029, 69031, 69061, 69067, 69073, 69109, 69119, 69127, 69143, 69149, 69151, 69163, 69191, 69193, 69197, 69203, 69221, 69233, 69239, 69247, 69257, 69259, 69263, 69313, 69317, 69337, 69341, 69371, 69379, 69383, 69389, 69401, 69403, 69427, 69431, 69439, 69457, 69463, 69467, 69473, 69481, 69491, 69493, 69497, 69499, 69539, 69557, 69593, 69623, 69653, 69661, 69677, 69691, 69697, 69709, 69737, 69739, 69761, 69763, 69767, 69779, 69809, 69821, 69827, 69829, 69833, 69847, 69857, 69859, 69877, 69899, 69911, 69929, 69931, 69941, 69959, 69991, 69997, 70001, 70003, 70009, 70019, 70039, 70051, 70061, 70067, 70079, 70099, 70111, 70117, 70121, 70123, 70139, 70141, 70157, 70163, 70177, 70181, 70183, 70199, 70201, 70207, 70223, 70229, 70237, 70241, 70249, 70271, 70289, 70297, 70309, 70313, 70321, 70327, 70351, 70373, 70379, 70381, 70393, 70423, 70429, 70439, 70451, 70457, 70459, 70481, 70487, 70489, 70501, 70507, 70529, 70537, 70549, 70571, 70573, 70583, 70589, 70607, 70619, 70621, 70627, 70639, 70657, 70663, 70667, 70687, 70709, 70717, 70729, 70753, 70769, 70783, 70793, 70823, 70841, 70843, 70849, 70853, 70867, 70877, 70879, 70891, 70901, 70913, 70919, 70921, 70937, 70949, 70951, 70957, 70969, 70979, 70981, 70991, 70997, 70999, 71011, 71023, 71039, 71059, 71069, 71081, 71089, 71119, 71129, 71143, 71147, 71153, 71161, 71167, 71171, 71191, 71209, 71233, 71237, 71249, 71257, 71261, 71263, 71287, 71293, 71317, 71327, 71329, 71333, 71339, 71341, 71347, 71353, 71359, 71363, 71387, 71389, 71399, 71411, 71413, 71419, 71429, 71437, 71443, 71453, 71471, 71473, 71479, 71483, 71503, 71527, 71537, 71549, 71551, 71563, 71569, 71593, 71597, 71633, 71647, 71663, 71671, 71693, 71699, 71707, 71711, 71713, 71719, 71741, 71761, 71777, 71789, 71807, 71809, 71821, 71837, 71843, 71849, 71861, 71867, 71879, 71881, 71887, 71899, 71909, 71917, 71933, 71941, 71947, 71963, 71971, 71983, 71987, 71993, 71999, 72019, 72031, 72043, 72047, 72053, 72073, 72077, 72089, 72091, 72101, 72103, 72109, 72139, 72161, 72167, 72169, 72173, 72211, 72221, 72223, 72227, 72229, 72251, 72253, 72269, 72271, 72277, 72287, 72307, 72313, 72337, 72341, 72353, 72367, 72379, 72383, 72421, 72431, 72461, 72467, 72469, 72481, 72493, 72497, 72503, 72533, 72547, 72551, 72559, 72577, 72613, 72617, 72623, 72643, 72647, 72649, 72661, 72671, 72673, 72679, 72689, 72701, 72707, 72719, 72727, 72733, 72739, 72763, 72767, 72797, 72817, 72823, 72859, 72869, 72871, 72883, 72889, 72893, 72901, 72907, 72911, 72923, 72931, 72937, 72949, 72953, 72959, 72973, 72977, 72997, 73009, 73013, 73019, 73037, 73039, 73043, 73061, 73063, 73079, 73091, 73121, 73127, 73133, 73141, 73181, 73189, 73237, 73243, 73259, 73277, 73291, 73303, 73309, 73327, 73331, 73351, 73361, 73363, 73369, 73379, 73387, 73417, 73421, 73433, 73453, 73459, 73471, 73477, 73483, 73517, 73523, 73529, 73547, 73553, 73561, 73571, 73583, 73589, 73597, 73607, 73609, 73613, 73637, 73643, 73651, 73673, 73679, 73681, 73693, 73699, 73709, 73721, 73727, 73751, 73757, 73771, 73783, 73819, 73823, 73847, 73849, 73859, 73867, 73877, 73883, 73897, 73907, 73939, 73943, 73951, 73961, 73973, 73999, 74017, 74021, 74027, 74047, 74051, 74071, 74077, 74093, 74099, 74101, 74131, 74143, 74149, 74159, 74161, 74167, 74177, 74189, 74197, 74201, 74203, 74209, 74219, 74231, 74257, 74279, 74287, 74293, 74297, 74311, 74317, 74323, 74353, 74357, 74363, 74377, 74381, 74383, 74411, 74413, 74419, 74441, 74449, 74453, 74471, 74489, 74507, 74509, 74521, 74527, 74531, 74551, 74561, 74567, 74573, 74587, 74597, 74609, 74611, 74623, 74653, 74687, 74699, 74707, 74713, 74717, 74719, 74729, 74731, 74747, 74759, 74761, 74771, 74779, 74797, 74821, 74827, 74831, 74843, 74857, 74861, 74869, 74873, 74887, 74891, 74897, 74903, 74923, 74929, 74933, 74941, 74959, 75011, 75013, 75017, 75029, 75037, 75041, 75079, 75083, 75109, 75133, 75149, 75161, 75167, 75169, 75181, 75193, 75209, 75211, 75217, 75223, 75227, 75239, 75253, 75269, 75277, 75289, 75307, 75323, 75329, 75337, 75347, 75353, 75367, 75377, 75389, 75391, 75401, 75403, 75407, 75431, 75437, 75479, 75503, 75511, 75521, 75527, 75533, 75539, 75541, 75553, 75557, 75571, 75577, 75583, 75611, 75617, 75619, 75629, 75641, 75653, 75659, 75679, 75683, 75689, 75703, 75707, 75709, 75721, 75731, 75743, 75767, 75773, 75781, 75787, 75793, 75797, 75821, 75833, 75853, 75869, 75883, 75913, 75931, 75937, 75941, 75967, 75979, 75983, 75989, 75991, 75997, 76001, 76003, 76031, 76039, 76079, 76081, 76091, 76099, 76103, 76123, 76129, 76147, 76157, 76159, 76163, 76207, 76213, 76231, 76243, 76249, 76253, 76259, 76261, 76283, 76289, 76303, 76333, 76343, 76367, 76369, 76379, 76387, 76403, 76421, 76423, 76441, 76463, 76471, 76481, 76487, 76493, 76507, 76511, 76519, 76537, 76541, 76543, 76561, 76579, 76597, 76603, 76607, 76631, 76649, 76651, 76667, 76673, 76679, 76697, 76717, 76733, 76753, 76757, 76771, 76777, 76781, 76801, 76819, 76829, 76831, 76837, 76847, 76871, 76873, 76883, 76907, 76913, 76919, 76943, 76949, 76961, 76963, 76991, 77003, 77017, 77023, 77029, 77041, 77047, 77069, 77081, 77093,

77101, 77137, 77141, 77153, 77167, 77171, 77191, 77201, 77213, 77237, 77239, 77243, 77249, 77261, 77263,
77267, 77269, 77279, 77291, 77317, 77323, 77339, 77347, 77351, 77359, 77369, 77377, 77383, 77417, 77419,
77431, 77447, 77471, 77477, 77479, 77489, 77491, 77509, 77513, 77521, 77527, 77543, 77549, 77551, 77557,
77563, 77569, 77573, 77587, 77591, 77611, 77617, 77621, 77641, 77647, 77659, 77681, 77687, 77689, 77699,
77711, 77713, 77719, 77723, 77731, 77743, 77747, 77761, 77773, 77783, 77797, 77801, 77813, 77839, 77849,
77863, 77867, 77893, 77899, 77929, 77933, 77951, 77969, 77977, 77983, 77999, 78007, 78017, 78031, 78041,
78049, 78059, 78079, 78101, 78121, 78137, 78139, 78157, 78163, 78167, 78173, 78179, 78191, 78193, 78203,
78229, 78233, 78241, 78259, 78277, 78283, 78301, 78307, 78311, 78317, 78341, 78347, 78367, 78401, 78427,
78437, 78439, 78467, 78479, 78487, 78497, 78509, 78511, 78517, 78539, 78541, 78553, 78569, 78571, 78577,
78583, 78593, 78607, 78623, 78643, 78649, 78653, 78691, 78697, 78707, 78713, 78721, 78737, 78779, 78781,
78787, 78791, 78797, 78803, 78809, 78823, 78839, 78853, 78857, 78877, 78887, 78889, 78893, 78901, 78919,
78929, 78941, 78977, 78979, 78989, 79031, 79039, 79043, 79063, 79087, 79103, 79111, 79133, 79139, 79147,
79151, 79153, 79159, 79181, 79187, 79193, 79201, 79229, 79231, 79241, 79259, 79273, 79279, 79283, 79301,
79309, 79319, 79333, 79337, 79349, 79357, 79367, 79379, 79393, 79397, 79399, 79411, 79423, 79427, 79433,
79451, 79481, 79493, 79531, 79537, 79549, 79559, 79561, 79579, 79589, 79601, 79609, 79613, 79621, 79627,
79631, 79633, 79657, 79669, 79687, 79691, 79693, 79697, 79699, 79757, 79769, 79777, 79801, 79811, 79813,
79817, 79823, 79829, 79841, 79843, 79847, 79861, 79867, 79873, 79889, 79901, 79903, 79907, 79939, 79943,
79967, 79973, 79979, 79987, 79997, 79999, 80021, 80039, 80051, 80071, 80077, 80107, 80111, 80141, 80147,
80149, 80153, 80167, 80173, 80177, 80191, 80207, 80209, 80221, 80231, 80233, 80239, 80251, 80263, 80273,
80279, 80287, 80309, 80317, 80329, 80341, 80347, 80363, 80369, 80387, 80407, 80429, 80447, 80449, 80471,
80473, 80489, 80491, 80513, 80527, 80537, 80557, 80567, 80599, 80603, 80611, 80621, 80627, 80629, 80651,
80657, 80669, 80671, 80677, 80681, 80683, 80687, 80701, 80713, 80737, 80747, 80749, 80761, 80777, 80779,
80783, 80789, 80803, 80809, 80819, 80831, 80833, 80849, 80863, 80897, 80909, 80911, 80917, 80923, 80929,
80933, 80953, 80963, 80989, 81001, 81013, 81017, 81019, 81023, 81031, 81041, 81043, 81047, 81049, 81071,
81077, 81083, 81097, 81101, 81119, 81131, 81157, 81163, 81173, 81181, 81197, 81199, 81203, 81223, 81233,
81239, 81281, 81283, 81293, 81299, 81307, 81331, 81343, 81349, 81353, 81359, 81371, 81373, 81401, 81409,
81421, 81439, 81457, 81463, 81509, 81517, 81527, 81533, 81547, 81551, 81553, 81559, 81563, 81569, 81611,
81619, 81629, 81637, 81647, 81649, 81667, 81671, 81677, 81689, 81701, 81703, 81707, 81727, 81737, 81749,
81761, 81769, 81773, 81799, 81817, 81839, 81847, 81853, 81869, 81883, 81899, 81901, 81919, 81929, 81931,
81937, 81943, 81953, 81967, 81971, 81973, 82003, 82007, 82009, 82013, 82021, 82031, 82037, 82039, 82051,
82067, 82073, 82129, 82139, 82141, 82153, 82163, 82171, 82183, 82189, 82193, 82207, 82217, 82219, 82223,
82231, 82237, 82241, 82261, 82267, 82279, 82301, 82307, 82339, 82349, 82351, 82361, 82373, 82387, 82393,
82421, 82457, 82463, 82469, 82471, 82483, 82487, 82493, 82499, 82507, 82529, 82531, 82549, 82559, 82561,
82567, 82571, 82591, 82601, 82609, 82613, 82619, 82633, 82651, 82657, 82699, 82721, 82723, 82727, 82729,
82757, 82759, 82763, 82781, 82787, 82793, 82799, 82811, 82813, 82837, 82847, 82883, 82889, 82891, 82903,
82913, 82939, 82963, 82981, 82997, 83003, 83009, 83023, 83047, 83059, 83063, 83071, 83077, 83089, 83093,
83101, 83117, 83137, 83177, 83203, 83207, 83219, 83221, 83227, 83231, 83233, 83243, 83257, 83267, 83269,
83273, 83299, 83311, 83339, 83341, 83357, 83383, 83389, 83399, 83401, 83407, 83417, 83423, 83431, 83437,
83443, 83449, 83459, 83471, 83477, 83497, 83537, 83557, 83561, 83563, 83579, 83591, 83597, 83609, 83617,
83621, 83639, 83641, 83653, 83663, 83689, 83701, 83717, 83719, 83737, 83761, 83773, 83777, 83791, 83813,
83833, 83843, 83857, 83869, 83873, 83891, 83903, 83911, 83921, 83933, 83939, 83969, 83983, 83987, 84011,
84017, 84047, 84053, 84059, 84061, 84067, 84089, 84121, 84127, 84131, 84137, 84143, 84163, 84179, 84181,
84191, 84199, 84211, 84221, 84223, 84229, 84239, 84247, 84263, 84299, 84307, 84313, 84317, 84319, 84347,
84349, 84377, 84389, 84391, 84401, 84407, 84421, 84431, 84437, 84443, 84449, 84457, 84463, 84467, 84481,
84499, 84503, 84509, 84521, 84523, 84533, 84551, 84559, 84589, 84629, 84631, 84649, 84653, 84659, 84673,
84691, 84697, 84701, 84713, 84719, 84731, 84737, 84751, 84761, 84787, 84793, 84809, 84811, 84827, 84857,
84859, 84869, 84871, 84913, 84919, 84947, 84961, 84967, 84977, 84979, 84991, 85009, 85021, 85027, 85037,
85049, 85061, 85081, 85087, 85091, 85093, 85103, 85109, 85121, 85133, 85147, 85159, 85193, 85199, 85201,
85213, 85223, 85229, 85237, 85243, 85247, 85259, 85297, 85303, 85313, 85331, 85333, 85361, 85363, 85369,

85381, 85411, 85427, 85429, 85439, 85447, 85451, 85453, 85469, 85487, 85513, 85517, 85523, 85531, 85549,
85571, 85577, 85597, 85601, 85607, 85619, 85621, 85627, 85639, 85643, 85661, 85667, 85669, 85691, 85703,
85711, 85717, 85733, 85751, 85781, 85793, 85817, 85819, 85829, 85831, 85837, 85843, 85847, 85853, 85889,
85903, 85909, 85931, 85933, 85991, 85999, 86011, 86017, 86027, 86029, 86069, 86077, 86083, 86111, 86113,
86117, 86131, 86137, 86143, 86161, 86171, 86179, 86183, 86197, 86201, 86209, 86239, 86243, 86249, 86257,
86263, 86269, 86287, 86291, 86293, 86297, 86311, 86323, 86341, 86351, 86353, 86357, 86369, 86371, 86381,
86389, 86399, 86413, 86423, 86441, 86453, 86461, 86467, 86477, 86491, 86501, 86509, 86531, 86533, 86539,
86561, 86573, 86579, 86587, 86599, 86627, 86629, 86677, 86689, 86693, 86711, 86719, 86729, 86743, 86753,
86767, 86771, 86783, 86813, 86837, 86843, 86851, 86857, 86861, 86869, 86923, 86927, 86929, 86939, 86951,
86959, 86969, 86981, 86993, 87011, 87013, 87037, 87041, 87049, 87071, 87083, 87103, 87107, 87119, 87121,
87133, 87149, 87151, 87179, 87181, 87187, 87211, 87221, 87223, 87251, 87253, 87257, 87277, 87281, 87293,
87299, 87313, 87317, 87323, 87337, 87359, 87383, 87403, 87407, 87421, 87427, 87433, 87443, 87473, 87481,
87491, 87509, 87511, 87517, 87523, 87539, 87541, 87547, 87553, 87557, 87559, 87583, 87587, 87589, 87613,
87623, 87629, 87631, 87641, 87643, 87649, 87671, 87679, 87683, 87691, 87697, 87701, 87719, 87721, 87739,
87743, 87751, 87767, 87793, 87797, 87803, 87811, 87833, 87853, 87869, 87877, 87881, 87887, 87911, 87917,
87931, 87943, 87959, 87961, 87973, 87977, 87991, 88001, 88003, 88007, 88019, 88037, 88069, 88079, 88093,
88117, 88129, 88169, 88177, 88211, 88223, 88237, 88241, 88259, 88261, 88289, 88301, 88321, 88327, 88337,
88339, 88379, 88397, 88411, 88423, 88427, 88463, 88469, 88471, 88493, 88499, 88513, 88523, 88547, 88589,
88591, 88607, 88609, 88643, 88651, 88657, 88661, 88663, 88667, 88681, 88721, 88729, 88741, 88747, 88771,
88789, 88793, 88799, 88801, 88807, 88811, 88813, 88817, 88819, 88843, 88853, 88861, 88867, 88873, 88883,
88897, 88903, 88919, 88937, 88951, 88969, 88993, 88997, 89003, 89009, 89017, 89021, 89041, 89051, 89057,
89069, 89071, 89083, 89087, 89101, 89107, 89113, 89119, 89123, 89137, 89153, 89189, 89203, 89209, 89213,
89227, 89231, 89237, 89261, 89269, 89273, 89293, 89303, 89317, 89329, 89363, 89371, 89381, 89387, 89393,
89399, 89413, 89417, 89431, 89443, 89449, 89459, 89477, 89491, 89501, 89513, 89519, 89521, 89527, 89533,
89561, 89563, 89567, 89591, 89597, 89599, 89603, 89611, 89627, 89633, 89653, 89657, 89659, 89669, 89671,
89681, 89689, 89753, 89759, 89767, 89779, 89783, 89797, 89809, 89819, 89821, 89833, 89839, 89849, 89867,
89891, 89897, 89899, 89909, 89917, 89923, 89939, 89959, 89963, 89977, 89983, 89989, 90001, 90007, 90011,
90017, 90019, 90023, 90031, 90053, 90059, 90067, 90071, 90073, 90089, 90107, 90121, 90127, 90149, 90163,
90173, 90187, 90191, 90197, 90199, 90203, 90217, 90227, 90239, 90247, 90263, 90271, 90281, 90289, 90313,
90353, 90359, 90371, 90373, 90379, 90397, 90401, 90403, 90407, 90437, 90439, 90469, 90473, 90481, 90499,
90511, 90523, 90527, 90529, 90533, 90547, 90583, 90599, 90617, 90619, 90631, 90641, 90647, 90659, 90677,
90679, 90697, 90703, 90709, 90731, 90749, 90787, 90793, 90803, 90821, 90823, 90833, 90841, 90847, 90863,
90887, 90901, 90907, 90911, 90917, 90931, 90947, 90971, 90977, 90989, 90997, 91009, 91019, 91033, 91079,
91081, 91097, 91099, 91121, 91127, 91129, 91139, 91141, 91151, 91153, 91159, 91163, 91183, 91193, 91199,
91229, 91237, 91243, 91249, 91253, 91283, 91291, 91297, 91303, 91309, 91331, 91367, 91369, 91373, 91381,
91387, 91393, 91397, 91411, 91423, 91433, 91453, 91457, 91459, 91463, 91493, 91499, 91513, 91529, 91541,
91571, 91573, 91577, 91583, 91591, 91621, 91631, 91639, 91673, 91691, 91703, 91711, 91733, 91753, 91757,
91771, 91781, 91801, 91807, 91811, 91813, 91823, 91837, 91841, 91867, 91873, 91909, 91921, 91939, 91943,
91951, 91957, 91961, 91967, 91969, 91997, 92003, 92009, 92033, 92041, 92051, 92077, 92083, 92107, 92111,
92119, 92143, 92153, 92173, 92177, 92179, 92189, 92203, 92219, 92221, 92227, 92233, 92237, 92243, 92251,
92269, 92297, 92311, 92317, 92333, 92347, 92353, 92357, 92363, 92369, 92377, 92381, 92383, 92387, 92399,
92401, 92413, 92419, 92431, 92459, 92461, 92467, 92479, 92489, 92503, 92507, 92551, 92557, 92567, 92569,
92581, 92593, 92623, 92627, 92639, 92641, 92647, 92657, 92669, 92671, 92681, 92683, 92693, 92699, 92707,
92717, 92723, 92737, 92753, 92761, 92767, 92779, 92789, 92791, 92801, 92809, 92821, 92831, 92849, 92857,
92861, 92863, 92867, 92893, 92899, 92921, 92927, 92941, 92951, 92957, 92959, 92987, 92993, 93001, 93047,
93053, 93059, 93077, 93083, 93089, 93097, 93103, 93113, 93131, 93133, 93139, 93151, 93169, 93179, 93187,
93199, 93229, 93239, 93241, 93251, 93253, 93257, 93263, 93281, 93283, 93287, 93307, 93319, 93323, 93329,
93337, 93371, 93377, 93383, 93407, 93419, 93427, 93463, 93479, 93481, 93487, 93491, 93493, 93497, 93503,
93523, 93529, 93553, 93557, 93559, 93563, 93581, 93601, 93607, 93629, 93637, 93683, 93701, 93703, 93719,

93739, 93761, 93763, 93787, 93809, 93811, 93827, 93851, 93871, 93887, 93889, 93893, 93901, 93911, 93913,
93923, 93937, 93941, 93949, 93967, 93971, 93979, 93983, 93997, 94007, 94009, 94033, 94049, 94057, 94063,
94079, 94099, 94109, 94111, 94117, 94121, 94151, 94153, 94169, 94201, 94207, 94219, 94229, 94253, 94261,
94273, 94291, 94307, 94309, 94321, 94327, 94331, 94343, 94349, 94351, 94379, 94397, 94399, 94421, 94427,
94433, 94439, 94441, 94447, 94463, 94477, 94483, 94513, 94529, 94531, 94541, 94543, 94547, 94559, 94561,
94573, 94583, 94597, 94603, 94613, 94621, 94649, 94651, 94687, 94693, 94709, 94723, 94727, 94747, 94771,
94777, 94781, 94789, 94793, 94811, 94819, 94823, 94837, 94841, 94847, 94849, 94873, 94889, 94903, 94907,
94933, 94949, 94951, 94961, 94993, 94999, 95003, 95009, 95021, 95027, 95063, 95071, 95083, 95087, 95089,
95093, 95101, 95107, 95111, 95131, 95143, 95153, 95177, 95189, 95191, 95203, 95213, 95219, 95231, 95233,
95239, 95257, 95261, 95267, 95273, 95279, 95287, 95311, 95317, 95327, 95339, 95369, 95383, 95393, 95401,
95413, 95419, 95429, 95441, 95443, 95461, 95467, 95471, 95479, 95483, 95507, 95527, 95531, 95539, 95549,
95561, 95569, 95581, 95597, 95603, 95617, 95621, 95629, 95633, 95651, 95701, 95707, 95713, 95717, 95723,
95731, 95737, 95747, 95773, 95783, 95789, 95791, 95801, 95803, 95813, 95819, 95857, 95869, 95873, 95881,
95891, 95911, 95917, 95923, 95929, 95947, 95957, 95959, 95971, 95987, 95989, 96001, 96013, 96017, 96043,
96053, 96059, 96079, 96097, 96137, 96149, 96157, 96167, 96179, 96181, 96199, 96211, 96221, 96223, 96233,
96259, 96263, 96269, 96281, 96289, 96293, 96323, 96329, 96331, 96337, 96353, 96377, 96401, 96419, 96431,
96443, 96451, 96457, 96461, 96469, 96479, 96487, 96493, 96497, 96517, 96527, 96553, 96557, 96581, 96587,
96589, 96601, 96643, 96661, 96667, 96671, 96697, 96703, 96731, 96737, 96739, 96749, 96757, 96763, 96769,
96779, 96787, 96797, 96799, 96821, 96823, 96827, 96847, 96851, 96857, 96893, 96907, 96911, 96931, 96953,
96959, 96973, 96979, 96989, 96997, 97001, 97003, 97007, 97021, 97039, 97073, 97081, 97103, 97117, 97127,
97151, 97157, 97159, 97169, 97171, 97177, 97187, 97213, 97231, 97241, 97259, 97283, 97301, 97303, 97327,
97367, 97369, 97373, 97379, 97381, 97387, 97397, 97423, 97429, 97441, 97453, 97459, 97463, 97499, 97501,
97511, 97523, 97547, 97549, 97553, 97561, 97571, 97577, 97579, 97583, 97607, 97609, 97613, 97649, 97651,
97673, 97687, 97711, 97729, 97771, 97777, 97787, 97789, 97813, 97829, 97841, 97843, 97847, 97849, 97859,
97861, 97871, 97879, 97883, 97919, 97927, 97931, 97943, 97961, 97967, 97973, 97987, 98009, 98011, 98017,
98041, 98047, 98057, 98081, 98101, 98123, 98129, 98143, 98179, 98207, 98213, 98221, 98227, 98251, 98257,
98269, 98297, 98299, 98317, 98321, 98323, 98327, 98347, 98369, 98377, 98387, 98389, 98407, 98411, 98419,
98429, 98443, 98453, 98459, 98467, 98473, 98479, 98491, 98507, 98519, 98533, 98543, 98561, 98563, 98573,
98597, 98621, 98627, 98639, 98641, 98663, 98669, 98689, 98711, 98713, 98717, 98729, 98731, 98737, 98773,
98779, 98801, 98807, 98809, 98837, 98849, 98867, 98869, 98873, 98887, 98893, 98897, 98899, 98909, 98911,
98927, 98929, 98939, 98947, 98953, 98963, 98981, 98993, 98999, 99013, 99017, 99023, 99041, 99053, 99079,
99083, 99089, 99103, 99109, 99119, 99131, 99133, 99137, 99139, 99149, 99173, 99181, 99191, 99223, 99233,
99241, 99251, 99257, 99259, 99277, 99289, 99317, 99347, 99349, 99367, 99371, 99377, 99391, 99397, 99401,
99409, 99431, 99439, 99469, 99487, 99497, 99523, 99527, 99529, 99551, 99559, 99563, 99571, 99577, 99581,
99607, 99611, 99623, 99643, 99661, 99667, 99679, 99689, 99707, 99709, 99713, 99719, 99721, 99733, 99761,
99767, 99787, 99793, 99809, 99817, 99823, 99829, 99833, 99839, 99859, 99871, 99877, 99881, 99901, 99907,
99923, 99929, 99961, 99971, 99989, 99991, 100003, 100019, 100043, 100049, 100057, 100069, 100103, 100109,
100129, 100151, 100153, 100169, 100183, 100189, 100193, 100207, 100213, 100237, 100267, 100271, 100279,
100291, 100297, 100313, 100333, 100343, 100357, 100361, 100363, 100379, 100391, 100393, 100403, 100411,
100417, 100447, 100459, 100469, 100483, 100493, 100501, 100511, 100517, 100519, 100523, 100537, 100547,
100549, 100559, 100591, 100609, 100613, 100621, 100649, 100669, 100673, 100693, 100699, 100703, 100733,
100741, 100747, 100769, 100787, 100799, 100801, 100811, 100823, 100829, 100847, 100853, 100907, 100913,
100927, 100931, 100937, 100943, 100957, 100981, 100987, 100999, 101009, 101021, 101027, 101051, 101063,
101081, 101089, 101107, 101111, 101113, 101117, 101119, 101141, 101149, 101159, 101161, 101173, 101183,
101197, 101203, 101207, 101209, 101221, 101267, 101273, 101279, 101281, 101287, 101293, 101323, 101333,
101341, 101347, 101359, 101363, 101377, 101383, 101399, 101411, 101419, 101429, 101449, 101467, 101477,
101483, 101489, 101501, 101503, 101513, 101527, 101531, 101533, 101537, 101561, 101573, 101581, 101599,
101603, 101611, 101627, 101641, 101653, 101663, 101681, 101693, 101701, 101719, 101723, 101737, 101741,
101747, 101749, 101771, 101789, 101797, 101807, 101833, 101837, 101839, 101863, 101869, 101873, 101879,

101891, 101917, 101921, 101929, 101939, 101957, 101963, 101977, 101987, 101999, 102001, 102013, 102019,
102023, 102031, 102043, 102059, 102061, 102071, 102077, 102079, 102101, 102103, 102107, 102121, 102139,
102149, 102161, 102181, 102191, 102197, 102199, 102203, 102217, 102229, 102233, 102241, 102251, 102253,
102259, 102293, 102299, 102301, 102317, 102329, 102337, 102359, 102367, 102397, 102407, 102409, 102433,
102437, 102451, 102461, 102481, 102497, 102499, 102503, 102523, 102533, 102539, 102547, 102551, 102559,
102563, 102587, 102593, 102607, 102611, 102643, 102647, 102653, 102667, 102673, 102677, 102679, 102701,
102761, 102763, 102769, 102793, 102797, 102811, 102829, 102841, 102859, 102871, 102877, 102881, 102911,
102913, 102929, 102931, 102953, 102967, 102983, 103001, 103007, 103043, 103049, 103067, 103069, 103079,
103087, 103091, 103093, 103099, 103123, 103141, 103171, 103177, 103183, 103217, 103231, 103237, 103289,
103291, 103307, 103319, 103333, 103349, 103357, 103387, 103391, 103393, 103399, 103409, 103421, 103423,
103451, 103457, 103471, 103483, 103511, 103529, 103549, 103553, 103561, 103567, 103573, 103577, 103583,
103591, 103613, 103619, 103643, 103651, 103657, 103669, 103681, 103687, 103699, 103703, 103723, 103769,
103787, 103801, 103811, 103813, 103837, 103841, 103843, 103867, 103889, 103903, 103913, 103919, 103951,
103963, 103967, 103969, 103979, 103981, 103991, 103993, 103997, 104003, 104009, 104021, 104033, 104047,
104053, 104059, 104087, 104089, 104107, 104113, 104119, 104123, 104147, 104149, 104161, 104173, 104179,
104183, 104207, 104231, 104233, 104239, 104243, 104281, 104287, 104297, 104309, 104311, 104323, 104327,
104347, 104369, 104381, 104383, 104393, 104399, 104417, 104459, 104471, 104473, 104479, 104491, 104513,
104527, 104537, 104543, 104549, 104551, 104561, 104579, 104593, 104597, 104623, 104639, 104651, 104659,
104677, 104681, 104683, 104693, 104701, 104707, 104711, 104717, 104723, 104729, 104743, 104759, 104761,
104773, 104779, 104789, 104801, 104803, 104827, 104831, 104849, 104851, 104869, 104879, 104891, 104911,
104917, 104933, 104947, 104953, 104959, 104971, 104987, 104999, 105019, 105023, 105031, 105037, 105071,
105097, 105107, 105137, 105143, 105167, 105173, 105199, 105211, 105227, 105229, 105239, 105251, 105253,
105263, 105269, 105277, 105319, 105323, 105331, 105337, 105341, 105359, 105361, 105367, 105373, 105379,
105389, 105397, 105401, 105407, 105437, 105449, 105467, 105491, 105499, 105503, 105509, 105517, 105527,
105529, 105533, 105541, 105557, 105563, 105601, 105607, 105613, 105619, 105649, 105653, 105667, 105673,
105683, 105691, 105701, 105727, 105733, 105751, 105761, 105767, 105769, 105817, 105829, 105863, 105871,
105883, 105899, 105907, 105913, 105929, 105943, 105953, 105967, 105971, 105977, 105983, 105997, 106013,
106019, 106031, 106033, 106087, 106103, 106109, 106121, 106123, 106129, 106163, 106181, 106187, 106189,
106207, 106213, 106217, 106219, 106243, 106261, 106273, 106277, 106279, 106291, 106297, 106303, 106307,
106319, 106321, 106331, 106349, 106357, 106363, 106367, 106373, 106391, 106397, 106411, 106417, 106427,
106433, 106441, 106451, 106453, 106487, 106501, 106531, 106537, 106541, 106543, 106591, 106619, 106621,
106627, 106637, 106649, 106657, 106661, 106663, 106669, 106681, 106693, 106699, 106703, 106721, 106727,
106739, 106747, 106751, 106753, 106759, 106781, 106783, 106787, 106801, 106823, 106853, 106859, 106861,
106867, 106871, 106877, 106903, 106907, 106921, 106937, 106949, 106957, 106961, 106963, 106979, 106993,
107021, 107033, 107053, 107057, 107069, 107071, 107077, 107089, 107099, 107101, 107119, 107123, 107137,
107171, 107183, 107197, 107201, 107209, 107227, 107243, 107251, 107269, 107273, 107279, 107309, 107323,
107339, 107347, 107351, 107357, 107377, 107441, 107449, 107453, 107467, 107473, 107507, 107509, 107563,
107581, 107599, 107603, 107609, 107621, 107641, 107647, 107671, 107687, 107693, 107699, 107713, 107717,
107719, 107741, 107747, 107761, 107773, 107777, 107791, 107827, 107837, 107839, 107843, 107857, 107867,
107873, 107881, 107897, 107903, 107923, 107927, 107941, 107951, 107971, 107981, 107999, 108007, 108011,
108013, 108023, 108037, 108041, 108061, 108079, 108089, 108107, 108109, 108127, 108131, 108139, 108161,
108179, 108187, 108191, 108193, 108203, 108211, 108217, 108223, 108233, 108247, 108263, 108271, 108287,
108289, 108293, 108301, 108343, 108347, 108359, 108377, 108379, 108401, 108413, 108421, 108439, 108457,
108461, 108463, 108497, 108499, 108503, 108517, 108529, 108533, 108541, 108553, 108557, 108571, 108587,
108631, 108637, 108643, 108649, 108677, 108707, 108709, 108727, 108739, 108751, 108761, 108769, 108791,
108793, 108799, 108803, 108821, 108827, 108863, 108869, 108877, 108881, 108883, 108887, 108893, 108907,
108917, 108923, 108929, 108943, 108947, 108949, 108959, 108961, 108967, 108971, 108991, 109001, 109013,
109037, 109049, 109063, 109073, 109097, 109103, 109111, 109121, 109133, 109139, 109141, 109147, 109159,
109169, 109171, 109199, 109201, 109211, 109229, 109253, 109267, 109279, 109297, 109303, 109313, 109321,

109331, 109357, 109363, 109367, 109379, 109387, 109391, 109397, 109423, 109433, 109441, 109451, 109453,
109469, 109471, 109481, 109507, 109517, 109519, 109537, 109541, 109547, 109567, 109579, 109583, 109589,
109597, 109609, 109619, 109621, 109639, 109661, 109663, 109673, 109717, 109721, 109741, 109751, 109789,
109793, 109807, 109819, 109829, 109831, 109841, 109843, 109847, 109849, 109859, 109873, 109883, 109891,
109897, 109903, 109913, 109919, 109937, 109943, 109961, 109987, 110017, 110023, 110039, 110051, 110059,
110063, 110069, 110083, 110119, 110129, 110161, 110183, 110221, 110233, 110237, 110251, 110261, 110269,
110273, 110281, 110291, 110311, 110321, 110323, 110339, 110359, 110419, 110431, 110437, 110441, 110459,
110477, 110479, 110491, 110501, 110503, 110527, 110533, 110543, 110557, 110563, 110567, 110569, 110573,
110581, 110587, 110597, 110603, 110609, 110623, 110629, 110641, 110647, 110651, 110681, 110711, 110729,
110731, 110749, 110753, 110771, 110777, 110807, 110813, 110819, 110821, 110849, 110863, 110879, 110881,
110899, 110909, 110917, 110921, 110923, 110927, 110933, 110939, 110947, 110951, 110969, 110977, 110989,
111029, 111031, 111043, 111049, 111053, 111091, 111103, 111109, 111119, 111121, 111127, 111143, 111149,
111187, 111191, 111211, 111217, 111227, 111229, 111253, 111263, 111269, 111271, 111301, 111317, 111323,
111337, 111341, 111347, 111373, 111409, 111427, 111431, 111439, 111443, 111467, 111487, 111491, 111493,
111497, 111509, 111521, 111533, 111539, 111577, 111581, 111593, 111599, 111611, 111623, 111637, 111641,
111653, 111659, 111667, 111697, 111721, 111731, 111733, 111751, 111767, 111773, 111779, 111781, 111791,
111799, 111821, 111827, 111829, 111833, 111847, 111857, 111863, 111869, 111871, 111893, 111913, 111919,
111949, 111953, 111959, 111973, 111977, 111997, 112019, 112031, 112061, 112067, 112069, 112087, 112097,
112103, 112111, 112121, 112129, 112139, 112153, 112163, 112181, 112199, 112207, 112213, 112223, 112237,
112241, 112247, 112249, 112253, 112261, 112279, 112289, 112291, 112297, 112303, 112327, 112331, 112337,
112339, 112349, 112361, 112363, 112397, 112403, 112429, 112459, 112481, 112501, 112507, 112543, 112559,
112571, 112573, 112577, 112583, 112589, 112601, 112603, 112621, 112643, 112657, 112663, 112687, 112691,
112741, 112757, 112759, 112771, 112787, 112799, 112807, 112831, 112843, 112859, 112877, 112901, 112909,
112913, 112919, 112921, 112927, 112939, 112951, 112967, 112979, 112997, 113011, 113017, 113021, 113023,
113027, 113039, 113041, 113051, 113063, 113081, 113083, 113089, 113093, 113111, 113117, 113123, 113131,
113143, 113147, 113149, 113153, 113159, 113161, 113167, 113171, 113173, 113177, 113189, 113209, 113213,
113227, 113233, 113279, 113287, 113327, 113329, 113341, 113357, 113359, 113363, 113371, 113381, 113383,
113417, 113437, 113453, 113467, 113489, 113497, 113501, 113513, 113537, 113539, 113557, 113567, 113591,
113621, 113623, 113647, 113657, 113683, 113717, 113719, 113723, 113731, 113749, 113759, 113761, 113777,
113779, 113783, 113797, 113809, 113819, 113837, 113843, 113891, 113899, 113903, 113909, 113921, 113933,
113947, 113957, 113963, 113969, 113983, 113989, 114001, 114013, 114031, 114041, 114043, 114067, 114073,
114077, 114083, 114089, 114113, 114143, 114157, 114161, 114167, 114193, 114197, 114199, 114203, 114217,
114221, 114229, 114259, 114269, 114277, 114281, 114299, 114311, 114319, 114329, 114343, 114371, 114377,
114407, 114419, 114451, 114467, 114473, 114479, 114487, 114493, 114547, 114553, 114571, 114577, 114593,
114599, 114601, 114613, 114617, 114641, 114643, 114649, 114659, 114661, 114671, 114679, 114689, 114691,
114713, 114743, 114749, 114757, 114761, 114769, 114773, 114781, 114797, 114799, 114809, 114827, 114833,
114847, 114859, 114883, 114889, 114901, 114913, 114941, 114967, 114973, 114997, 115001, 115013, 115019,
115021, 115057, 115061, 115067, 115079, 115099, 115117, 115123, 115127, 115133, 115151, 115153, 115163,
115183, 115201, 115211, 115223, 115237, 115249, 115259, 115279, 115301, 115303, 115309, 115319, 115321,
115327, 115331, 115337, 115343, 115361, 115363, 115399, 115421, 115429, 115459, 115469, 115471, 115499,
115513, 115523, 115547, 115553, 115561, 115571, 115589, 115597, 115601, 115603, 115613, 115631, 115637,
115657, 115663, 115679, 115693, 115727, 115733, 115741, 115751, 115757, 115763, 115769, 115771, 115777,
115781, 115783, 115793, 115807, 115811, 115823, 115831, 115837, 115849, 115853, 115859, 115861, 115873,
115877, 115879, 115883, 115891, 115901, 115903, 115931, 115933, 115963, 115979, 115981, 115987, 116009,
116027, 116041, 116047, 116089, 116099, 116101, 116107, 116113, 116131, 116141, 116159, 116167, 116177,
116189, 116191, 116201, 116239, 116243, 116257, 116269, 116273, 116279, 116293, 116329, 116341, 116351,
116359, 116371, 116381, 116387, 116411, 116423, 116437, 116443, 116447, 116461, 116471, 116483, 116491,
116507, 116531, 116533, 116537, 116539, 116549, 116579, 116593, 116639, 116657, 116663, 116681, 116687,
116689, 116707, 116719, 116731, 116741, 116747, 116789, 116791, 116797, 116803, 116819, 116827, 116833,

116849, 116867, 116881, 116903, 116911, 116923, 116927, 116929, 116933, 116953, 116959, 116969, 116981,
116989, 116993, 117017, 117023, 117037, 117041, 117043, 117053, 117071, 117101, 117109, 117119, 117127,
117133, 117163, 117167, 117191, 117193, 117203, 117209, 117223, 117239, 117241, 117251, 117259, 117269,
117281, 117307, 117319, 117329, 117331, 117353, 117361, 117371, 117373, 117389, 117413, 117427, 117431,
117437, 117443, 117497, 117499, 117503, 117511, 117517, 117529, 117539, 117541, 117563, 117571, 117577,
117617, 117619, 117643, 117659, 117671, 117673, 117679, 117701, 117703, 117709, 117721, 117727, 117731,
117751, 117757, 117763, 117773, 117779, 117787, 117797, 117809, 117811, 117833, 117839, 117841, 117851,
117877, 117881, 117883, 117889, 117899, 117911, 117917, 117937, 117959, 117973, 117977, 117979, 117989,
117991, 118033, 118037, 118043, 118051, 118057, 118061, 118081, 118093, 118127, 118147, 118163, 118169,
118171, 118189, 118211, 118213, 118219, 118247, 118249, 118253, 118259, 118273, 118277, 118297, 118343,
118361, 118369, 118373, 118387, 118399, 118409, 118411, 118423, 118429, 118453, 118457, 118463, 118471,
118493, 118529, 118543, 118549, 118571, 118583, 118589, 118603, 118619, 118621, 118633, 118661, 118669,
118673, 118681, 118687, 118691, 118709, 118717, 118739, 118747, 118751, 118757, 118787, 118799, 118801,
118819, 118831, 118843, 118861, 118873, 118891, 118897, 118901, 118903, 118907, 118913, 118927, 118931,
118967, 118973, 119027, 119033, 119039, 119047, 119057, 119069, 119083, 119087, 119089, 119099, 119101,
119107, 119129, 119131, 119159, 119173, 119179, 119183, 119191, 119227, 119233, 119237, 119243, 119267,
119291, 119293, 119297, 119299, 119311, 119321, 119359, 119363, 119389, 119417, 119419, 119429, 119447,
119489, 119503, 119513, 119533, 119549, 119551, 119557, 119563, 119569, 119591, 119611, 119617, 119627,
119633, 119653, 119657, 119659, 119671, 119677, 119687, 119689, 119699, 119701, 119723, 119737, 119747,
119759, 119771, 119773, 119783, 119797, 119809, 119813, 119827, 119831, 119839, 119849, 119851, 119869,
119881, 119891, 119921, 119923, 119929, 119953, 119963, 119971, 119981, 119983, 119993, 120011, 120017,
120041, 120047, 120049, 120067, 120077, 120079, 120091, 120097, 120103, 120121, 120157, 120163, 120167,
120181, 120193, 120199, 120209, 120223, 120233, 120247, 120277, 120283, 120293, 120299, 120319, 120331,
120349, 120371, 120383, 120391, 120397, 120401, 120413, 120427, 120431, 120473, 120503, 120511, 120539,
120551, 120557, 120563, 120569, 120577, 120587, 120607, 120619, 120623, 120641, 120647, 120661, 120671,
120677, 120689, 120691, 120709, 120713, 120721, 120737, 120739, 120749, 120763, 120767, 120779, 120811,
120817, 120823, 120829, 120833, 120847, 120851, 120863, 120871, 120877, 120889, 120899, 120907, 120917,
120919, 120929, 120937, 120941, 120943, 120947, 120977, 120997, 121001, 121007, 121013, 121019, 121021,
121039, 121061, 121063, 121067, 121081, 121123, 121139, 121151, 121157, 121169, 121171, 121181, 121189,
121229, 121259, 121267, 121271, 121283, 121291, 121309, 121313, 121321, 121327, 121333, 121343, 121349,
121351, 121357, 121367, 121369, 121379, 121403, 121421, 121439, 121441, 121447, 121453, 121469, 121487,
121493, 121501, 121507, 121523, 121531, 121547, 121553, 121559, 121571, 121577, 121579, 121591, 121607,
121609, 121621, 121631, 121633, 121637, 121661, 121687, 121697, 121711, 121721, 121727, 121763, 121787,
121789, 121843, 121853, 121867, 121883, 121889, 121909, 121921, 121931, 121937, 121949, 121951, 121963,
121967, 121993, 121997, 122011, 122021, 122027, 122029, 122033, 122039, 122041, 122051, 122053, 122069,
122081, 122099, 122117, 122131, 122147, 122149, 122167, 122173, 122201, 122203, 122207, 122209, 122219,
122231, 122251, 122263, 122267, 122273, 122279, 122299, 122321, 122323, 122327, 122347, 122363, 122387,
122389, 122393, 122399, 122401, 122443, 122449, 122453, 122471, 122477, 122489, 122497, 122501, 122503,
122509, 122527, 122533, 122557, 122561, 122579, 122597, 122599, 122609, 122611, 122651, 122653, 122663,
122693, 122701, 122719, 122741, 122743, 122753, 122761, 122777, 122789, 122819, 122827, 122833, 122839,
122849, 122861, 122867, 122869, 122887, 122891, 122921, 122929, 122939, 122953, 122957, 122963, 122971,
123001, 123007, 123017, 123031, 123049, 123059, 123077, 123083, 123091, 123113, 123121, 123127, 123143,
123169, 123191, 123203, 123209, 123217, 123229, 123239, 123259, 123269, 123289, 123307, 123311, 123323,
123341, 123373, 123377, 123379, 123397, 123401, 123407, 123419, 123427, 123433, 123439, 123449, 123457,
123479, 123491, 123493, 123499, 123503, 123517, 123527, 123547, 123551, 123553, 123581, 123583, 123593,
123601, 123619, 123631, 123637, 123653, 123661, 123667, 123677, 123701, 123707, 123719, 123727, 123731,
123733, 123737, 123757, 123787, 123791, 123803, 123817, 123821, 123829, 123833, 123853, 123863, 123887,
123911, 123923, 123931, 123941, 123953, 123973, 123979, 123983, 123989, 123997, 124001, 124021, 124067,
124087, 124097, 124121, 124123, 124133, 124139, 124147, 124153, 124171, 124181, 124183, 124193, 124199,

124213, 124231, 124247, 124249, 124277, 124291, 124297, 124301, 124303, 124309, 124337, 124339, 124343,
124349, 124351, 124363, 124367, 124427, 124429, 124433, 124447, 124459, 124471, 124477, 124489, 124493,
124513, 124529, 124541, 124543, 124561, 124567, 124577, 124601, 124633, 124643, 124669, 124673, 124679,
124693, 124699, 124703, 124717, 124721, 124739, 124753, 124759, 124769, 124771, 124777, 124781, 124783,
124793, 124799, 124819, 124823, 124847, 124853, 124897, 124907, 124909, 124919, 124951, 124979, 124981,
124987, 124991, 125003, 125017, 125029, 125053, 125063, 125093, 125101, 125107, 125113, 125117, 125119,
125131, 125141, 125149, 125183, 125197, 125201, 125207, 125219, 125221, 125231, 125243, 125261, 125269,
125287, 125299, 125303, 125311, 125329, 125339, 125353, 125371, 125383, 125387, 125399, 125407, 125423,
125429, 125441, 125453, 125471, 125497, 125507, 125509, 125527, 125539, 125551, 125591, 125597, 125617,
125621, 125627, 125639, 125641, 125651, 125659, 125669, 125683, 125687, 125693, 125707, 125711, 125717,
125731, 125737, 125743, 125753, 125777, 125789, 125791, 125803, 125813, 125821, 125863, 125887, 125897,
125899, 125921, 125927, 125929, 125933, 125941, 125959, 125963, 126001, 126011, 126013, 126019, 126023,
126031, 126037, 126041, 126047, 126067, 126079, 126097, 126107, 126127, 126131, 126143, 126151, 126173,
126199, 126211, 126223, 126227, 126229, 126233, 126241, 126257, 126271, 126307, 126311, 126317, 126323,
126337, 126341, 126349, 126359, 126397, 126421, 126433, 126443, 126457, 126461, 126473, 126481, 126487,
126491, 126493, 126499, 126517, 126541, 126547, 126551, 126583, 126601, 126611, 126613, 126631, 126641,
126653, 126683, 126691, 126703, 126713, 126719, 126733, 126739, 126743, 126751, 126757, 126761, 126781,
126823, 126827, 126839, 126851, 126857, 126859, 126913, 126923, 126943, 126949, 126961, 126967, 126989,
127031, 127033, 127037, 127051, 127079, 127081, 127103, 127123, 127133, 127139, 127157, 127163, 127189,
127207, 127217, 127219, 127241, 127247, 127249, 127261, 127271, 127277, 127289, 127291, 127297, 127301,
127321, 127331, 127343, 127363, 127373, 127399, 127403, 127423, 127447, 127453, 127481, 127487, 127493,
127507, 127529, 127541, 127549, 127579, 127583, 127591, 127597, 127601, 127607, 127609, 127637, 127643,
127649, 127657, 127663, 127669, 127679, 127681, 127691, 127703, 127709, 127711, 127717, 127727, 127733,
127739, 127747, 127763, 127781, 127807, 127817, 127819, 127837, 127843, 127849, 127859, 127867, 127873,
127877, 127913, 127921, 127931, 127951, 127973, 127979, 127997, 128021, 128033, 128047, 128053, 128099,
128111, 128113, 128119, 128147, 128153, 128159, 128173, 128189, 128201, 128203, 128213, 128221, 128237,
128239, 128257, 128273, 128287, 128291, 128311, 128321, 128327, 128339, 128341, 128347, 128351, 128377,
128389, 128393, 128399, 128411, 128413, 128431, 128437, 128449, 128461, 128467, 128473, 128477, 128483,
128489, 128509, 128519, 128521, 128549, 128551, 128563, 128591, 128599, 128603, 128621, 128629, 128657,
128659, 128663, 128669, 128677, 128683, 128693, 128717, 128747, 128749, 128761, 128767, 128813, 128819,
128831, 128833, 128837, 128857, 128861, 128873, 128879, 128903, 128923, 128939, 128941, 128951, 128959,
128969, 128971, 128981, 128983, 128987, 128993, 129001, 129011, 129023, 129037, 129049, 129061, 129083,
129089, 129097, 129113, 129119, 129121, 129127, 129169, 129187, 129193, 129197, 129209, 129221, 129223,
129229, 129263, 129277, 129281, 129287, 129289, 129293, 129313, 129341, 129347, 129361, 129379, 129401,
129403, 129419, 129439, 129443, 129449, 129457, 129461, 129469, 129491, 129497, 129499, 129509, 129517,
129527, 129529, 129533, 129539, 129553, 129581, 129587, 129589, 129593, 129607, 129629, 129631, 129641,
129643, 129671, 129707, 129719, 129733, 129737, 129749, 129757, 129763, 129769, 129793, 129803, 129841,
129853, 129887, 129893, 129901, 129917, 129919, 129937, 129953, 129959, 129967, 129971, 130003, 130021,
130027, 130043, 130051, 130057, 130069, 130073, 130079, 130087, 130099, 130121, 130127, 130147, 130171,
130183, 130199, 130201, 130211, 130223, 130241, 130253, 130259, 130261, 130267, 130279, 130303, 130307,
130337, 130343, 130349, 130363, 130367, 130369, 130379, 130399, 130409, 130411, 130423, 130439, 130447,
130457, 130469, 130477, 130483, 130489, 130513, 130517, 130523, 130531, 130547, 130553, 130579, 130589,
130619, 130621, 130631, 130633, 130639, 130643, 130649, 130651, 130657, 130681, 130687, 130693, 130699,
130729, 130769, 130783, 130787, 130807, 130811, 130817, 130829, 130841, 130843, 130859, 130873, 130927,
130957, 130969, 130973, 130981, 130987, 131009, 131011, 131023, 131041, 131059, 131063, 131071, 131101,
131111, 131113, 131129, 131143, 131149, 131171, 131203, 131213, 131221, 131231, 131249, 131251, 131267,
131293, 131297, 131303, 131311, 131317, 131321, 131357, 131363, 131371, 131381, 131413, 131431, 131437,
131441, 131447, 131449, 131477, 131479, 131489, 131497, 131501, 131507, 131519, 131543, 131561, 131581,
131591, 131611, 131617, 131627, 131639, 131641, 131671, 131687, 131701, 131707, 131711, 131713, 131731,

131743, 131749, 131759, 131771, 131777, 131779, 131783, 131797, 131837, 131839, 131849, 131861, 131891,
131893, 131899, 131909, 131927, 131933, 131939, 131941, 131947, 131959, 131969, 132001, 132019, 132047,
132049, 132059, 132071, 132103, 132109, 132113, 132137, 132151, 132157, 132169, 132173, 132199, 132229,
132233, 132241, 132247, 132257, 132263, 132283, 132287, 132299, 132313, 132329, 132331, 132347, 132361,
132367, 132371, 132383, 132403, 132409, 132421, 132437, 132439, 132469, 132491, 132499, 132511, 132523,
132527, 132529, 132533, 132541, 132547, 132589, 132607, 132611, 132619, 132623, 132631, 132637, 132647,
132661, 132667, 132679, 132689, 132697, 132701, 132707, 132709, 132721, 132739, 132749, 132751, 132757,
132761, 132763, 132817, 132833, 132851, 132857, 132859, 132863, 132887, 132893, 132911, 132929, 132947,
132949, 132953, 132961, 132967, 132971, 132989, 133013, 133033, 133039, 133051, 133069, 133073, 133087,
133097, 133103, 133109, 133117, 133121, 133153, 133157, 133169, 133183, 133187, 133201, 133213, 133241,
133253, 133261, 133271, 133277, 133279, 133283, 133303, 133319, 133321, 133327, 133337, 133349, 133351,
133379, 133387, 133391, 133403, 133417, 133439, 133447, 133451, 133481, 133493, 133499, 133519, 133541,
133543, 133559, 133571, 133583, 133597, 133631, 133633, 133649, 133657, 133669, 133673, 133691, 133697,
133709, 133711, 133717, 133723, 133733, 133769, 133781, 133801, 133811, 133813, 133831, 133843, 133853,
133873, 133877, 133919, 133949, 133963, 133967, 133979, 133981, 133993, 133999, 134033, 134039, 134047,
134053, 134059, 134077, 134081, 134087, 134089, 134093, 134129, 134153, 134161, 134171, 134177, 134191,
134207, 134213, 134219, 134227, 134243, 134257, 134263, 134269, 134287, 134291, 134293, 134327, 134333,
134339, 134341, 134353, 134359, 134363, 134369, 134371, 134399, 134401, 134417, 134437, 134443, 134471,
134489, 134503, 134507, 134513, 134581, 134587, 134591, 134593, 134597, 134609, 134639, 134669, 134677,
134681, 134683, 134699, 134707, 134731, 134741, 134753, 134777, 134789, 134807, 134837, 134839, 134851,
134857, 134867, 134873, 134887, 134909, 134917, 134921, 134923, 134947, 134951, 134989, 134999, 135007,
135017, 135019, 135029, 135043, 135049, 135059, 135077, 135089, 135101, 135119, 135131, 135151, 135173,
135181, 135193, 135197, 135209, 135211, 135221, 135241, 135257, 135271, 135277, 135281, 135283, 135301,
135319, 135329, 135347, 135349, 135353, 135367, 135389, 135391, 135403, 135409, 135427, 135431, 135433,
135449, 135461, 135463, 135467, 135469, 135479, 135497, 135511, 135533, 135559, 135571, 135581, 135589,
135593, 135599, 135601, 135607, 135613, 135617, 135623, 135637, 135647, 135649, 135661, 135671, 135697,
135701, 135719, 135721, 135727, 135731, 135743, 135757, 135781, 135787, 135799, 135829, 135841, 135851,
135859, 135887, 135893, 135899, 135911, 135913, 135929, 135937, 135977, 135979, 136013, 136027, 136033,
136043, 136057, 136067, 136069, 136093, 136099, 136111, 136133, 136139, 136163, 136177, 136189, 136193,
136207, 136217, 136223, 136237, 136247, 136261, 136273, 136277, 136303, 136309, 136319, 136327, 136333,
136337, 136343, 136351, 136361, 136373, 136379, 136393, 136397, 136399, 136403, 136417, 136421, 136429,
136447, 136453, 136463, 136471, 136481, 136483, 136501, 136511, 136519, 136523, 136531, 136537, 136541,
136547, 136559, 136573, 136601, 136603, 136607, 136621, 136649, 136651, 136657, 136691, 136693, 136709,
136711, 136727, 136733, 136739, 136751, 136753, 136769, 136777, 136811, 136813, 136841, 136849, 136859,
136861, 136879, 136883, 136889, 136897, 136943, 136949, 136951, 136963, 136973, 136979, 136987, 136991,
136993, 136999, 137029, 137077, 137087, 137089, 137117, 137119, 137131, 137143, 137147, 137153, 137177,
137183, 137191, 137197, 137201, 137209, 137219, 137239, 137251, 137273, 137279, 137303, 137321, 137339,
137341, 137353, 137359, 137363, 137369, 137383, 137387, 137393, 137399, 137413, 137437, 137443, 137447,
137453, 137477, 137483, 137491, 137507, 137519, 137537, 137567, 137573, 137587, 137593, 137597, 137623,
137633, 137639, 137653, 137659, 137699, 137707, 137713, 137723, 137737, 137743, 137771, 137777, 137791,
137803, 137827, 137831, 137849, 137867, 137869, 137873, 137909, 137911, 137927, 137933, 137941, 137947,
137957, 137983, 137993, 137999, 138007, 138041, 138053, 138059, 138071, 138077, 138079, 138101, 138107,
138113, 138139, 138143, 138157, 138163, 138179, 138181, 138191, 138197, 138209, 138239, 138241, 138247,
138251, 138283, 138289, 138311, 138319, 138323, 138337, 138349, 138371, 138373, 138389, 138401, 138403,
138407, 138427, 138433, 138449, 138451, 138461, 138469, 138493, 138497, 138511, 138517, 138547, 138559,
138563, 138569, 138571, 138577, 138581, 138587, 138599, 138617, 138629, 138637, 138641, 138647, 138661,
138679, 138683, 138727, 138731, 138739, 138763, 138793, 138797, 138799, 138821, 138829, 138841, 138863,
138869, 138883, 138889, 138893, 138899, 138917, 138923, 138937, 138959, 138967, 138977, 139021, 139033,
139067, 139079, 139091, 139109, 139121, 139123, 139133, 139169, 139177, 139187, 139199, 139201, 139241,

139267, 139273, 139291, 139297, 139301, 139303, 139309, 139313, 139333, 139339, 139343, 139361, 139367,
139369, 139387, 139393, 139397, 139409, 139423, 139429, 139439, 139457, 139459, 139483, 139487, 139493,
139501, 139511, 139537, 139547, 139571, 139589, 139591, 139597, 139609, 139619, 139627, 139661, 139663,
139681, 139697, 139703, 139709, 139721, 139729, 139739, 139747, 139753, 139759, 139787, 139801, 139813,
139831, 139837, 139861, 139871, 139883, 139891, 139901, 139907, 139921, 139939, 139943, 139967, 139969,
139981, 139987, 139991, 139999, 140009, 140053, 140057, 140069, 140071, 140111, 140123, 140143, 140159,
140167, 140171, 140177, 140191, 140197, 140207, 140221, 140227, 140237, 140249, 140263, 140269, 140281,
140297, 140317, 140321, 140333, 140339, 140351, 140363, 140381, 140401, 140407, 140411, 140417, 140419,
140423, 140443, 140449, 140453, 140473, 140477, 140521, 140527, 140533, 140549, 140551, 140557, 140587,
140593, 140603, 140611, 140617, 140627, 140629, 140639, 140659, 140663, 140677, 140681, 140683, 140689,
140717, 140729, 140731, 140741, 140759, 140761, 140773, 140779, 140797, 140813, 140827, 140831, 140837,
140839, 140863, 140867, 140869, 140891, 140893, 140897, 140909, 140929, 140939, 140977, 140983, 140989,
141023, 141041, 141061, 141067, 141073, 141079, 141101, 141107, 141121, 141131, 141157, 141161, 141179,
141181, 141199, 141209, 141221, 141223, 141233, 141241, 141257, 141263, 141269, 141277, 141283, 141301,
141307, 141311, 141319, 141353, 141359, 141371, 141397, 141403, 141413, 141439, 141443, 141461, 141481,
141497, 141499, 141509, 141511, 141529, 141539, 141551, 141587, 141601, 141613, 141619, 141623, 141629,
141637, 141649, 141653, 141667, 141671, 141677, 141679, 141689, 141697, 141707, 141709, 141719, 141731,
141761, 141767, 141769, 141773, 141793, 141803, 141811, 141829, 141833, 141851, 141853, 141863, 141871,
141907, 141917, 141931, 141937, 141941, 141959, 141961, 141971, 141991, 142007, 142019, 142031, 142039,
142049, 142057, 142061, 142067, 142097, 142099, 142111, 142123, 142151, 142157, 142159, 142169, 142183,
142189, 142193, 142211, 142217, 142223, 142231, 142237, 142271, 142297, 142319, 142327, 142357, 142369,
142381, 142391, 142403, 142421, 142427, 142433, 142453, 142469, 142501, 142529, 142537, 142543, 142547,
142553, 142559, 142567, 142573, 142589, 142591, 142601, 142607, 142609, 142619, 142657, 142673, 142697,
142699, 142711, 142733, 142757, 142759, 142771, 142787, 142789, 142799, 142811, 142837, 142841, 142867,
142871, 142873, 142897, 142903, 142907, 142939, 142949, 142963, 142969, 142973, 142979, 142981, 142993,
143053, 143063, 143093, 143107, 143111, 143113, 143137, 143141, 143159, 143177, 143197, 143239, 143243,
143249, 143257, 143261, 143263, 143281, 143287, 143291, 143329, 143333, 143357, 143387, 143401, 143413,
143419, 143443, 143461, 143467, 143477, 143483, 143489, 143501, 143503, 143509, 143513, 143519, 143527,
143537, 143551, 143567, 143569, 143573, 143593, 143609, 143617, 143629, 143651, 143653, 143669, 143677,
143687, 143699, 143711, 143719, 143729, 143743, 143779, 143791, 143797, 143807, 143813, 143821, 143827,
143831, 143833, 143873, 143879, 143881, 143909, 143947, 143953, 143971, 143977, 143981, 143999, 144013,
144031, 144037, 144061, 144071, 144073, 144103, 144139, 144161, 144163, 144167, 144169, 144173, 144203,
144223, 144241, 144247, 144253, 144259, 144271, 144289, 144299, 144307, 144311, 144323, 144341, 144349,
144379, 144383, 144407, 144409, 144413, 144427, 144439, 144451, 144461, 144479, 144481, 144497, 144511,
144539, 144541, 144563, 144569, 144577, 144583, 144589, 144593, 144611, 144629, 144659, 144667, 144671,
144701, 144709, 144719, 144731, 144737, 144751, 144757, 144763, 144773, 144779, 144791, 144817, 144829,
144839, 144847, 144883, 144887, 144889, 144899, 144917, 144931, 144941, 144961, 144967, 144973, 144983,
145007, 145009, 145021, 145031, 145037, 145043, 145063, 145069, 145091, 145109, 145121, 145133, 145139,
145177, 145193, 145207, 145213, 145219, 145253, 145259, 145267, 145283, 145289, 145303, 145307, 145349,
145361, 145381, 145391, 145399, 145417, 145423, 145433, 145441, 145451, 145459, 145463, 145471, 145477,
145487, 145501, 145511, 145513, 145517, 145531, 145543, 145547, 145549, 145577, 145589, 145601, 145603,
145633, 145637, 145643, 145661, 145679, 145681, 145687, 145703, 145709, 145721, 145723, 145753, 145757,
145759, 145771, 145777, 145799, 145807, 145819, 145823, 145829, 145861, 145879, 145897, 145903, 145931,
145933, 145949, 145963, 145967, 145969, 145987, 145991, 146009, 146011, 146021, 146023, 146033, 146051,
146057, 146059, 146063, 146077, 146093, 146099, 146117, 146141, 146161, 146173, 146191, 146197, 146203,
146213, 146221, 146239, 146249, 146273, 146291, 146297, 146299, 146309, 146317, 146323, 146347, 146359,
146369, 146381, 146383, 146389, 146407, 146417, 146423, 146437, 146449, 146477, 146513, 146519, 146521,
146527, 146539, 146543, 146563, 146581, 146603, 146609, 146617, 146639, 146647, 146669, 146677, 146681,
146683, 146701, 146719, 146743, 146749, 146767, 146777, 146801, 146807, 146819, 146833, 146837, 146843,

146849, 146857, 146891, 146893, 146917, 146921, 146933, 146941, 146953, 146977, 146983, 146987, 146989,
147011, 147029, 147031, 147047, 147073, 147083, 147089, 147097, 147107, 147137, 147139, 147151, 147163,
147179, 147197, 147209, 147211, 147221, 147227, 147229, 147253, 147263, 147283, 147289, 147293, 147299,
147311, 147319, 147331, 147341, 147347, 147353, 147377, 147391, 147397, 147401, 147409, 147419, 147449,
147451, 147457, 147481, 147487, 147503, 147517, 147541, 147547, 147551, 147557, 147571, 147583, 147607,
147613, 147617, 147629, 147647, 147661, 147671, 147673, 147689, 147703, 147709, 147727, 147739, 147743,
147761, 147769, 147773, 147779, 147787, 147793, 147799, 147811, 147827, 147853, 147859, 147863, 147881,
147919, 147937, 147949, 147977, 147997, 148013, 148021, 148061, 148063, 148073, 148079, 148091, 148123,
148139, 148147, 148151, 148153, 148157, 148171, 148193, 148199, 148201, 148207, 148229, 148243, 148249,
148279, 148301, 148303, 148331, 148339, 148361, 148367, 148381, 148387, 148399, 148403, 148411, 148429,
148439, 148457, 148469, 148471, 148483, 148501, 148513, 148517, 148531, 148537, 148549, 148573, 148579,
148609, 148627, 148633, 148639, 148663, 148667, 148669, 148691, 148693, 148711, 148721, 148723, 148727,
148747, 148763, 148781, 148783, 148793, 148817, 148829, 148853, 148859, 148861, 148867, 148873, 148891,
148913, 148921, 148927, 148931, 148933, 148949, 148957, 148961, 148991, 148997, 149011, 149021, 149027,
149033, 149053, 149057, 149059, 149069, 149077, 149087, 149099, 149101, 149111, 149113, 149119, 149143,
149153, 149159, 149161, 149173, 149183, 149197, 149213, 149239, 149249, 149251, 149257, 149269, 149287,
149297, 149309, 149323, 149333, 149341, 149351, 149371, 149377, 149381, 149393, 149399, 149411, 149417,
149419, 149423, 149441, 149459, 149489, 149491, 149497, 149503, 149519, 149521, 149531, 149533, 149543,
149551, 149561, 149563, 149579, 149603, 149623, 149627, 149629, 149689, 149711, 149713, 149717, 149729,
149731, 149749, 149759, 149767, 149771, 149791, 149803, 149827, 149837, 149839, 149861, 149867, 149873,
149893, 149899, 149909, 149911, 149921, 149939, 149953, 149969, 149971, 149993, 150001, 150011, 150041,
150053, 150061, 150067, 150077, 150083, 150089, 150091, 150097, 150107, 150131, 150151, 150169, 150193,
150197, 150203, 150209, 150211, 150217, 150221, 150223, 150239, 150247, 150287, 150299, 150301, 150323,
150329, 150343, 150373, 150377, 150379, 150383, 150401, 150407, 150413, 150427, 150431, 150439, 150473,
150497, 150503, 150517, 150523, 150533, 150551, 150559, 150571, 150583, 150587, 150589, 150607, 150611,
150617, 150649, 150659, 150697, 150707, 150721, 150743, 150767, 150769, 150779, 150791, 150797, 150827,
150833, 150847, 150869, 150881, 150883, 150889, 150893, 150901, 150907, 150919, 150929, 150959, 150961,
150967, 150979, 150989, 150991, 151007, 151009, 151013, 151027, 151049, 151051, 151057, 151091, 151121,
151141, 151153, 151157, 151163, 151169, 151171, 151189, 151201, 151213, 151237, 151241, 151243, 151247,
151253, 151273, 151279, 151289, 151303, 151337, 151339, 151343, 151357, 151379, 151381, 151391, 151397,
151423, 151429, 151433, 151451, 151471, 151477, 151483, 151499, 151507, 151517, 151523, 151531, 151537,
151549, 151553, 151561, 151573, 151579, 151597, 151603, 151607, 151609, 151631, 151637, 151643, 151651,
151667, 151673, 151681, 151687, 151693, 151703, 151717, 151729, 151733, 151769, 151771, 151783, 151787,
151799, 151813, 151817, 151841, 151847, 151849, 151871, 151883, 151897, 151901, 151903, 151909, 151937,
151939, 151967, 151969, 152003, 152017, 152027, 152029, 152039, 152041, 152063, 152077, 152081, 152083,
152093, 152111, 152123, 152147, 152183, 152189, 152197, 152203, 152213, 152219, 152231, 152239, 152249,
152267, 152287, 152293, 152297, 152311, 152363, 152377, 152381, 152389, 152393, 152407, 152417, 152419,
152423, 152429, 152441, 152443, 152459, 152461, 152501, 152519, 152531, 152533, 152539, 152563, 152567,
152597, 152599, 152617, 152623, 152629, 152639, 152641, 152657, 152671, 152681, 152717, 152723, 152729,
152753, 152767, 152777, 152783, 152791, 152809, 152819, 152821, 152833, 152837, 152839, 152843, 152851,
152857, 152879, 152897, 152899, 152909, 152939, 152941, 152947, 152953, 152959, 152981, 152989, 152993,
153001, 153059, 153067, 153071, 153073, 153077, 153089, 153107, 153113, 153133, 153137, 153151, 153191,
153247, 153259, 153269, 153271, 153277, 153281, 153287, 153313, 153319, 153337, 153343, 153353, 153359,
153371, 153379, 153407, 153409, 153421, 153427, 153437, 153443, 153449, 153457, 153469, 153487, 153499,
153509, 153511, 153521, 153523, 153529, 153533, 153557, 153563, 153589, 153607, 153611, 153623, 153641,
153649, 153689, 153701, 153719, 153733, 153739, 153743, 153749, 153757, 153763, 153817, 153841, 153871,
153877, 153887, 153889, 153911, 153913, 153929, 153941, 153947, 153949, 153953, 153991, 153997, 154001,
154027, 154043, 154057, 154061, 154067, 154073, 154079, 154081, 154087, 154097, 154111, 154127, 154153,
154157, 154159, 154181, 154183, 154211, 154213, 154229, 154243, 154247, 154267, 154277, 154279, 154291,

154303, 154313, 154321, 154333, 154339, 154351, 154369, 154373, 154387, 154409, 154417, 154423, 154439,
154459, 154487, 154493, 154501, 154523, 154543, 154571, 154573, 154579, 154589, 154591, 154613, 154619,
154621, 154643, 154667, 154669, 154681, 154691, 154699, 154723, 154727, 154733, 154747, 154753, 154769,
154787, 154789, 154799, 154807, 154823, 154841, 154849, 154871, 154873, 154877, 154883, 154897, 154927,
154933, 154937, 154943, 154981, 154991, 155003, 155009, 155017, 155027, 155047, 155069, 155081, 155083,
155087, 155119, 155137, 155153, 155161, 155167, 155171, 155191, 155201, 155203, 155209, 155219, 155231,
155251, 155269, 155291, 155299, 155303, 155317, 155327, 155333, 155371, 155377, 155381, 155383, 155387,
155399, 155413, 155423, 155443, 155453, 155461, 155473, 155501, 155509, 155521, 155537, 155539, 155557,
155569, 155579, 155581, 155593, 155599, 155609, 155621, 155627, 155653, 155657, 155663, 155671, 155689,
155693, 155699, 155707, 155717, 155719, 155723, 155731, 155741, 155747, 155773, 155777, 155783, 155797,
155801, 155809, 155821, 155833, 155849, 155851, 155861, 155863, 155887, 155891, 155893, 155921, 156007,
156011, 156019, 156041, 156059, 156061, 156071, 156089, 156109, 156119, 156127, 156131, 156139, 156151,
156157, 156217, 156227, 156229, 156241, 156253, 156257, 156259, 156269, 156307, 156319, 156329, 156347,
156353, 156361, 156371, 156419, 156421, 156437, 156467, 156487, 156491, 156493, 156511, 156521, 156539,
156577, 156589, 156593, 156601, 156619, 156623, 156631, 156641, 156659, 156671, 156677, 156679, 156683,
156691, 156703, 156707, 156719, 156727, 156733, 156749, 156781, 156797, 156799, 156817, 156823, 156833,
156841, 156887, 156899, 156901, 156913, 156941, 156943, 156967, 156971, 156979, 157007, 157013, 157019,
157037, 157049, 157051, 157057, 157061, 157081, 157103, 157109, 157127, 157133, 157141, 157163, 157177,
157181, 157189, 157207, 157211, 157217, 157219, 157229, 157231, 157243, 157247, 157253, 157259, 157271,
157273, 157277, 157279, 157291, 157303, 157307, 157321, 157327, 157349, 157351, 157363, 157393, 157411,
157427, 157429, 157433, 157457, 157477, 157483, 157489, 157513, 157519, 157523, 157543, 157559, 157561,
157571, 157579, 157627, 157637, 157639, 157649, 157667, 157669, 157679, 157721, 157733, 157739, 157747,
157769, 157771, 157793, 157799, 157813, 157823, 157831, 157837, 157841, 157867, 157877, 157889, 157897,
157901, 157907, 157931, 157933, 157951, 157991, 157999, 158003, 158009, 158017, 158029, 158047, 158071,
158077, 158113, 158129, 158141, 158143, 158161, 158189, 158201, 158209, 158227, 158231, 158233, 158243,
158261, 158269, 158293, 158303, 158329, 158341, 158351, 158357, 158359, 158363, 158371, 158393, 158407,
158419, 158429, 158443, 158449, 158489, 158507, 158519, 158527, 158537, 158551, 158563, 158567, 158573,
158581, 158591, 158597, 158611, 158617, 158621, 158633, 158647, 158657, 158663, 158699, 158731, 158747,
158749, 158759, 158761, 158771, 158777, 158791, 158803, 158843, 158849, 158863, 158867, 158881, 158909,
158923, 158927, 158941, 158959, 158981, 158993, 159013, 159017, 159023, 159059, 159073, 159079, 159097,
159113, 159119, 159157, 159161, 159167, 159169, 159179, 159191, 159193, 159199, 159209, 159223, 159227,
159233, 159287, 159293, 159311, 159319, 159337, 159347, 159349, 159361, 159389, 159403, 159407, 159421,
159431, 159437, 159457, 159463, 159469, 159473, 159491, 159499, 159503, 159521, 159539, 159541, 159553,
159563, 159569, 159571, 159589, 159617, 159623, 159629, 159631, 159667, 159671, 159673, 159683, 159697,
159701, 159707, 159721, 159737, 159739, 159763, 159769, 159773, 159779, 159787, 159791, 159793, 159799,
159811, 159833, 159839, 159853, 159857, 159869, 159871, 159899, 159911, 159931, 159937, 159977, 159979,
160001, 160009, 160019, 160031, 160033, 160049, 160073, 160079, 160081, 160087, 160091, 160093, 160117,
160141, 160159, 160163, 160169, 160183, 160201, 160207, 160217, 160231, 160243, 160253, 160309, 160313,
160319, 160343, 160357, 160367, 160373, 160387, 160397, 160403, 160409, 160423, 160441, 160453, 160481,
160483, 160499, 160507, 160541, 160553, 160579, 160583, 160591, 160603, 160619, 160621, 160627, 160637,
160639, 160649, 160651, 160663, 160669, 160681, 160687, 160697, 160709, 160711, 160723, 160739, 160751,
160753, 160757, 160781, 160789, 160807, 160813, 160817, 160829, 160841, 160861, 160877, 160879, 160883,
160903, 160907, 160933, 160967, 160969, 160981, 160997, 161009, 161017, 161033, 161039, 161047, 161053,
161059, 161071, 161087, 161093, 161123, 161137, 161141, 161149, 161159, 161167, 161201, 161221, 161233,
161237, 161263, 161267, 161281, 161303, 161309, 161323, 161333, 161339, 161341, 161363, 161377, 161387,
161407, 161411, 161453, 161459, 161461, 161471, 161503, 161507, 161521, 161527, 161531, 161543, 161561,
161563, 161569, 161573, 161591, 161599, 161611, 161627, 161639, 161641, 161659, 161683, 161717, 161729,
161731, 161741, 161743, 161753, 161761, 161771, 161773, 161779, 161783, 161807, 161831, 161839, 161869,
161873, 161879, 161881, 161911, 161921, 161923, 161947, 161957, 161969, 161971, 161977, 161983, 161999,

162007, 162011, 162017, 162053, 162059, 162079, 162091, 162109, 162119, 162143, 162209, 162221, 162229,
162251, 162257, 162263, 162269, 162277, 162287, 162289, 162293, 162343, 162359, 162389, 162391, 162413,
162419, 162439, 162451, 162457, 162473, 162493, 162499, 162517, 162523, 162527, 162529, 162553, 162557,
162563, 162577, 162593, 162601, 162611, 162623, 162629, 162641, 162649, 162671, 162677, 162683, 162691,
162703, 162709, 162713, 162727, 162731, 162739, 162749, 162751, 162779, 162787, 162791, 162821, 162823,
162829, 162839, 162847, 162853, 162859, 162881, 162889, 162901, 162907, 162917, 162937, 162947, 162971,
162973, 162989, 162997, 163003, 163019, 163021, 163027, 163061, 163063, 163109, 163117, 163127, 163129,
163147, 163151, 163169, 163171, 163181, 163193, 163199, 163211, 163223, 163243, 163249, 163259, 163307,
163309, 163321, 163327, 163337, 163351, 163363, 163367, 163393, 163403, 163409, 163411, 163417, 163433,
163469, 163477, 163481, 163483, 163487, 163517, 163543, 163561, 163567, 163573, 163601, 163613, 163621,
163627, 163633, 163637, 163643, 163661, 163673, 163679, 163697, 163729, 163733, 163741, 163753, 163771,
163781, 163789, 163811, 163819, 163841, 163847, 163853, 163859, 163861, 163871, 163883, 163901, 163909,
163927, 163973, 163979, 163981, 163987, 163991, 163993, 163997, 164011, 164023, 164039, 164051, 164057,
164071, 164089, 164093, 164113, 164117, 164147, 164149, 164173, 164183, 164191, 164201, 164209, 164231,
164233, 164239, 164249, 164251, 164267, 164279, 164291, 164299, 164309, 164321, 164341, 164357, 164363,
164371, 164377, 164387, 164413, 164419, 164429, 164431, 164443, 164447, 164449, 164471, 164477, 164503,
164513, 164531, 164569, 164581, 164587, 164599, 164617, 164621, 164623, 164627, 164653, 164663, 164677,
164683, 164701, 164707, 164729, 164743, 164767, 164771, 164789, 164809, 164821, 164831, 164837, 164839,
164881, 164893, 164911, 164953, 164963, 164987, 164999, 165001, 165037, 165041, 165047, 165049, 165059,
165079, 165083, 165089, 165103, 165133, 165161, 165173, 165181, 165203, 165211, 165229, 165233, 165247,
165287, 165293, 165311, 165313, 165317, 165331, 165343, 165349, 165367, 165379, 165383, 165391, 165397,
165437, 165443, 165449, 165457, 165463, 165469, 165479, 165511, 165523, 165527, 165533, 165541, 165551,
165553, 165559, 165569, 165587, 165589, 165601, 165611, 165617, 165653, 165667, 165673, 165701, 165703,
165707, 165709, 165713, 165719, 165721, 165749, 165779, 165799, 165811, 165817, 165829, 165833, 165857,
165877, 165883, 165887, 165901, 165931, 165941, 165947, 165961, 165983, 166013, 166021, 166027, 166031,
166043, 166063, 166081, 166099, 166147, 166151, 166157, 166169, 166183, 166189, 166207, 166219, 166237,
166247, 166259, 166273, 166289, 166297, 166301, 166303, 166319, 166349, 166351, 166357, 166363, 166393,
166399, 166403, 166409, 166417, 166429, 166457, 166471, 166487, 166541, 166561, 166567, 166571, 166597,
166601, 166603, 166609, 166613, 166619, 166627, 166631, 166643, 166657, 166667, 166669, 166679, 166693,
166703, 166723, 166739, 166741, 166781, 166783, 166799, 166807, 166823, 166841, 166843, 166847, 166849,
166853, 166861, 166867, 166871, 166909, 166919, 166931, 166949, 166967, 166973, 166979, 166987, 167009,
167017, 167021, 167023, 167033, 167039, 167047, 167051, 167071, 167077, 167081, 167087, 167099, 167107,
167113, 167117, 167119, 167149, 167159, 167173, 167177, 167191, 167197, 167213, 167221, 167249, 167261,
167267, 167269, 167309, 167311, 167317, 167329, 167339, 167341, 167381, 167393, 167407, 167413, 167423,
167429, 167437, 167441, 167443, 167449, 167471, 167483, 167491, 167521, 167537, 167543, 167593, 167597,
167611, 167621, 167623, 167627, 167633, 167641, 167663, 167677, 167683, 167711, 167729, 167747, 167759,
167771, 167777, 167779, 167801, 167809, 167861, 167863, 167873, 167879, 167887, 167891, 167899, 167911,
167917, 167953, 167971, 167987, 168013, 168023, 168029, 168037, 168043, 168067, 168071, 168083, 168089,
168109, 168127, 168143, 168151, 168193, 168197, 168211, 168227, 168247, 168253, 168263, 168269, 168277,
168281, 168293, 168323, 168331, 168347, 168353, 168391, 168409, 168433, 168449, 168451, 168457, 168463,
168481, 168491, 168499, 168523, 168527, 168533, 168541, 168559, 168599, 168601, 168617, 168629, 168631,
168643, 168673, 168677, 168697, 168713, 168719, 168731, 168737, 168743, 168761, 168769, 168781, 168803,
168851, 168863, 168869, 168887, 168893, 168899, 168901, 168913, 168937, 168943, 168977, 168991, 169003,
169007, 169009, 169019, 169049, 169063, 169067, 169069, 169079, 169093, 169097, 169111, 169129, 169151,
169159, 169177, 169181, 169199, 169217, 169219, 169241, 169243, 169249, 169259, 169283, 169307, 169313,
169319, 169321, 169327, 169339, 169343, 169361, 169369, 169373, 169399, 169409, 169427, 169457, 169471,
169483, 169489, 169493, 169501, 169523, 169531, 169553, 169567, 169583, 169591, 169607, 169627, 169633,
169639, 169649, 169657, 169661, 169667, 169681, 169691, 169693, 169709, 169733, 169751, 169753, 169769,
169777, 169783, 169789, 169817, 169823, 169831, 169837, 169843, 169859, 169889, 169891, 169909, 169913,

169919, 169933, 169937, 169943, 169951, 169957, 169987, 169991, 170003, 170021, 170029, 170047, 170057,
170063, 170081, 170099, 170101, 170111, 170123, 170141, 170167, 170179, 170189, 170197, 170207, 170213,
170227, 170231, 170239, 170243, 170249, 170263, 170267, 170279, 170293, 170299, 170327, 170341, 170347,
170351, 170353, 170363, 170369, 170371, 170383, 170389, 170393, 170413, 170441, 170447, 170473, 170483,
170497, 170503, 170509, 170537, 170539, 170551, 170557, 170579, 170603, 170609, 170627, 170633, 170641,
170647, 170669, 170689, 170701, 170707, 170711, 170741, 170749, 170759, 170761, 170767, 170773, 170777,
170801, 170809, 170813, 170827, 170837, 170843, 170851, 170857, 170873, 170881, 170887, 170899, 170921,
170927, 170953, 170957, 170971, 171007, 171023, 171029, 171043, 171047, 171049, 171053, 171077, 171079,
171091, 171103, 171131, 171161, 171163, 171167, 171169, 171179, 171203, 171233, 171251, 171253, 171263,
171271, 171293, 171299, 171317, 171329, 171341, 171383, 171401, 171403, 171427, 171439, 171449, 171467,
171469, 171473, 171481, 171491, 171517, 171529, 171539, 171541, 171553, 171559, 171571, 171583, 171617,
171629, 171637, 171641, 171653, 171659, 171671, 171673, 171679, 171697, 171707, 171713, 171719, 171733,
171757, 171761, 171763, 171793, 171799, 171803, 171811, 171823, 171827, 171851, 171863, 171869, 171877,
171881, 171889, 171917, 171923, 171929, 171937, 171947, 172001, 172009, 172021, 172027, 172031, 172049,
172069, 172079, 172093, 172097, 172127, 172147, 172153, 172157, 172169, 172171, 172181, 172199, 172213,
172217, 172219, 172223, 172243, 172259, 172279, 172283, 172297, 172307, 172313, 172321, 172331, 172343,
172351, 172357, 172373, 172399, 172411, 172421, 172423, 172427, 172433, 172439, 172441, 172489, 172507,
172517, 172519, 172541, 172553, 172561, 172573, 172583, 172589, 172597, 172603, 172607, 172619, 172633,
172643, 172649, 172657, 172663, 172673, 172681, 172687, 172709, 172717, 172721, 172741, 172751, 172759,
172787, 172801, 172807, 172829, 172849, 172853, 172859, 172867, 172871, 172877, 172883, 172933, 172969,
172973, 172981, 172987, 172993, 172999, 173021, 173023, 173039, 173053, 173059, 173081, 173087, 173099,
173137, 173141, 173149, 173177, 173183, 173189, 173191, 173207, 173209, 173219, 173249, 173263, 173267,
173273, 173291, 173293, 173297, 173309, 173347, 173357, 173359, 173429, 173431, 173473, 173483, 173491,
173497, 173501, 173531, 173539, 173543, 173549, 173561, 173573, 173599, 173617, 173629, 173647, 173651,
173659, 173669, 173671, 173683, 173687, 173699, 173707, 173713, 173729, 173741, 173743, 173773, 173777,
173779, 173783, 173807, 173819, 173827, 173839, 173851, 173861, 173867, 173891, 173897, 173909, 173917,
173923, 173933, 173969, 173977, 173981, 173993, 174007, 174017, 174019, 174047, 174049, 174061, 174067,
174071, 174077, 174079, 174091, 174101, 174121, 174137, 174143, 174149, 174157, 174169, 174197, 174221,
174241, 174257, 174259, 174263, 174281, 174289, 174299, 174311, 174329, 174331, 174337, 174347, 174367,
174389, 174407, 174413, 174431, 174443, 174457, 174467, 174469, 174481, 174487, 174491, 174527, 174533,
174569, 174571, 174583, 174599, 174613, 174617, 174631, 174637, 174649, 174653, 174659, 174673, 174679,
174703, 174721, 174737, 174749, 174761, 174763, 174767, 174773, 174799, 174821, 174829, 174851, 174859,
174877, 174893, 174901, 174907, 174917, 174929, 174931, 174943, 174959, 174989, 174991, 175003, 175013,
175039, 175061, 175067, 175069, 175081, 175103, 175129, 175141, 175211, 175229, 175261, 175267,
175277, 175291, 175303, 175309, 175327, 175333, 175349, 175361, 175391, 175393, 175403, 175411, 175433,
175447, 175453, 175463, 175481, 175493, 175499, 175519, 175523, 175543, 175573, 175601, 175621, 175631,
175633, 175649, 175663, 175673, 175687, 175691, 175699, 175709, 175723, 175727, 175753, 175757, 175759,
175781, 175783, 175811, 175829, 175837, 175843, 175853, 175859, 175873, 175891, 175897, 175909, 175919,
175937, 175939, 175949, 175961, 175963, 175979, 175991, 175993, 176017, 176021, 176023, 176041, 176047,
176051, 176053, 176063, 176081, 176087, 176089, 176123, 176129, 176153, 176159, 176161, 176179, 176191,
176201, 176207, 176213, 176221, 176227, 176237, 176243, 176261, 176299, 176303, 176317, 176321, 176327,
176329, 176333, 176347, 176353, 176357, 176369, 176383, 176389, 176401, 176413, 176417, 176419, 176431,
176459, 176461, 176467, 176489, 176497, 176503, 176507, 176509, 176521, 176531, 176537, 176549, 176551,
176557, 176573, 176591, 176597, 176599, 176609, 176611, 176629, 176641, 176651, 176677, 176699, 176711,
176713, 176741, 176747, 176753, 176777, 176779, 176789, 176791, 176797, 176807, 176809, 176819, 176849,
176857, 176887, 176899, 176903, 176921, 176923, 176927, 176933, 176951, 176977, 176983, 176989, 177007,
177011, 177013, 177019, 177043, 177091, 177101, 177109, 177113, 177127, 177131, 177167, 177173, 177209,
177211, 177217, 177223, 177239, 177257, 177269, 177283, 177301, 177319, 177323, 177337, 177347, 177379,
177383, 177409, 177421, 177427, 177431, 177433, 177467, 177473, 177481, 177487, 177493, 177511, 177533,

177539, 177553, 177589, 177601, 177623, 177647, 177677, 177679, 177691, 177739, 177743, 177761, 177763,
177787, 177791, 177797, 177811, 177823, 177839, 177841, 177883, 177887, 177889, 177893, 177907, 177913,
177917, 177929, 177943, 177949, 177953, 177967, 177979, 178001, 178021, 178037, 178039, 178067, 178069,
178091, 178093, 178103, 178117, 178127, 178141, 178151, 178169, 178183, 178187, 178207, 178223, 178231,
178247, 178249, 178259, 178261, 178289, 178301, 178307, 178327, 178333, 178349, 178351, 178361, 178393,
178397, 178403, 178417, 178439, 178441, 178447, 178469, 178481, 178487, 178489, 178501, 178513, 178531,
178537, 178559, 178561, 178567, 178571, 178597, 178601, 178603, 178609, 178613, 178621, 178627, 178639,
178643, 178681, 178691, 178693, 178697, 178753, 178757, 178781, 178793, 178799, 178807, 178813, 178817,
178819, 178831, 178853, 178859, 178873, 178877, 178889, 178897, 178903, 178907, 178909, 178921, 178931,
178933, 178939, 178951, 178973, 178987, 179021, 179029, 179033, 179041, 179051, 179057, 179083, 179089,
179099, 179107, 179111, 179119, 179143, 179161, 179167, 179173, 179203, 179209, 179213, 179233, 179243,
179261, 179269, 179281, 179287, 179317, 179321, 179327, 179351, 179357, 179369, 179381, 179383, 179393,
179407, 179411, 179429, 179437, 179441, 179453, 179461, 179471, 179479, 179483, 179497, 179519, 179527,
179533, 179549, 179563, 179573, 179579, 179581, 179591, 179593, 179603, 179623, 179633, 179651, 179657,
179659, 179671, 179687, 179689, 179693, 179717, 179719, 179737, 179743, 179749, 179779, 179801, 179807,
179813, 179819, 179821, 179827, 179833, 179849, 179897, 179899, 179903, 179909, 179917, 179923, 179939,
179947, 179951, 179953, 179957, 179969, 179981, 179989, 179999, 180001, 180007, 180023, 180043, 180053,
180071, 180073, 180077, 180097, 180137, 180161, 180179, 180181, 180211, 180221, 180233, 180239, 180241,
180247, 180259, 180263, 180281, 180287, 180289, 180307, 180311, 180317, 180331, 180337, 180347, 180361,
180371, 180379, 180391, 180413, 180419, 180437, 180463, 180473, 180491, 180497, 180503, 180511, 180533,
180539, 180541, 180547, 180563, 180569, 180617, 180623, 180629, 180647, 180667, 180679, 180701, 180731,
180749, 180751, 180773, 180779, 180793, 180797, 180799, 180811, 180847, 180871, 180883, 180907, 180949,
180959, 181001, 181003, 181019, 181031, 181039, 181061, 181063, 181081, 181087, 181123, 181141, 181157,
181183, 181193, 181199, 181201, 181211, 181213, 181219, 181243, 181253, 181273, 181277, 181283, 181297,
181301, 181303, 181361, 181387, 181397, 181399, 181409, 181421, 181439, 181457, 181459, 181499, 181501,
181513, 181523, 181537, 181549, 181553, 181603, 181607, 181609, 181619, 181639, 181667, 181669, 181693,
181711, 181717, 181721, 181729, 181739, 181751, 181757, 181759, 181763, 181777, 181787, 181789, 181813,
181837, 181871, 181873, 181889, 181891, 181903, 181913, 181919, 181927, 181931, 181943, 181957, 181967,
181981, 181997, 182009, 182011, 182027, 182029, 182041, 182047, 182057, 182059, 182089, 182099, 182101,
182107, 182111, 182123, 182129, 182131, 182141, 182159, 182167, 182177, 182179, 182201, 182209, 182233,
182239, 182243, 182261, 182279, 182297, 182309, 182333, 182339, 182341, 182353, 182387, 182389, 182417,
182423, 182431, 182443, 182453, 182467, 182471, 182473, 182489, 182503, 182509, 182519, 182537, 182549,
182561, 182579, 182587, 182593, 182599, 182603, 182617, 182627, 182639, 182641, 182653, 182657, 182659,
182681, 182687, 182701, 182711, 182713, 182747, 182773, 182779, 182789, 182803, 182813, 182821, 182839,
182851, 182857, 182867, 182887, 182893, 182899, 182921, 182927, 182929, 182933, 182953, 182957, 182969,
182981, 182999, 183023, 183037, 183041, 183047, 183059, 183067, 183089, 183091, 183119, 183151, 183167,
183191, 183203, 183247, 183259, 183263, 183283, 183289, 183299, 183301, 183307, 183317, 183319, 183329,
183343, 183349, 183361, 183373, 183377, 183383, 183389, 183397, 183437, 183439, 183451, 183461, 183473,
183479, 183487, 183497, 183499, 183503, 183509, 183511, 183523, 183527, 183569, 183571, 183577, 183581,
183587, 183593, 183611, 183637, 183661, 183683, 183691, 183697, 183707, 183709, 183713, 183761, 183763,
183797, 183809, 183823, 183829, 183871, 183877, 183881, 183907, 183917, 183919, 183943, 183949, 183959,
183971, 183973, 183979, 184003, 184007, 184013, 184031, 184039, 184043, 184057, 184073, 184081, 184087,
184111, 184117, 184133, 184153, 184157, 184181, 184187, 184189, 184199, 184211, 184231, 184241, 184259,
184271, 184273, 184279, 184291, 184309, 184321, 184333, 184337, 184351, 184369, 184409, 184417, 184441,
184447, 184463, 184477, 184487, 184489, 184511, 184517, 184523, 184553, 184559, 184567, 184571, 184577,
184607, 184609, 184627, 184631, 184633, 184649, 184651, 184669, 184687, 184693, 184703, 184711, 184721,
184727, 184733, 184753, 184777, 184823, 184829, 184831, 184837, 184843, 184859, 184879, 184901, 184903,
184913, 184949, 184957, 184967, 184969, 184993, 184997, 184999, 185021, 185027, 185051, 185057, 185063,
185069, 185071, 185077, 185089, 185099, 185123, 185131, 185137, 185149, 185153, 185161, 185167, 185177,

185183, 185189, 185221, 185233, 185243, 185267, 185291, 185299, 185303, 185309, 185323, 185327, 185359,
185363, 185369, 185371, 185401, 185429, 185441, 185467, 185477, 185483, 185491, 185519, 185527, 185531,
185533, 185539, 185543, 185551, 185557, 185567, 185569, 185593, 185599, 185621, 185641, 185651, 185677,
185681, 185683, 185693, 185699, 185707, 185711, 185723, 185737, 185747, 185749, 185753, 185767, 185789,
185797, 185813, 185819, 185821, 185831, 185833, 185849, 185869, 185873, 185893, 185897, 185903, 185917,
185923, 185947, 185951, 185957, 185959, 185971, 185987, 185993, 186007, 186013, 186019, 186023, 186037,
186041, 186049, 186071, 186097, 186103, 186107, 186113, 186119, 186149, 186157, 186161, 186163, 186187,
186191, 186211, 186227, 186229, 186239, 186247, 186253, 186259, 186271, 186283, 186299, 186301, 186311,
186317, 186343, 186377, 186379, 186391, 186397, 186419, 186437, 186451, 186469, 186479, 186481, 186551,
186569, 186581, 186583, 186587, 186601, 186619, 186629, 186647, 186649, 186653, 186671, 186679, 186689,
186701, 186707, 186709, 186727, 186733, 186743, 186757, 186761, 186763, 186773, 186793, 186799, 186841,
186859, 186869, 186871, 186877, 186883, 186889, 186917, 186947, 186959, 187003, 187009, 187027, 187043,
187049, 187067, 187069, 187073, 187081, 187091, 187111, 187123, 187127, 187129, 187133, 187139, 187141,
187163, 187171, 187177, 187181, 187189, 187193, 187211, 187217, 187219, 187223, 187237, 187273, 187277,
187303, 187337, 187339, 187349, 187361, 187367, 187373, 187379, 187387, 187393, 187409, 187417, 187423,
187433, 187441, 187463, 187469, 187471, 187477, 187507, 187513, 187531, 187547, 187559, 187573, 187597,
187631, 187633, 187637, 187639, 187651, 187661, 187669, 187687, 187699, 187711, 187721, 187751, 187763,
187787, 187793, 187823, 187843, 187861, 187871, 187877, 187883, 187897, 187907, 187909, 187921, 187927,
187931, 187951, 187963, 187973, 187987, 188011, 188017, 188021, 188029, 188107, 188137, 188143, 188147,
188159, 188171, 188179, 188189, 188197, 188249, 188261, 188273, 188281, 188291, 188299, 188303, 188311,
188317, 188323, 188333, 188351, 188359, 188369, 188389, 188401, 188407, 188417, 188431, 188437, 188443,
188459, 188473, 188483, 188491, 188519, 188527, 188533, 188563, 188579, 188603, 188609, 188621, 188633,
188653, 188677, 188681, 188687, 188693, 188701, 188707, 188711, 188719, 188729, 188753, 188767, 188779,
188791, 188801, 188827, 188831, 188833, 188843, 188857, 188861, 188863, 188869, 188891, 188911, 188927,
188933, 188939, 188941, 188953, 188957, 188983, 188999, 189011, 189017, 189019, 189041, 189043, 189061,
189067, 189127, 189139, 189149, 189151, 189169, 189187, 189199, 189223, 189229, 189239, 189251, 189253,
189257, 189271, 189307, 189311, 189337, 189347, 189349, 189353, 189361, 189377, 189389, 189391, 189401,
189407, 189421, 189433, 189437, 189439, 189463, 189467, 189473, 189479, 189491, 189493, 189509, 189517,
189523, 189529, 189547, 189559, 189583, 189593, 189599, 189613, 189617, 189619, 189643, 189653, 189661,
189671, 189691, 189697, 189701, 189713, 189733, 189743, 189757, 189767, 189797, 189799, 189817, 189823,
189851, 189853, 189859, 189877, 189881, 189887, 189901, 189913, 189929, 189947, 189949, 189961, 189967,
189977, 189983, 189989, 189997, 190027, 190031, 190051, 190063, 190093, 190097, 190121, 190129, 190147,
190159, 190181, 190207, 190243, 190249, 190261, 190271, 190283, 190297, 190301, 190313, 190321, 190331,
190339, 190357, 190367, 190369, 190387, 190391, 190403, 190409, 190471, 190507, 190523, 190529, 190537,
190543, 190573, 190577, 190579, 190583, 190591, 190607, 190613, 190633, 190639, 190649, 190657, 190667,
190669, 190699, 190709, 190711, 190717, 190753, 190759, 190763, 190769, 190783, 190787, 190793, 190807,
190811, 190823, 190829, 190837, 190843, 190871, 190889, 190891, 190901, 190909, 190913, 190921, 190979,
190997, 191021, 191027, 191033, 191039, 191047, 191057, 191071, 191089, 191099, 191119, 191123, 191137,
191141, 191143, 191161, 191173, 191189, 191227, 191231, 191237, 191249, 191251, 191281, 191297, 191299,
191339, 191341, 191353, 191413, 191441, 191447, 191449, 191453, 191459, 191461, 191467, 191473, 191491,
191497, 191507, 191509, 191519, 191531, 191533, 191537, 191551, 191561, 191563, 191579, 191599, 191621,
191627, 191657, 191669, 191671, 191677, 191689, 191693, 191699, 191707, 191717, 191747, 191749, 191773,
191783, 191791, 191801, 191803, 191827, 191831, 191833, 191837, 191861, 191899, 191903, 191911, 191929,
191953, 191969, 191977, 191999, 192007, 192013, 192029, 192037, 192043, 192047, 192053, 192091, 192097,
192103, 192113, 192121, 192133, 192149, 192161, 192173, 192187, 192191, 192193, 192229, 192233, 192239,
192251, 192259, 192263, 192271, 192307, 192317, 192319, 192323, 192341, 192343, 192347, 192373, 192377,
192383, 192391, 192407, 192431, 192461, 192463, 192497, 192499, 192529, 192539, 192547, 192553, 192557,
192571, 192581, 192583, 192587, 192601, 192611, 192613, 192617, 192629, 192631, 192637, 192667, 192677,
192697, 192737, 192743, 192749, 192757, 192767, 192781, 192791, 192799, 192811, 192817, 192833, 192847,

192853, 192859, 192877, 192883, 192887, 192889, 192917, 192923, 192931, 192949, 192961, 192971, 192977,
192979, 192991, 193003, 193009, 193013, 193031, 193043, 193051, 193057, 193073, 193093, 193133, 193139,
193147, 193153, 193163, 193181, 193183, 193189, 193201, 193243, 193247, 193261, 193283, 193301, 193327,
193337, 193357, 193367, 193373, 193379, 193381, 193387, 193393, 193423, 193433, 193441, 193447, 193451,
193463, 193469, 193493, 193507, 193513, 193541, 193549, 193559, 193573, 193577, 193597, 193601, 193603,
193607, 193619, 193649, 193663, 193679, 193703, 193723, 193727, 193741, 193751, 193757, 193763, 193771,
193789, 193793, 193799, 193811, 193813, 193841, 193847, 193859, 193861, 193871, 193873, 193877, 193883,
193891, 193937, 193939, 193943, 193951, 193957, 193979, 193993, 194003, 194017, 194027, 194057, 194069,
194071, 194083, 194087, 194093, 194101, 194113, 194119, 194141, 194149, 194167, 194179, 194197, 194203,
194239, 194263, 194267, 194269, 194309, 194323, 194353, 194371, 194377, 194413, 194431, 194443, 194471,
194479, 194483, 194507, 194521, 194527, 194543, 194569, 194581, 194591, 194609, 194647, 194653, 194659,
194671, 194681, 194683, 194687, 194707, 194713, 194717, 194723, 194729, 194749, 194767, 194771, 194809,
194813, 194819, 194827, 194839, 194861, 194863, 194867, 194869, 194891, 194899, 194911, 194917, 194933,
194963, 194977, 194981, 194989, 195023, 195029, 195043, 195047, 195049, 195053, 195071, 195077, 195089,
195103, 195121, 195127, 195131, 195137, 195157, 195161, 195163, 195193, 195197, 195203, 195229, 195241,
195253, 195259, 195271, 195277, 195281, 195311, 195319, 195329, 195341, 195343, 195353, 195359, 195389,
195401, 195407, 195413, 195427, 195443, 195457, 195469, 195479, 195493, 195497, 195511, 195527, 195539,
195541, 195581, 195593, 195599, 195659, 195677, 195691, 195697, 195709, 195731, 195733, 195737, 195739,
195743, 195751, 195761, 195781, 195787, 195791, 195809, 195817, 195863, 195869, 195883, 195887, 195893,
195907, 195913, 195919, 195929, 195931, 195967, 195971, 195973, 195977, 195991, 195997, 196003, 196033,
196039, 196043, 196051, 196073, 196081, 196087, 196111, 196117, 196139, 196159, 196169, 196171, 196177,
196181, 196187, 196193, 196201, 196247, 196271, 196277, 196279, 196291, 196303, 196307, 196331, 196337,
196379, 196387, 196429, 196439, 196453, 196459, 196477, 196499, 196501, 196519, 196523, 196541, 196543,
196549, 196561, 196579, 196583, 196597, 196613, 196643, 196657, 196661, 196663, 196681, 196687, 196699,
196709, 196717, 196727, 196739, 196751, 196769, 196771, 196799, 196817, 196831, 196837, 196853, 196871,
196873, 196879, 196901, 196907, 196919, 196927, 196961, 196991, 196993, 197003, 197009, 197023, 197033,
197059, 197063, 197077, 197083, 197089, 197101, 197117, 197123, 197137, 197147, 197159, 197161, 197203,
197207, 197221, 197233, 197243, 197257, 197261, 197269, 197273, 197279, 197293, 197297, 197299, 197311,
197339, 197341, 197347, 197359, 197369, 197371, 197381, 197383, 197389, 197419, 197423, 197441, 197453,
197479, 197507, 197521, 197539, 197551, 197567, 197569, 197573, 197597, 197599, 197609, 197621, 197641,
197647, 197651, 197677, 197683, 197689, 197699, 197711, 197713, 197741, 197753, 197759, 197767, 197773,
197779, 197803, 197807, 197831, 197837, 197887, 197891, 197893, 197909, 197921, 197927, 197933, 197947,
197957, 197959, 197963, 197969, 197971, 198013, 198017, 198031, 198043, 198047, 198073, 198083, 198091,
198097, 198109, 198127, 198139, 198173, 198179, 198193, 198197, 198221, 198223, 198241, 198251, 198257,
198259, 198277, 198281, 198301, 198313, 198323, 198337, 198347, 198349, 198377, 198391, 198397, 198409,
198413, 198427, 198437, 198439, 198461, 198463, 198469, 198479, 198491, 198503, 198529, 198533, 198553,
198571, 198589, 198593, 198599, 198613, 198623, 198637, 198641, 198647, 198659, 198673, 198689, 198701,
198719, 198733, 198761, 198769, 198811, 198817, 198823, 198827, 198829, 198833, 198839, 198841, 198851,
198859, 198899, 198901, 198929, 198937, 198941, 198943, 198953, 198959, 198967, 198971, 198977, 198997,
199021, 199033, 199037, 199039, 199049, 199081, 199103, 199109, 199151, 199153, 199181, 199193, 199207,
199211, 199247, 199261, 199267, 199289, 199313, 199321, 199337, 199343, 199357, 199373, 199379, 199399,
199403, 199411, 199429, 199447, 199453, 199457, 199483, 199487, 199489, 199499, 199501, 199523,
199559, 199567, 199583, 199601, 199603, 199621, 199637, 199657, 199669, 199673, 199679, 199687, 199697,
199721, 199729, 199739, 199741, 199751, 199753, 199777, 199783, 199799, 199807, 199811, 199813, 199819,
199831, 199853, 199873, 199877, 199889, 199909, 199921, 199931, 199933, 199961, 199967, 199999, 200003,
200009, 200017, 200023, 200029, 200033, 200041, 200063, 200087, 200117, 200131, 200153, 200159, 200171,
200177, 200183, 200191, 200201, 200227, 200231, 200237, 200257, 200273, 200293, 200297, 200323, 200329,
200341, 200351, 200357, 200363, 200371, 200381, 200383, 200401, 200407, 200437, 200443, 200461, 200467,
200483, 200513, 200569, 200573, 200579, 200587, 200591, 200597, 200609, 200639, 200657, 200671, 200689,

200699, 200713, 200723, 200731, 200771, 200779, 200789, 200797, 200807, 200843, 200861, 200867, 200869, 200881, 200891, 200899, 200903, 200909, 200927, 200929, 200971, 200983, 200987, 200989, 201007, 201011, 201031, 201037, 201049, 201073, 201101, 201107, 201119, 201121, 201139, 201151, 201163, 201167, 201193, 201203, 201209, 201211, 201233, 201247, 201251, 201281, 201287, 201307, 201329, 201337, 201359, 201389, 201401, 201403, 201413, 201437, 201449, 201451, 201473, 201491, 201493, 201497, 201499, 201511, 201517, 201547, 201557, 201577, 201581, 201589, 201599, 201611, 201623, 201629, 201653, 201661, 201667, 201673, 201683, 201701, 201709, 201731, 201743, 201757, 201767, 201769, 201781, 201787, 201791, 201797, 201809, 201821, 201823, 201827, 201829, 201833, 201847, 201881, 201889, 201893, 201907, 201911, 201919, 201923, 201937, 201947, 201953, 201961, 201973, 201979, 201997, 202001, 202021, 202031, 202049, 202061, 202063, 202067, 202087, 202099, 202109, 202121, 202127, 202129, 202183, 202187, 202201, 202219, 202231, 202243, 202277, 202289, 202291, 202309, 202327, 202339, 202343, 202357, 202361, 202381, 202387, 202393, 202403, 202409, 202441, 202471, 202481, 202493, 202519, 202529, 202549, 202567, 202577, 202591, 202613, 202621, 202627, 202637, 202639, 202661, 202667, 202679, 202693, 202717, 202729, 202733, 202747, 202751, 202753, 202757, 202777, 202799, 202817, 202823, 202841, 202859, 202877, 202879, 202889, 202907, 202921, 202931, 202933, 202949, 202967, 202973, 202981, 202987, 202999, 203011, 203017, 203023, 203039, 203051, 203057, 203117, 203141, 203173, 203183, 203207, 203209, 203213, 203221, 203227, 203233, 203249, 203279, 203293, 203309, 203311, 203317, 203321, 203323, 203339, 203341, 203351, 203353, 203363, 203381, 203383, 203387, 203393, 203417, 203419, 203429, 203431, 203449, 203459, 203461, 203531, 203549, 203563, 203569, 203579, 203591, 203617, 203627, 203641, 203653, 203657, 203659, 203663, 203669, 203713, 203761, 203767, 203771, 203773, 203789, 203807, 203809, 203821, 203843, 203857, 203869, 203873, 203897, 203909, 203911, 203921, 203947, 203953, 203969, 203971, 203977, 203989, 203999, 204007, 204013, 204019, 204023, 204047, 204059, 204067, 204101, 204107, 204133, 204137, 204143, 204151, 204161, 204163, 204173, 204233, 204251, 204299, 204301, 204311, 204319, 204329, 204331, 204353, 204359, 204361, 204367, 204371, 204377, 204397, 204427, 204431, 204437, 204439, 204443, 204461, 204481, 204487, 204509, 204511, 204517, 204521, 204557, 204563, 204583, 204587, 204599, 204601, 204613, 204623, 204641, 204667, 204679, 204707, 204719, 204733, 204749, 204751, 204781, 204791, 204793, 204797, 204803, 204821, 204857, 204859, 204871, 204887, 204913, 204917, 204923, 204931, 204947, 204973, 204979, 204983, 205019, 205031, 205033, 205043, 205063, 205069, 205081, 205097, 205103, 205111, 205129, 205133, 205141, 205151, 205157, 205171, 205187, 205201, 205211, 205213, 205223, 205237, 205253, 205267, 205297, 205307, 205319, 205327, 205339, 205357, 205391, 205397, 205399, 205417, 205421, 205423, 205427, 205433, 205441, 205453, 205463, 205477, 205483, 205487, 205493, 205507, 205519, 205529, 205537, 205549, 205553, 205559, 205589, 205603, 205607, 205619, 205627, 205633, 205651, 205657, 205661, 205663, 205703, 205721, 205759, 205763, 205783, 205817, 205823, 205837, 205847, 205879, 205883, 205913, 205937, 205949, 205951, 205957, 205963, 205967, 205981, 205991, 205993, 206009, 206021, 206027, 206033, 206039, 206047, 206051, 206069, 206077, 206081, 206083, 206123, 206153, 206177, 206179, 206183, 206191, 206197, 206203, 206209, 206221, 206233, 206237, 206249, 206251, 206263, 206273, 206279, 206281, 206291, 206299, 206303, 206341, 206347, 206351, 206369, 206383, 206399, 206407, 206411, 206413, 206419, 206447, 206461, 206467, 206477, 206483, 206489, 206501, 206519, 206527, 206543, 206551, 206593, 206597, 206603, 206623, 206627, 206639, 206641, 206651, 206699, 206749, 206779, 206783, 206803, 206807, 206813, 206819, 206821, 206827, 206879, 206887, 206897, 206909, 206911, 206917, 206923, 206933, 206939, 206951, 206953, 206993, 207013, 207017, 207029, 207037, 207041, 207061, 207073, 207079, 207113, 207121, 207127, 207139, 207169, 207187, 207191, 207197, 207199, 207227, 207239, 207241, 207257, 207269, 207287, 207293, 207301, 207307, 207329, 207331, 207341, 207343, 207367, 207371, 207377, 207401, 207409, 207433, 207443, 207457, 207463, 207469, 207479, 207481, 207491, 207497, 207509, 207511, 207517, 207521, 207523, 207541, 207547, 207551, 207563, 207569, 207589, 207593, 207619, 207629, 207643, 207653, 207661, 207671, 207673, 207679, 207709, 207719, 207721, 207743, 207763, 207769, 207797, 207799, 207811, 207821, 207833, 207847, 207869, 207877, 207923, 207931, 207941, 207947, 207953, 207967, 207971, 207973, 207997, 208001, 208003, 208009, 208037, 208049, 208057, 208067, 208073, 208099, 208111, 208121, 208129, 208139, 208141, 208147, 208189, 208207, 208213, 208217, 208223, 208231, 208253, 208261, 208277, 208279, 208283, 208291, 208309, 208319, 208333, 208337, 208367, 208379, 208387, 208391, 208393, 208409, 208433, 208441, 208457,

208459, 208463, 208469, 208489, 208493, 208499, 208501, 208511, 208513, 208519, 208529, 208553, 208577,
208589, 208591, 208609, 208627, 208631, 208657, 208667, 208673, 208687, 208697, 208699, 208721, 208729,
208739, 208759, 208787, 208799, 208807, 208837, 208843, 208877, 208889, 208891, 208907, 208927, 208931,
208933, 208961, 208963, 208991, 208993, 208997, 209021, 209029, 209039, 209063, 209071, 209089, 209123,
209147, 209159, 209173, 209179, 209189, 209201, 209203, 209213, 209221, 209227, 209233, 209249, 209257,
209263, 209267, 209269, 209299, 209311, 209317, 209327, 209333, 209347, 209353, 209357, 209359, 209371,
209381, 209393, 209401, 209431, 209441, 209449, 209459, 209471, 209477, 209497, 209519, 209533, 209543,
209549, 209563, 209567, 209569, 209579, 209581, 209597, 209621, 209623, 209639, 209647, 209659, 209669,
209687, 209701, 209707, 209717, 209719, 209743, 209767, 209771, 209789, 209801, 209809, 209813, 209819,
209821, 209837, 209851, 209857, 209861, 209887, 209917, 209927, 209929, 209939, 209953, 209959, 209971,
209977, 209983, 209987, 210011, 210019, 210031, 210037, 210053, 210071, 210097, 210101, 210109, 210113,
210127, 210131, 210139, 210143, 210157, 210169, 210173, 210187, 210191, 210193, 210209, 210229, 210233,
210241, 210247, 210257, 210263, 210277, 210283, 210299, 210317, 210319, 210323, 210347, 210359, 210361,
210391, 210401, 210403, 210407, 210421, 210437, 210461, 210467, 210481, 210487, 210491, 210499, 210523,
210527, 210533, 210557, 210599, 210601, 210619, 210631, 210643, 210659, 210671, 210709, 210713, 210719,
210731, 210739, 210761, 210773, 210803, 210809, 210811, 210823, 210827, 210839, 210853, 210857, 210869,
210901, 210907, 210911, 210913, 210923, 210929, 210943, 210961, 210967, 211007, 211039, 211049, 211051,
211061, 211063, 211067, 211073, 211093, 211097, 211129, 211151, 211153, 211177, 211187, 211193, 211199,
211213, 211217, 211219, 211229, 211231, 211241, 211247, 211271, 211283, 211291, 211297, 211313, 211319,
211333, 211339, 211349, 211369, 211373, 211403, 211427, 211433, 211441, 211457, 211469, 211493, 211499,
211501, 211507, 211543, 211559, 211571, 211573, 211583, 211597, 211619, 211639, 211643, 211657, 211661,
211663, 211681, 211691, 211693, 211711, 211723, 211727, 211741, 211747, 211777, 211781, 211789, 211801,
211811, 211817, 211859, 211867, 211873, 211877, 211879, 211889, 211891, 211927, 211931, 211933, 211943,
211949, 211969, 211979, 211997, 212029, 212039, 212057, 212081, 212099, 212117, 212123, 212131, 212141,
212161, 212167, 212183, 212203, 212207, 212209, 212227, 212239, 212243, 212281, 212293, 212297, 212353,
212369, 212383, 212411, 212419, 212423, 212437, 212447, 212453, 212461, 212467, 212479, 212501, 212507,
212557, 212561, 212573, 212579, 212587, 212593, 212627, 212633, 212651, 212669, 212671, 212677, 212683,
212701, 212777, 212791, 212801, 212827, 212837, 212843, 212851, 212867, 212869, 212873, 212881, 212897,
212903, 212909, 212917, 212923, 212969, 212981, 212987, 212999, 213019, 213023, 213029, 213043, 213067,
213079, 213091, 213097, 213119, 213131, 213133, 213139, 213149, 213173, 213181, 213193, 213203, 213209,
213217, 213223, 213229, 213247, 213253, 213263, 213281, 213287, 213289, 213307, 213319, 213329, 213337,
213349, 213359, 213361, 213383, 213391, 213397, 213407, 213449, 213461, 213467, 213481, 213491, 213523,
213533, 213539, 213553, 213557, 213589, 213599, 213611, 213613, 213623, 213637, 213641, 213649, 213659,
213713, 213721, 213727, 213737, 213751, 213791, 213799, 213821, 213827, 213833, 213847, 213859, 213881,
213887, 213901, 213919, 213929, 213943, 213947, 213949, 213953, 213973, 213977, 213989, 214003, 214007,
214009, 214021, 214031, 214033, 214043, 214051, 214063, 214069, 214087, 214091, 214129, 214133, 214141,
214147, 214163, 214177, 214189, 214211, 214213, 214219, 214237, 214243, 214259, 214283, 214297, 214309,
214351, 214363, 214373, 214381, 214391, 214399, 214433, 214439, 214451, 214457, 214463, 214469, 214481,
214483, 214499, 214507, 214517, 214519, 214531, 214541, 214559, 214561, 214589, 214603, 214607, 214631,
214639, 214651, 214657, 214663, 214667, 214673, 214691, 214723, 214729, 214733, 214741, 214759, 214763,
214771, 214783, 214787, 214789, 214807, 214811, 214817, 214831, 214849, 214853, 214867, 214883, 214891,
214913, 214939, 214943, 214967, 214987, 214993, 215051, 215063, 215077, 215087, 215123, 215141, 215143,
215153, 215161, 215179, 215183, 215191, 215197, 215239, 215249, 215261, 215273, 215279, 215297, 215309,
215317, 215329, 215351, 215353, 215359, 215381, 215389, 215393, 215399, 215417, 215443, 215447, 215459,
215461, 215471, 215483, 215497, 215503, 215507, 215521, 215531, 215563, 215573, 215587, 215617, 215653,
215659, 215681, 215687, 215689, 215693, 215723, 215737, 215753, 215767, 215771, 215797, 215801, 215827,
215833, 215843, 215851, 215857, 215863, 215893, 215899, 215909, 215921, 215927, 215939, 215953, 215959,
215981, 215983, 216023, 216037, 216061, 216071, 216091, 216103, 216107, 216113, 216119, 216127, 216133,
216149, 216157, 216173, 216179, 216211, 216217, 216233, 216259, 216263, 216289, 216317, 216319, 216329,

216347, 216371, 216373, 216379, 216397, 216401, 216421, 216431, 216451, 216481, 216493, 216509, 216523,
216551, 216553, 216569, 216571, 216577, 216607, 216617, 216641, 216647, 216649, 216653, 216661, 216679,
216703, 216719, 216731, 216743, 216751, 216757, 216761, 216779, 216781, 216787, 216791, 216803, 216829,
216841, 216851, 216859, 216877, 216899, 216901, 216911, 216917, 216919, 216947, 216967, 216973, 216991,
217001, 217003, 217027, 217033, 217057, 217069, 217081, 217111, 217117, 217121, 217157, 217163, 217169,
217199, 217201, 217207, 217219, 217223, 217229, 217241, 217253, 217271, 217307, 217309, 217313, 217319,
217333, 217337, 217339, 217351, 217361, 217363, 217367, 217369, 217387, 217397, 217409, 217411, 217421,
217429, 217439, 217457, 217463, 217489, 217499, 217517, 217519, 217559, 217561, 217573, 217577, 217579,
217619, 217643, 217661, 217667, 217681, 217687, 217691, 217697, 217717, 217727, 217733, 217739, 217747,
217771, 217781, 217793, 217823, 217829, 217849, 217859, 217901, 217907, 217909, 217933, 217937, 217969,
217979, 217981, 218003, 218021, 218047, 218069, 218077, 218081, 218083, 218087, 218107, 218111, 218117,
218131, 218137, 218143, 218149, 218171, 218191, 218213, 218227, 218233, 218249, 218279, 218287, 218357,
218363, 218371, 218381, 218389, 218401, 218417, 218419, 218423, 218437, 218447, 218453, 218459, 218461,
218479, 218509, 218513, 218521, 218527, 218531, 218549, 218551, 218579, 218591, 218599, 218611, 218623,
218627, 218629, 218641, 218651, 218657, 218677, 218681, 218711, 218717, 218719, 218723, 218737, 218749,
218761, 218783, 218797, 218809, 218819, 218833, 218839, 218843, 218849, 218857, 218873, 218887, 218923,
218941, 218947, 218963, 218969, 218971, 218987, 218989, 218993, 219001, 219017, 219019, 219031, 219041,
219053, 219059, 219071, 219083, 219091, 219097, 219103, 219119, 219133, 219143, 219169, 219187, 219217,
219223, 219251, 219277, 219281, 219293, 219301, 219311, 219313, 219353, 219361, 219371, 219377, 219389,
219407, 219409, 219433, 219437, 219451, 219463, 219467, 219491, 219503, 219517, 219523, 219529, 219533,
219547, 219577, 219587, 219599, 219607, 219613, 219619, 219629, 219647, 219649, 219677, 219679, 219683,
219689, 219707, 219721, 219727, 219731, 219749, 219757, 219761, 219763, 219767, 219787, 219797, 219799,
219809, 219823, 219829, 219839, 219847, 219851, 219871, 219881, 219889, 219911, 219917, 219931, 219937,
219941, 219943, 219953, 219959, 219971, 219977, 219979, 219983, 220009, 220013, 220019, 220021, 220057,
220063, 220123, 220141, 220147, 220151, 220163, 220169, 220177, 220189, 220217, 220243, 220279, 220291,
220301, 220307, 220327, 220333, 220351, 220357, 220361, 220369, 220373, 220391, 220399, 220403, 220411,
220421, 220447, 220469, 220471, 220511, 220513, 220529, 220537, 220543, 220553, 220559, 220573, 220579,
220589, 220613, 220663, 220667, 220673, 220681, 220687, 220699, 220709, 220721, 220747, 220757, 220771,
220783, 220789, 220793, 220807, 220811, 220841, 220859, 220861, 220873, 220877, 220879, 220889, 220897,
220901, 220903, 220907, 220919, 220931, 220933, 220939, 220973, 221021, 221047, 221059, 221069, 221071,
221077, 221083, 221087, 221093, 221101, 221159, 221171, 221173, 221197, 221201, 221203, 221209, 221219,
221227, 221233, 221239, 221251, 221261, 221281, 221303, 221311, 221317, 221327, 221393, 221399, 221401,
221411, 221413, 221447, 221453, 221461, 221471, 221477, 221489, 221497, 221509, 221537, 221539, 221549,
221567, 221581, 221587, 221603, 221621, 221623, 221653, 221657, 221659, 221671, 221677, 221707, 221713,
221717, 221719, 221723, 221729, 221737, 221747, 221773, 221797, 221807, 221813, 221827, 221831, 221849,
221873, 221891, 221909, 221941, 221951, 221953, 221957, 221987, 221989, 221999, 222007, 222011, 222023,
222029, 222041, 222043, 222059, 222067, 222073, 222107, 222109, 222113, 222127, 222137, 222149, 222151,
222161, 222163, 222193, 222197, 222199, 222247, 222269, 222289, 222293, 222311, 222317, 222323, 222329,
222337, 222347, 222349, 222361, 222367, 222379, 222389, 222403, 222419, 222437, 222461, 222493, 222499,
222511, 222527, 222533, 222553, 222557, 222587, 222601, 222613, 222619, 222643, 222647, 222659, 222679,
222707, 222713, 222731, 222773, 222779, 222787, 222791, 222793, 222799, 222823, 222839, 222841, 222857,
222863, 222877, 222883, 222913, 222919, 222931, 222941, 222947, 222953, 222967, 222977, 222979, 222991,
223007, 223009, 223019, 223037, 223049, 223051, 223061, 223063, 223087, 223099, 223103, 223129, 223133,
223151, 223207, 223211, 223217, 223219, 223229, 223241, 223243, 223247, 223253, 223259, 223273, 223277,
223283, 223291, 223303, 223313, 223319, 223331, 223337, 223339, 223361, 223367, 223381, 223403, 223423,
223429, 223439, 223441, 223463, 223469, 223481, 223493, 223507, 223529, 223543, 223547, 223549, 223577,
223589, 223621, 223633, 223637, 223667, 223679, 223681, 223697, 223711, 223747, 223753, 223757, 223759,
223781, 223823, 223829, 223831, 223837, 223841, 223843, 223849, 223903, 223919, 223921, 223939, 223963,
223969, 223999, 224011, 224027, 224033, 224041, 224047, 224057, 224069, 224071, 224101, 224113, 224129,

224131, 224149, 224153, 224171, 224177, 224197, 224201, 224209, 224221, 224233, 224239, 224251, 224261,
224267, 224291, 224299, 224303, 224309, 224317, 224327, 224351, 224359, 224363, 224401, 224423, 224429,
224443, 224449, 224461, 224467, 224473, 224491, 224501, 224513, 224527, 224563, 224569, 224579, 224591,
224603, 224611, 224617, 224629, 224633, 224669, 224677, 224683, 224699, 224711, 224717, 224729, 224737,
224743, 224759, 224771, 224797, 224813, 224831, 224863, 224869, 224881, 224891, 224897, 224909, 224911,
224921, 224929, 224947, 224951, 224969, 224977, 224993, 225023, 225037, 225061, 225067, 225077, 225079,
225089, 225109, 225119, 225133, 225143, 225149, 225157, 225161, 225163, 225167, 225217, 225221, 225223,
225227, 225241, 225257, 225263, 225287, 225289, 225299, 225307, 225341, 225343, 225347, 225349, 225353,
225371, 225373, 225383, 225427, 225431, 225457, 225461, 225479, 225493, 225499, 225503, 225509, 225523,
225527, 225529, 225569, 225581, 225583, 225601, 225611, 225613, 225619, 225629, 225637, 225671, 225683,
225689, 225697, 225721, 225733, 225749, 225751, 225767, 225769, 225779, 225781, 225809, 225821, 225829,
225839, 225859, 225871, 225889, 225919, 225931, 225941, 225943, 225949, 225961, 225977, 225983, 225989,
226001, 226007, 226013, 226027, 226063, 226087, 226099, 226103, 226123, 226129, 226133, 226141, 226169,
226183, 226189, 226199, 226201, 226217, 226231, 226241, 226267, 226283, 226307, 226313, 226337, 226357,
226367, 226379, 226381, 226397, 226409, 226427, 226433, 226451, 226453, 226463, 226483, 226487, 226511,
226531, 226547, 226549, 226553, 226571, 226601, 226609, 226621, 226631, 226637, 226643, 226649, 226657,
226663, 226669, 226691, 226697, 226741, 226753, 226769, 226777, 226783, 226789, 226799, 226813, 226817,
226819, 226823, 226843, 226871, 226901, 226903, 226907, 226913, 226937, 226943, 226991, 227011, 227027,
227053, 227081, 227089, 227093, 227111, 227113, 227131, 227147, 227153, 227159, 227167, 227177, 227189,
227191, 227207, 227219, 227231, 227233, 227251, 227257, 227267, 227281, 227299, 227303, 227363, 227371,
227377, 227387, 227393, 227399, 227407, 227419, 227431, 227453, 227459, 227467, 227471, 227473, 227489,
227497, 227501, 227519, 227531, 227533, 227537, 227561, 227567, 227569, 227581, 227593, 227597, 227603,
227609, 227611, 227627, 227629, 227651, 227653, 227663, 227671, 227693, 227699, 227707, 227719, 227729,
227743, 227789, 227797, 227827, 227849, 227869, 227873, 227893, 227947, 227951, 227977, 227989, 227993,
228013, 228023, 228049, 228061, 228077, 228097, 228103, 228113, 228127, 228131, 228139, 228181, 228197,
228199, 228203, 228211, 228223, 228233, 228251, 228257, 228281, 228289, 228301, 228307, 228311, 228331,
228337, 228341, 228353, 228359, 228383, 228409, 228419, 228421, 228427, 228443, 228451, 228457, 228461,
228469, 228479, 228509, 228511, 228517, 228521, 228523, 228539, 228559, 228577, 228581, 228587, 228593,
228601, 228611, 228617, 228619, 228637, 228647, 228677, 228707, 228713, 228731, 228733, 228737, 228751,
228757, 228773, 228793, 228797, 228799, 228829, 228841, 228847, 228853, 228859, 228869, 228881, 228883,
228887, 228901, 228911, 228913, 228923, 228929, 228953, 228959, 228961, 228983, 228989, 229003, 229027,
229037, 229081, 229093, 229123, 229127, 229133, 229139, 229153, 229157, 229171, 229181, 229189, 229199,
229213, 229217, 229223, 229237, 229247, 229249, 229253, 229261, 229267, 229283, 229309, 229321, 229343,
229351, 229373, 229393, 229399, 229403, 229409, 229423, 229433, 229459, 229469, 229487, 229499, 229507,
229519, 229529, 229547, 229549, 229553, 229561, 229583, 229589, 229591, 229601, 229613, 229627, 229631,
229637, 229639, 229681, 229693, 229699, 229703, 229711, 229717, 229727, 229739, 229751, 229753, 229759,
229763, 229769, 229771, 229777, 229781, 229799, 229813, 229819, 229837, 229841, 229847, 229849, 229897,
229903, 229937, 229939, 229949, 229961, 229963, 229979, 229981, 230003, 230017, 230047, 230059, 230063,
230077, 230081, 230089, 230101, 230107, 230117, 230123, 230137, 230143, 230149, 230189, 230203, 230213,
230221, 230227, 230233, 230239, 230257, 230273, 230281, 230291, 230303, 230309, 230311, 230327, 230339,
230341, 230353, 230357, 230369, 230383, 230387, 230389, 230393, 230431, 230449, 230453, 230467, 230471,
230479, 230501, 230507, 230539, 230551, 230561, 230563, 230567, 230597, 230611, 230647, 230653, 230663,
230683, 230693, 230719, 230729, 230743, 230761, 230767, 230771, 230773, 230779, 230807, 230819, 230827,
230833, 230849, 230861, 230863, 230873, 230891, 230929, 230933, 230939, 230941, 230959, 230969, 230977,
230999, 231001, 231017, 231019, 231031, 231041, 231053, 231067, 231079, 231107, 231109, 231131, 231169,
231197, 231223, 231241, 231269, 231271, 231277, 231289, 231293, 231299, 231317, 231323, 231331, 231347,
231349, 231359, 231367, 231379, 231409, 231419, 231431, 231433, 231443, 231461, 231463, 231479, 231481,
231493, 231503, 231529, 231533, 231547, 231551, 231559, 231563, 231571, 231589, 231599, 231607, 231611,
231613, 231631, 231643, 231661, 231677, 231701, 231709, 231719, 231779, 231799, 231809, 231821, 231823,

231827, 231839, 231841, 231859, 231871, 231877, 231893, 231901, 231919, 231923, 231943, 231947, 231961,
231967, 232003, 232007, 232013, 232049, 232051, 232073, 232079, 232081, 232091, 232103, 232109, 232117,
232129, 232153, 232171, 232187, 232189, 232207, 232217, 232259, 232303, 232307, 232333, 232357, 232363,
232367, 232381, 232391, 232409, 232411, 232417, 232433, 232439, 232451, 232457, 232459, 232487, 232499,
232513, 232523, 232549, 232567, 232571, 232591, 232597, 232607, 232621, 232633, 232643, 232663, 232669,
232681, 232699, 232709, 232711, 232741, 232751, 232753, 232777, 232801, 232811, 232819, 232823, 232847,
232853, 232861, 232871, 232877, 232891, 232901, 232907, 232919, 232937, 232961, 232963, 232987, 233021,
233069, 233071, 233083, 233113, 233117, 233141, 233143, 233159, 233161, 233173, 233183, 233201, 233221,
233231, 233239, 233251, 233267, 233279, 233293, 233297, 233323, 233327, 233329, 233341, 233347, 233353,
233357, 233371, 233407, 233417, 233419, 233423, 233437, 233477, 233489, 233509, 233549, 233551, 233557,
233591, 233599, 233609, 233617, 233621, 233641, 233663, 233669, 233683, 233687, 233689, 233693, 233713,
233743, 233747, 233759, 233777, 233837, 233851, 233861, 233879, 233881, 233911, 233917, 233921, 233923,
233939, 233941, 233969, 233983, 233993, 234007, 234029, 234043, 234067, 234083, 234089, 234103, 234121,
234131, 234139, 234149, 234161, 234167, 234181, 234187, 234191, 234193, 234197, 234203, 234211, 234217,
234239, 234259, 234271, 234281, 234287, 234293, 234317, 234319, 234323, 234331, 234341, 234343, 234361,
234383, 234431, 234457, 234461, 234463, 234467, 234473, 234499, 234511, 234527, 234529, 234539, 234541,
234547, 234571, 234587, 234589, 234599, 234613, 234629, 234653, 234659, 234673, 234683, 234713, 234721,
234727, 234733, 234743, 234749, 234769, 234781, 234791, 234799, 234803, 234809, 234811, 234833, 234847,
234851, 234863, 234869, 234893, 234907, 234917, 234931, 234947, 234959, 234961, 234967, 234977, 234979,
234989, 235003, 235007, 235009, 235013, 235043, 235051, 235057, 235069, 235091, 235099, 235111, 235117,
235159, 235171, 235177, 235181, 235199, 235211, 235231, 235241, 235243, 235273, 235289, 235307, 235309,
235337, 235349, 235369, 235397, 235439, 235441, 235447, 235483, 235489, 235493, 235513, 235519, 235523,
235537, 235541, 235553, 235559, 235577, 235591, 235601, 235607, 235621, 235661, 235663, 235673, 235679,
235699, 235723, 235747, 235751, 235783, 235787, 235789, 235793, 235811, 235813, 235849, 235871, 235877,
235889, 235891, 235901, 235919, 235927, 235951, 235967, 235979, 235997, 236017, 236021, 236053, 236063,
236069, 236077, 236087, 236107, 236111, 236129, 236143, 236153, 236167, 236207, 236209, 236219, 236231,
236261, 236287, 236293, 236297, 236323, 236329, 236333, 236339, 236377, 236381, 236387, 236399, 236407,
236429, 236449, 236461, 236471, 236477, 236479, 236503, 236507, 236519, 236527, 236549, 236563, 236573,
236609, 236627, 236641, 236653, 236659, 236681, 236699, 236701, 236707, 236713, 236723, 236729, 236737,
236749, 236771, 236773, 236779, 236783, 236807, 236813, 236867, 236869, 236879, 236881, 236891, 236893,
236897, 236909, 236917, 236947, 236981, 236983, 236993, 237011, 237019, 237043, 237053, 237067, 237071,
237073, 237089, 237091, 237137, 237143, 237151, 237157, 237161, 237163, 237173, 237179, 237203, 237217,
237233, 237257, 237271, 237277, 237283, 237287, 237301, 237313, 237319, 237331, 237343, 237361, 237373,
237379, 237401, 237409, 237467, 237487, 237509, 237547, 237563, 237571, 237581, 237607, 237619, 237631,
237673, 237683, 237689, 237691, 237701, 237707, 237733, 237737, 237749, 237763, 237767, 237781, 237791,
237821, 237851, 237857, 237859, 237877, 237883, 237901, 237911, 237929, 237959, 237967, 237971, 237973,
237977, 237997, 238001, 238009, 238019, 238031, 238037, 238039, 238079, 238081, 238093, 238099, 238103,
238109, 238141, 238151, 238157, 238159, 238163, 238171, 238181, 238201, 238207, 238213, 238223, 238229,
238237, 238247, 238261, 238267, 238291, 238307, 238313, 238321, 238331, 238339, 238361, 238363, 238369,
238373, 238397, 238417, 238423, 238439, 238451, 238463, 238471, 238477, 238481, 238499, 238519, 238529,
238531, 238547, 238573, 238591, 238627, 238639, 238649, 238657, 238673, 238681, 238691, 238703, 238709,
238723, 238727, 238729, 238747, 238759, 238781, 238789, 238801, 238829, 238837, 238841, 238853, 238859,
238877, 238879, 238883, 238897, 238919, 238921, 238939, 238943, 238949, 238967, 238991, 239017, 239023,
239027, 239053, 239069, 239081, 239087, 239119, 239137, 239147, 239167, 239171, 239179, 239201, 239231,
239233, 239237, 239243, 239251, 239263, 239273, 239287, 239297, 239329, 239333, 239347, 239357, 239383,
239387, 239389, 239417, 239423, 239429, 239431, 239441, 239461, 239489, 239509, 239521, 239527, 239531,
239539, 239543, 239557, 239567, 239579, 239587, 239597, 239611, 239623, 239633, 239641, 239671, 239689,
239699, 239711, 239713, 239731, 239737, 239753, 239779, 239783, 239803, 239807, 239831, 239843, 239849,
239851, 239857, 239873, 239879, 239893, 239929, 239933, 239947, 239957, 239963, 239977, 239999, 240007,

240011, 240017, 240041, 240043, 240047, 240049, 240059, 240073, 240089, 240101, 240109, 240113, 240131,
240139, 240151, 240169, 240173, 240197, 240203, 240209, 240257, 240259, 240263, 240271, 240283, 240287,
240319, 240341, 240347, 240349, 240353, 240371, 240379, 240421, 240433, 240437, 240473, 240479, 240491,
240503, 240509, 240517, 240551, 240571, 240587, 240589, 240599, 240607, 240623, 240631, 240641, 240659,
240677, 240701, 240707, 240719, 240727, 240733, 240739, 240743, 240763, 240769, 240797, 240811, 240829,
240841, 240853, 240859, 240869, 240881, 240883, 240893, 240899, 240913, 240943, 240953, 240959, 240967,
240997, 241013, 241027, 241037, 241049, 241051, 241061, 241067, 241069, 241079, 241093, 241117, 241127,
241141, 241169, 241177, 241183, 241207, 241229, 241249, 241253, 241259, 241261, 241271, 241291, 241303,
241313, 241321, 241327, 241333, 241337, 241343, 241361, 241363, 241391, 241393, 241421, 241429, 241441,
241453, 241463, 241469, 241489, 241511, 241513, 241517, 241537, 241543, 241559, 241561, 241567, 241589,
241597, 241601, 241603, 241639, 241643, 241651, 241663, 241667, 241679, 241687, 241691, 241711, 241727,
241739, 241771, 241781, 241783, 241793, 241807, 241811, 241817, 241823, 241847, 241861, 241867, 241873,
241877, 241883, 241903, 241907, 241919, 241921, 241931, 241939, 241951, 241963, 241973, 241979, 241981,
241993, 242009, 242057, 242059, 242069, 242083, 242093, 242101, 242119, 242129, 242147, 242161, 242171,
242173, 242197, 242201, 242227, 242243, 242257, 242261, 242273, 242279, 242309, 242329, 242357, 242371,
242377, 242393, 242399, 242413, 242419, 242441, 242447, 242449, 242453, 242467, 242479, 242483, 242491,
242509, 242519, 242521, 242533, 242551, 242591, 242603, 242617, 242621, 242629, 242633, 242639, 242647,
242659, 242677, 242681, 242689, 242713, 242729, 242731, 242747, 242773, 242779, 242789, 242797, 242807,
242813, 242819, 242863, 242867, 242873, 242887, 242911, 242923, 242927, 242971, 242989, 242999, 243011,
243031, 243073, 243077, 243091, 243101, 243109, 243119, 243121, 243137, 243149, 243157, 243161, 243167,
243197, 243203, 243209, 243227, 243233, 243239, 243259, 243263, 243301, 243311, 243343, 243367, 243391,
243401, 243403, 243421, 243431, 243433, 243437, 243461, 243469, 243473, 243479, 243487, 243517, 243521,
243527, 243533, 243539, 243553, 243577, 243583, 243587, 243589, 243613, 243623, 243631, 243643, 243647,
243671, 243673, 243701, 243703, 243707, 243709, 243769, 243781, 243787, 243799, 243809, 243829, 243839,
243851, 243857, 243863, 243871, 243889, 243911, 243917, 243931, 243953, 243973, 243989, 244003, 244009,
244021, 244033, 244043, 244087, 244091, 244109, 244121, 244129, 244141, 244147, 244157, 244159, 244177,
244199, 244217, 244219, 244243, 244247, 244253, 244261, 244291, 244297, 244301, 244303, 244313, 244333,
244339, 244351, 244357, 244367, 244379, 244381, 244393, 244399, 244403, 244411, 244423, 244429, 244451,
244457, 244463, 244471, 244481, 244493, 244507, 244529, 244547, 244553, 244561, 244567, 244583, 244589,
244597, 244603, 244619, 244633, 244637, 244639, 244667, 244669, 244687, 244691, 244703, 244711, 244721,
244733, 244747, 244753, 244759, 244781, 244787, 244813, 244837, 244841, 244843, 244859, 244861, 244873,
244877, 244889, 244897, 244901, 244939, 244943, 244957, 244997, 245023, 245029, 245033, 245039, 245071,
245083, 245087, 245107, 245129, 245131, 245149, 245171, 245173, 245177, 245183, 245209, 245251, 245257,
245261, 245269, 245279, 245291, 245299, 245317, 245321, 245339, 245383, 245389, 245407, 245411, 245417,
245419, 245437, 245471, 245473, 245477, 245501, 245513, 245519, 245521, 245527, 245533, 245561, 245563,
245587, 245591, 245593, 245621, 245627, 245629, 245639, 245653, 245671, 245681, 245683, 245711, 245719,
245723, 245741, 245747, 245753, 245759, 245771, 245783, 245789, 245821, 245849, 245851, 245863, 245881,
245897, 245899, 245909, 245911, 245941, 245963, 245977, 245981, 245983, 245989, 246011, 246017, 246049,
246073, 246097, 246119, 246121, 246131, 246133, 246151, 246167, 246173, 246187, 246193, 246203, 246209,
246217, 246223, 246241, 246247, 246251, 246271, 246277, 246289, 246317, 246319, 246329, 246343, 246349,
246361, 246371, 246391, 246403, 246439, 246469, 246473, 246497, 246509, 246511, 246523, 246527, 246539,
246557, 246569, 246577, 246599, 246607, 246611, 246613, 246637, 246641, 246643, 246661, 246683, 246689,
246707, 246709, 246713, 246731, 246739, 246769, 246773, 246781, 246787, 246793, 246803, 246809, 246811,
246817, 246833, 246839, 246889, 246899, 246907, 246913, 246919, 246923, 246929, 246931, 246937, 246941,
246947, 246971, 246979, 247001, 247007, 247031, 247067, 247069, 247073, 247087, 247099, 247141, 247183,
247193, 247201, 247223, 247229, 247241, 247249, 247259, 247279, 247301, 247309, 247337, 247339, 247343,
247363, 247369, 247381, 247391, 247393, 247409, 247421, 247433, 247439, 247451, 247463, 247501, 247519,
247529, 247531, 247547, 247553, 247579, 247591, 247601, 247603, 247607, 247609, 247613, 247633, 247649,
247651, 247691, 247693, 247697, 247711, 247717, 247729, 247739, 247759, 247769, 247771, 247781, 247799,

247811, 247813, 247829, 247847, 247853, 247873, 247879, 247889, 247901, 247913, 247939, 247943, 247957,
247991, 247993, 247997, 247999, 248021, 248033, 248041, 248051, 248057, 248063, 248071, 248077, 248089,
248099, 248117, 248119, 248137, 248141, 248161, 248167, 248177, 248179, 248189, 248201, 248203, 248231,
248243, 248257, 248267, 248291, 248293, 248299, 248309, 248317, 248323, 248351, 248357, 248371, 248389,
248401, 248407, 248431, 248441, 248447, 248461, 248473, 248477, 248483, 248509, 248533, 248537, 248543,
248569, 248579, 248587, 248593, 248597, 248609, 248621, 248627, 248639, 248641, 248657, 248683, 248701,
248707, 248719, 248723, 248737, 248749, 248753, 248779, 248783, 248789, 248797, 248813, 248821, 248827,
248839, 248851, 248861, 248867, 248869, 248879, 248887, 248891, 248893, 248903, 248909, 248971, 248981,
248987, 249017, 249037, 249059, 249079, 249089, 249097, 249103, 249107, 249127, 249131, 249133, 249143,
249181, 249187, 249199, 249211, 249217, 249229, 249233, 249253, 249257, 249287, 249311, 249317, 249329,
249341, 249367, 249377, 249383, 249397, 249419, 249421, 249427, 249433, 249437, 249439, 249449, 249463,
249497, 249499, 249503, 249517, 249521, 249533, 249539, 249541, 249563, 249583, 249589, 249593, 249607,
249647, 249659, 249671, 249677, 249703, 249721, 249727, 249737, 249749, 249763, 249779, 249797, 249811,
249827, 249833, 249853, 249857, 249859, 249863, 249871, 249881, 249911, 249923, 249943, 249947, 249967,
249971, 249973, 249989, 250007, 250013, 250027, 250031, 250037, 250043, 250049, 250051, 250057, 250073,
250091, 250109, 250123, 250147, 250153, 250169, 250199, 250253, 250259, 250267, 250279, 250301, 250307,
250343, 250361, 250403, 250409, 250423, 250433, 250441, 250451, 250489, 250499, 250501, 250543, 250583,
250619, 250643, 250673, 250681, 250687, 250693, 250703, 250709, 250721, 250727, 250739, 250741, 250751,
250753, 250777, 250787, 250793, 250799, 250807, 250813, 250829, 250837, 250841, 250853, 250867, 250871,
250889, 250919, 250949, 250951, 250963, 250967, 250969, 250979, 250993, 251003, 251033, 251051, 251057,
251059, 251063, 251071, 251081, 251087, 251099, 251117, 251143, 251149, 251159, 251171, 251177, 251179,
251191, 251197, 251201, 251203, 251219, 251221, 251231, 251233, 251257, 251261, 251263, 251287, 251291,
251297, 251323, 251347, 251353, 251359, 251387, 251393, 251417, 251429, 251431, 251437, 251443, 251467,
251473, 251477, 251483, 251491, 251501, 251513, 251519, 251527, 251533, 251539, 251543, 251561, 251567,
251609, 251611, 251621, 251623, 251639, 251653, 251663, 251677, 251701, 251707, 251737, 251761, 251789,
251791, 251809, 251831, 251833, 251843, 251857, 251861, 251879, 251887, 251893, 251897, 251903, 251917,
251939, 251941, 251947, 251969, 251971, 251983, 252001, 252013, 252017, 252029, 252037, 252079, 252101,
252139, 252143, 252151, 252157, 252163, 252169, 252173, 252181, 252193, 252209, 252223, 252233, 252253,
252277, 252283, 252289, 252293, 252313, 252319, 252323, 252341, 252359, 252383, 252391, 252401, 252409,
252419, 252431, 252443, 252449, 252457, 252463, 252481, 252509, 252533, 252541, 252559, 252583, 252589,
252607, 252611, 252617, 252641, 252667, 252691, 252709, 252713, 252727, 252731, 252737, 252761, 252767,
252779, 252817, 252823, 252827, 252829, 252869, 252877, 252881, 252887, 252893, 252899, 252911, 252913,
252919, 252937, 252949, 252971, 252979, 252983, 253003, 253013, 253049, 253063, 253081, 253103, 253109,
253133, 253153, 253157, 253159, 253229, 253243, 253247, 253273, 253307, 253321, 253343, 253349, 253361,
253367, 253369, 253381, 253387, 253417, 253423, 253427, 253433, 253439, 253447, 253469, 253481, 253493,
253501, 253507, 253531, 253537, 253543, 253553, 253567, 253573, 253601, 253607, 253609, 253613, 253633,
253637, 253639, 253651, 253661, 253679, 253681, 253703, 253717, 253733, 253741, 253751, 253763, 253769,
253777, 253787, 253789, 253801, 253811, 253819, 253823, 253853, 253867, 253871, 253879, 253901, 253907,
253909, 253919, 253937, 253949, 253951, 253969, 253987, 253993, 253999, 254003, 254021, 254027, 254039,
254041, 254047, 254053, 254071, 254083, 254119, 254141, 254147, 254161, 254179, 254197, 254207, 254209,
254213, 254249, 254257, 254279, 254281, 254291, 254299, 254329, 254369, 254377, 254383, 254389, 254407,
254413, 254437, 254447, 254461, 254489, 254491, 254519, 254537, 254557, 254593, 254623, 254627, 254647,
254659, 254663, 254699, 254713, 254729, 254731, 254741, 254747, 254753, 254773, 254777, 254783, 254791,
254803, 254827, 254831, 254833, 254857, 254869, 254873, 254879, 254887, 254899, 254911, 254927, 254929,
254941, 254959, 254963, 254971, 254977, 254987, 254993, 255007, 255019, 255023, 255043, 255049, 255053,
255071, 255077, 255083, 255097, 255107, 255121, 255127, 255133, 255137, 255149, 255173, 255179, 255181,
255191, 255193, 255197, 255209, 255217, 255239, 255247, 255251, 255253, 255259, 255313, 255329, 255349,
255361, 255371, 255383, 255413, 255419, 255443, 255457, 255467, 255469, 255473, 255487, 255499, 255503,
255511, 255517, 255523, 255551, 255571, 255587, 255589, 255613, 255617, 255637, 255641, 255649, 255653,

255659, 255667, 255679, 255709, 255713, 255733, 255743, 255757, 255763, 255767, 255803, 255839, 255841,
255847, 255851, 255859, 255869, 255877, 255887, 255907, 255917, 255919, 255923, 255947, 255961, 255971,
255973, 255977, 255989, 256019, 256021, 256031, 256033, 256049, 256057, 256079, 256093, 256117, 256121,
256129, 256133, 256147, 256163, 256169, 256181, 256187, 256189, 256199, 256211, 256219, 256279, 256301,
256307, 256313, 256337, 256349, 256363, 256369, 256391, 256393, 256423, 256441, 256469, 256471, 256483,
256489, 256493, 256499, 256517, 256541, 256561, 256567, 256577, 256579, 256589, 256603, 256609, 256639,
256643, 256651, 256661, 256687, 256699, 256721, 256723, 256757, 256771, 256799, 256801, 256813, 256831,
256873, 256877, 256889, 256901, 256903, 256931, 256939, 256957, 256967, 256981, 257003, 257017, 257053,
257069, 257077, 257093, 257099, 257107, 257123, 257141, 257161, 257171, 257177, 257189, 257219, 257221,
257239, 257249, 257263, 257273, 257281, 257287, 257293, 257297, 257311, 257321, 257339, 257351, 257353,
257371, 257381, 257399, 257401, 257407, 257437, 257443, 257447, 257459, 257473, 257489, 257497, 257501,
257503, 257519, 257539, 257561, 257591, 257611, 257627, 257639, 257657, 257671, 257687, 257689, 257707,
257711, 257713, 257717, 257731, 257783, 257791, 257797, 257837, 257857, 257861, 257863, 257867, 257869,
257879, 257893, 257903, 257921, 257947, 257953, 257981, 257987, 257989, 257993, 258019, 258023, 258031,
258061, 258067, 258101, 258107, 258109, 258113, 258119, 258127, 258131, 258143, 258157, 258161, 258173,
258197, 258211, 258233, 258241, 258253, 258277, 258283, 258299, 258317, 258319, 258329, 258331, 258337,
258353, 258373, 258389, 258403, 258407, 258413, 258421, 258437, 258443, 258449, 258469, 258487, 258491,
258499, 258521, 258527, 258539, 258551, 258563, 258569, 258581, 258607, 258611, 258613, 258617, 258623,
258631, 258637, 258659, 258673, 258677, 258691, 258697, 258703, 258707, 258721, 258733, 258737, 258743,
258763, 258779, 258787, 258803, 258809, 258827, 258847, 258871, 258887, 258917, 258919, 258949, 258959,
258967, 258971, 258977, 258983, 258991, 259001, 259009, 259019, 259033, 259099, 259121, 259123, 259151,
259157, 259159, 259163, 259169, 259177, 259183, 259201, 259211, 259213, 259219, 259229, 259271, 259277,
259309, 259321, 259339, 259379, 259381, 259387, 259397, 259411, 259421, 259429, 259451, 259453, 259459,
259499, 259507, 259517, 259531, 259537, 259547, 259577, 259583, 259603, 259619, 259621, 259627, 259631,
259639, 259643, 259657, 259667, 259681, 259691, 259697, 259717, 259723, 259733, 259751, 259771, 259781,
259783, 259801, 259813, 259823, 259829, 259837, 259841, 259867, 259907, 259933, 259937, 259943, 259949,
259967, 259991, 259993, 260003, 260009, 260011, 260017, 260023, 260047, 260081, 260089, 260111, 260137,
260171, 260179, 260189, 260191, 260201, 260207, 260209, 260213, 260231, 260263, 260269, 260317, 260329,
260339, 260363, 260387, 260399, 260411, 260413, 260417, 260419, 260441, 260453, 260461, 260467, 260483,
260489, 260527, 260539, 260543, 260549, 260551, 260569, 260573, 260581, 260587, 260609, 260629, 260647,
260651, 260671, 260677, 260713, 260717, 260723, 260747, 260753, 260761, 260773, 260791, 260807, 260809,
260849, 260857, 260861, 260863, 260873, 260879, 260893, 260921, 260941, 260951, 260959, 260969, 260983,
260987, 260999, 261011, 261013, 261017, 261031, 261043, 261059, 261061, 261071, 261077, 261089, 261101,
261127, 261167, 261169, 261223, 261229, 261241, 261251, 261271, 261281, 261301, 261323, 261329, 261337,
261347, 261353, 261379, 261389, 261407, 261427, 261431, 261433, 261439, 261451, 261463, 261467, 261509,
261523, 261529, 261557, 261563, 261577, 261581, 261587, 261593, 261601, 261619, 261631, 261637, 261641,
261643, 261673, 261697, 261707, 261713, 261721, 261739, 261757, 261761, 261773, 261787, 261791, 261799,
261823, 261847, 261881, 261887, 261917, 261959, 261971, 261973, 261977, 261983, 262007, 262027, 262049,
262051, 262069, 262079, 262103, 262109, 262111, 262121, 262127, 262133, 262139, 262147, 262151, 262153,
262187, 262193, 262217, 262231, 262237, 262253, 262261, 262271, 262303, 262313, 262321, 262331, 262337,
262349, 262351, 262369, 262387, 262391, 262399, 262411, 262433, 262459, 262469, 262489, 262501, 262511,
262513, 262519, 262541, 262543, 262553, 262567, 262583, 262597, 262621, 262627, 262643, 262649, 262651,
262657, 262681, 262693, 262697, 262709, 262723, 262733, 262739, 262741, 262747, 262781, 262783, 262807,
262819, 262853, 262877, 262883, 262897, 262901, 262909, 262927, 262949, 262957, 262981, 263009, 263023,
263047, 263063, 263071, 263077, 263083, 263089, 263101, 263111, 263119, 263129, 263167, 263171, 263183,
263191, 263201, 263209, 263213, 263227, 263239, 263257, 263267, 263269, 263273, 263287, 263293, 263303,
263323, 263369, 263383, 263387, 263399, 263401, 263411, 263423, 263429, 263437, 263443, 263489, 263491,
263503, 263513, 263519, 263521, 263533, 263537, 263561, 263567, 263573, 263591, 263597, 263609, 263611,
263621, 263647, 263651, 263657, 263677, 263723, 263729, 263737, 263759, 263761, 263803, 263819, 263821,

263827, 263843, 263849, 263863, 263867, 263869, 263881, 263899, 263909, 263911, 263927, 263933, 263941,
263951, 263953, 263957, 263983, 264007, 264013, 264029, 264031, 264053, 264059, 264071, 264083, 264091,
264101, 264113, 264127, 264133, 264137, 264139, 264167, 264169, 264179, 264211, 264221, 264263, 264269,
264283, 264289, 264301, 264323, 264331, 264343, 264349, 264353, 264359, 264371, 264391, 264403, 264437,
264443, 264463, 264487, 264527, 264529, 264553, 264559, 264577, 264581, 264599, 264601, 264619, 264631,
264637, 264643, 264659, 264697, 264731, 264739, 264743, 264749, 264757, 264763, 264769, 264779, 264787,
264791, 264793, 264811, 264827, 264829, 264839, 264871, 264881, 264889, 264893, 264899, 264919, 264931,
264949, 264959, 264961, 264977, 264991, 264997, 265003, 265007, 265021, 265037, 265079, 265091, 265093,
265117, 265123, 265129, 265141, 265151, 265157, 265163, 265169, 265193, 265207, 265231, 265241, 265247,
265249, 265261, 265271, 265273, 265277, 265313, 265333, 265337, 265339, 265381, 265399, 265403, 265417,
265423, 265427, 265451, 265459, 265471, 265483, 265493, 265511, 265513, 265541, 265543, 265547, 265561,
265567, 265571, 265579, 265607, 265613, 265619, 265621, 265703, 265709, 265711, 265717, 265729, 265739,
265747, 265757, 265781, 265787, 265807, 265813, 265819, 265831, 265841, 265847, 265861, 265871, 265873,
265883, 265891, 265921, 265957, 265961, 265987, 266003, 266009, 266023, 266027, 266029, 266047, 266051,
266053, 266059, 266081, 266083, 266089, 266093, 266099, 266111, 266117, 266129, 266137, 266153, 266159,
266177, 266183, 266221, 266239, 266261, 266269, 266281, 266291, 266293, 266297, 266333, 266351, 266353,
266359, 266369, 266381, 266401, 266411, 266417, 266447, 266449, 266477, 266479, 266489, 266491, 266521,
266549, 266587, 266599, 266603, 266633, 266641, 266647, 266663, 266671, 266677, 266681, 266683, 266687,
266689, 266701, 266711, 266719, 266759, 266767, 266797, 266801, 266821, 266837, 266839, 266863, 266867,
266891, 266897, 266899, 266909, 266921, 266927, 266933, 266947, 266953, 266957, 266971, 266977, 266983,
266993, 266999, 267017, 267037, 267049, 267097, 267131, 267133, 267139, 267143, 267167, 267187, 267193,
267199, 267203, 267217, 267227, 267229, 267233, 267259, 267271, 267277, 267299, 267301, 267307, 267317,
267341, 267353, 267373, 267389, 267391, 267401, 267403, 267413, 267419, 267431, 267433, 267439, 267451,
267469, 267479, 267481, 267493, 267497, 267511, 267517, 267521, 267523, 267541, 267551, 267557, 267569,
267581, 267587, 267593, 267601, 267611, 267613, 267629, 267637, 267643, 267647, 267649, 267661, 267667,
267671, 267677, 267679, 267713, 267719, 267721, 267727, 267737, 267739, 267749, 267763, 267781, 267791,
267797, 267803, 267811, 267829, 267833, 267857, 267863, 267877, 267887, 267893, 267899, 267901, 267907,
267913, 267929, 267941, 267959, 267961, 268003, 268013, 268043, 268049, 268063, 268069, 268091, 268123,
268133, 268153, 268171, 268189, 268199, 268207, 268211, 268237, 268253, 268267, 268271, 268283, 268291,
268297, 268343, 268403, 268439, 268459, 268487, 268493, 268501, 268507, 268517, 268519, 268529, 268531,
268537, 268547, 268573, 268607, 268613, 268637, 268643, 268661, 268693, 268721, 268729, 268733, 268747,
268757, 268759, 268771, 268777, 268781, 268783, 268789, 268811, 268813, 268817, 268819, 268823, 268841,
268843, 268861, 268883, 268897, 268909, 268913, 268921, 268927, 268937, 268969, 268973, 268979, 268993,
268997, 268999, 269023, 269029, 269039, 269041, 269057, 269063, 269069, 269089, 269117, 269131, 269141,
269167, 269177, 269179, 269183, 269189, 269201, 269209, 269219, 269221, 269231, 269237, 269251, 269257,
269281, 269317, 269327, 269333, 269341, 269351, 269377, 269383, 269387, 269389, 269393, 269413, 269419,
269429, 269431, 269441, 269461, 269473, 269513, 269519, 269527, 269539, 269543, 269561, 269573, 269579,
269597, 269617, 269623, 269641, 269651, 269663, 269683, 269701, 269713, 269719, 269723, 269741, 269749,
269761, 269779, 269783, 269791, 269851, 269879, 269887, 269891, 269897, 269923, 269939, 269947, 269953,
269981, 269987, 270001, 270029, 270031, 270037, 270059, 270071, 270073, 270097, 270121, 270131, 270133,
270143, 270157, 270163, 270167, 270191, 270209, 270217, 270223, 270229, 270239, 270241, 270269, 270271,
270287, 270299, 270307, 270311, 270323, 270329, 270337, 270343, 270371, 270379, 270407, 270421, 270437,
270443, 270451, 270461, 270463, 270493, 270509, 270527, 270539, 270547, 270551, 270553, 270563, 270577,
270583, 270587, 270593, 270601, 270619, 270631, 270653, 270659, 270667, 270679, 270689, 270701, 270709,
270719, 270737, 270749, 270761, 270763, 270791, 270797, 270799, 270821, 270833, 270841, 270859, 270899,
270913, 270923, 270931, 270937, 270953, 270961, 270967, 270973, 271003, 271013, 271021, 271027, 271043,
271057, 271067, 271079, 271097, 271109, 271127, 271129, 271163, 271169, 271177, 271181, 271211, 271217,
271231, 271241, 271253, 271261, 271273, 271277, 271279, 271289, 271333, 271351, 271357, 271363, 271367,
271393, 271409, 271429, 271451, 271463, 271471, 271483, 271489, 271499, 271501, 271517, 271549, 271553,

271571, 271573, 271597, 271603, 271619, 271637, 271639, 271651, 271657, 271693, 271703, 271723, 271729, 271753, 271769, 271771, 271787, 271807, 271811, 271829, 271841, 271849, 271853, 271861, 271867, 271879, 271897, 271903, 271919, 271927, 271939, 271967, 271969, 271981, 272003, 272009, 272011, 272029, 272039, 272053, 272059, 272093, 272131, 272141, 272171, 272179, 272183, 272189, 272191, 272201, 272203, 272227, 272231, 272249, 272257, 272263, 272267, 272269, 272287, 272299, 272317, 272329, 272333, 272341, 272347, 272351, 272353, 272359, 272369, 272381, 272383, 272399, 272407, 272411, 272417, 272423, 272449, 272453, 272477, 272507, 272533, 272537, 272539, 272549, 272563, 272567, 272581, 272603, 272621, 272651, 272659, 272683, 272693, 272717, 272719, 272737, 272759, 272761, 272771, 272777, 272807, 272809, 272813, 272863, 272879, 272887, 272903, 272911, 272917, 272927, 272933, 272959, 272971, 272981, 272983, 272989, 272999, 273001, 273029, 273043, 273047, 273059, 273061, 273067, 273073, 273083, 273107, 273113, 273127, 273131, 273149, 273157, 273181, 273187, 273193, 273233, 273253, 273269, 273271, 273281, 273283, 273289, 273311, 273313, 273323, 273349, 273359, 273367, 273433, 273457, 273473, 273503, 273517, 273521, 273527, 273551, 273569, 273601, 273613, 273617, 273629, 273641, 273643, 273653, 273697, 273709, 273719, 273727, 273739, 273773, 273787, 273797, 273803, 273821, 273827, 273857, 273881, 273899, 273901, 273913, 273919, 273929, 273941, 273943, 273967, 273971, 273979, 273997, 274007, 274019, 274033, 274061, 274069, 274081, 274093, 274103, 274117, 274121, 274123, 274139, 274147, 274163, 274171, 274177, 274187, 274199, 274201, 274213, 274223, 274237, 274243, 274259, 274271, 274277, 274283, 274301, 274333, 274349, 274357, 274361, 274403, 274423, 274441, 274451, 274453, 274457, 274471, 274489, 274517, 274529, 274579, 274583, 274591, 274609, 274627, 274661, 274667, 274679, 274693, 274697, 274709, 274711, 274723, 274739, 274751, 274777, 274783, 274787, 274811, 274817, 274829, 274831, 274837, 274843, 274847, 274853, 274861, 274867, 274871, 274889, 274909, 274931, 274943, 274951, 274957, 274961, 274973, 274993, 275003, 275027, 275039, 275047, 275053, 275059, 275083, 275087, 275129, 275131, 275147, 275153, 275159, 275161, 275167, 275183, 275201, 275207, 275227, 275251, 275263, 275269, 275299, 275309, 275321, 275323, 275339, 275357, 275371, 275389, 275393, 275399, 275419, 275423, 275447, 275449, 275453, 275459, 275461, 275489, 275491, 275503, 275521, 275531, 275543, 275549, 275573, 275579, 275581, 275591, 275593, 275599, 275623, 275641, 275651, 275657, 275669, 275677, 275699, 275711, 275719, 275729, 275741, 275767, 275773, 275783, 275813, 275827, 275837, 275881, 275897, 275911, 275917, 275921, 275923, 275929, 275939, 275941, 275963, 275969, 275981, 275987, 275999, 276007, 276011, 276019, 276037, 276041, 276043, 276047, 276049, 276079, 276083, 276091, 276113, 276137, 276151, 276173, 276181, 276187, 276191, 276209, 276229, 276239, 276247, 276251, 276257, 276277, 276293, 276319, 276323, 276337, 276343, 276347, 276359, 276371, 276373, 276389, 276401, 276439, 276443, 276449, 276461, 276467, 276487, 276499, 276503, 276517, 276527, 276553, 276557, 276581, 276587, 276589, 276593, 276599, 276623, 276629, 276637, 276671, 276673, 276707, 276721, 276739, 276763, 276767, 276779, 276781, 276817, 276821, 276823, 276827, 276833, 276839, 276847, 276869, 276883, 276901, 276907, 276917, 276919, 276929, 276949, 276953, 276961, 276977, 277003, 277007, 277021, 277051, 277063, 277073, 277087, 277097, 277099, 277157, 277163, 277169, 277177, 277183, 277213, 277217, 277223, 277231, 277247, 277259, 277261, 277273, 277279, 277297, 277301, 277309, 277331, 277363, 277373, 277411, 277421, 277427, 277429, 277483, 277493, 277499, 277513, 277531, 277547, 277549, 277567, 277577, 277579, 277597, 277601, 277603, 277637, 277639, 277643, 277657, 277663, 277687, 277691, 277703, 277741, 277747, 277751, 277757, 277787, 277789, 277793, 277813, 277829, 277847, 277859, 277883, 277889, 277891, 277897, 277903, 277919, 277961, 277993, 277999, 278017, 278029, 278041, 278051, 278063, 278071, 278087, 278111, 278119, 278123, 278143, 278147, 278149, 278177, 278191, 278207, 278209, 278219, 278227, 278233, 278237, 278261, 278269, 278279, 278321, 278329, 278347, 278353, 278363, 278387, 278393, 278413, 278437, 278459, 278479, 278489, 278491, 278497, 278501, 278503, 278543, 278549, 278557, 278561, 278563, 278581, 278591, 278609, 278611, 278617, 278623, 278627, 278639, 278651, 278671, 278687, 278689, 278701, 278717, 278741, 278743, 278753, 278767, 278801, 278807, 278809, 278813, 278819, 278827, 278843, 278849, 278867, 278879, 278881, 278891, 278903, 278909, 278911, 278917, 278947, 278981, 279001, 279007, 279023, 279029, 279047, 279073, 279109, 279119, 279121, 279127, 279131, 279137, 279143, 279173, 279179, 279187, 279203, 279211, 279221, 279269, 279311, 279317, 279329, 279337, 279353, 279397, 279407, 279413, 279421, 279431, 279443, 279451, 279479, 279481, 279511, 279523, 279541, 279551, 279553, 279557, 279571, 279577, 279583, 279593, 279607, 279613, 279619, 279637,

279641, 279649, 279659, 279679, 279689, 279707, 279709, 279731, 279751, 279761, 279767, 279779, 279817,
279823, 279847, 279857, 279863, 279883, 279913, 279919, 279941, 279949, 279967, 279977, 279991, 280001,
280009, 280013, 280031, 280037, 280061, 280069, 280097, 280099, 280103, 280121, 280129, 280139, 280183,
280187, 280199, 280207, 280219, 280223, 280229, 280243, 280249, 280253, 280277, 280297, 280303, 280321,
280327, 280337, 280339, 280351, 280373, 280409, 280411, 280451, 280463, 280487, 280499, 280507, 280513,
280537, 280541, 280547, 280549, 280561, 280583, 280589, 280591, 280597, 280603, 280607, 280613, 280627,
280639, 280673, 280681, 280697, 280699, 280703, 280711, 280717, 280729, 280751, 280759, 280769, 280771,
280811, 280817, 280837, 280843, 280859, 280871, 280879, 280883, 280897, 280909, 280913, 280921, 280927,
280933, 280939, 280949, 280957, 280963, 280967, 280979, 280997, 281023, 281033, 281053, 281063, 281069,
281081, 281117, 281131, 281153, 281159, 281167, 281189, 281191, 281207, 281227, 281233, 281243, 281249,
281251, 281273, 281279, 281291, 281297, 281317, 281321, 281327, 281339, 281353, 281357, 281363, 281381,
281419, 281423, 281429, 281431, 281509, 281527, 281531, 281539, 281549, 281551, 281557, 281563, 281579,
281581, 281609, 281621, 281623, 281627, 281641, 281647, 281651, 281653, 281663, 281669, 281683, 281717,
281719, 281737, 281747, 281761, 281767, 281777, 281783, 281791, 281797, 281803, 281807, 281833, 281837,
281839, 281849, 281857, 281867, 281887, 281893, 281921, 281923, 281927, 281933, 281947, 281959, 281971,
281989, 281993, 282001, 282011, 282019, 282053, 282059, 282071, 282089, 282091, 282097, 282101, 282103,
282127, 282143, 282157, 282167, 282221, 282229, 282239, 282241, 282253, 282281, 282287, 282299, 282307,
282311, 282313, 282349, 282377, 282383, 282389, 282391, 282407, 282409, 282413, 282427, 282439, 282461,
282481, 282487, 282493, 282559, 282563, 282571, 282577, 282589, 282599, 282617, 282661, 282671, 282677,
282679, 282683, 282691, 282697, 282703, 282707, 282713, 282767, 282769, 282773, 282797, 282809, 282827,
282833, 282847, 282851, 282869, 282881, 282889, 282907, 282911, 282913, 282917, 282959, 282973, 282977,
282991, 283001, 283007, 283009, 283027, 283051, 283079, 283093, 283097, 283099, 283111, 283117, 283121,
283133, 283139, 283159, 283163, 283181, 283183, 283193, 283207, 283211, 283267, 283277, 283289, 283303,
283369, 283397, 283403, 283411, 283447, 283463, 283487, 283489, 283501, 283511, 283519, 283541, 283553,
283571, 283573, 283579, 283583, 283601, 283607, 283609, 283631, 283637, 283639, 283669, 283687, 283697,
283721, 283741, 283763, 283769, 283771, 283793, 283799, 283807, 283813, 283817, 283831, 283837, 283859,
283861, 283873, 283909, 283937, 283949, 283957, 283961, 283979, 284003, 284023, 284041, 284051, 284057,
284059, 284083, 284093, 284111, 284117, 284129, 284131, 284149, 284153, 284159, 284161, 284173, 284191,
284201, 284227, 284231, 284233, 284237, 284243, 284261, 284267, 284269, 284293, 284311, 284341, 284357,
284369, 284377, 284387, 284407, 284413, 284423, 284429, 284447, 284467, 284477, 284483, 284489, 284507,
284509, 284521, 284527, 284539, 284551, 284561, 284573, 284587, 284591, 284593, 284623, 284633, 284651,
284657, 284659, 284681, 284689, 284701, 284707, 284723, 284729, 284731, 284737, 284741, 284743, 284747,
284749, 284759, 284777, 284783, 284803, 284807, 284813, 284819, 284831, 284833, 284839, 284857, 284881,
284897, 284899, 284917, 284927, 284957, 284969, 284989, 285007, 285023, 285031, 285049, 285071, 285079,
285091, 285101, 285113, 285119, 285121, 285139, 285151, 285161, 285179, 285191, 285199, 285221, 285227,
285251, 285281, 285283, 285287, 285289, 285301, 285317, 285343, 285377, 285421, 285433, 285451, 285457,
285463, 285469, 285473, 285497, 285517, 285521, 285533, 285539, 285553, 285557, 285559, 285569, 285599,
285611, 285613, 285629, 285631, 285641, 285643, 285661, 285667, 285673, 285697, 285707, 285709, 285721,
285731, 285749, 285757, 285763, 285767, 285773, 285781, 285823, 285827, 285839, 285841, 285871, 285937,
285949, 285953, 285977, 285979, 285983, 285997, 286001, 286009, 286019, 286043, 286049, 286061, 286063,
286073, 286103, 286129, 286163, 286171, 286199, 286243, 286249, 286289, 286301, 286333, 286367, 286369,
286381, 286393, 286397, 286411, 286421, 286427, 286453, 286457, 286459, 286469, 286477, 286483, 286487,
286493, 286499, 286513, 286519, 286541, 286543, 286547, 286553, 286589, 286591, 286609, 286613, 286619,
286633, 286651, 286673, 286687, 286697, 286703, 286711, 286721, 286733, 286751, 286753, 286763, 286771,
286777, 286789, 286801, 286813, 286831, 286859, 286873, 286927, 286973, 286981, 286987, 286999, 287003,
287047, 287057, 287059, 287087, 287093, 287099, 287107, 287117, 287137, 287141, 287149, 287159, 287167,
287173, 287179, 287191, 287219, 287233, 287237, 287239, 287251, 287257, 287269, 287279, 287281, 287291,
287297, 287321, 287327, 287333, 287341, 287347, 287383, 287387, 287393, 287437, 287449, 287491, 287501,
287503, 287537, 287549, 287557, 287579, 287597, 287611, 287629, 287669, 287671, 287681, 287689, 287701,

287731, 287747, 287783, 287789, 287801, 287813, 287821, 287849, 287851, 287857, 287863, 287867, 287873,
287887, 287921, 287933, 287939, 287977, 288007, 288023, 288049, 288053, 288061, 288077, 288089, 288109,
288137, 288179, 288181, 288191, 288199, 288203, 288209, 288227, 288241, 288247, 288257, 288283, 288293,
288307, 288313, 288317, 288349, 288359, 288361, 288383, 288389, 288403, 288413, 288427, 288433, 288461,
288467, 288481, 288493, 288499, 288527, 288529, 288539, 288551, 288559, 288571, 288577, 288583, 288647,
288649, 288653, 288661, 288679, 288683, 288689, 288697, 288731, 288733, 288751, 288767, 288773, 288803,
288817, 288823, 288833, 288839, 288851, 288853, 288877, 288907, 288913, 288929, 288931, 288947, 288973,
288979, 288989, 288991, 288997, 289001, 289019, 289021, 289031, 289033, 289039, 289049, 289063, 289067,
289099, 289103, 289109, 289111, 289127, 289129, 289139, 289141, 289151, 289169, 289171, 289181, 289189,
289193, 289213, 289241, 289243, 289249, 289253, 289273, 289283, 289291, 289297, 289309, 289319, 289343,
289349, 289361, 289369, 289381, 289397, 289417, 289423, 289439, 289453, 289463, 289469, 289477, 289489,
289511, 289543, 289559, 289573, 289577, 289589, 289603, 289607, 289637, 289643, 289657, 289669, 289717,
289721, 289727, 289733, 289741, 289759, 289763, 289771, 289789, 289837, 289841, 289843, 289847, 289853,
289859, 289871, 289889, 289897, 289937, 289951, 289957, 289967, 289973, 289987, 289999, 290011, 290021,
290023, 290027, 290033, 290039, 290041, 290047, 290057, 290083, 290107, 290113, 290119, 290137, 290141,
290161, 290183, 290189, 290201, 290209, 290219, 290233, 290243, 290249, 290317, 290327, 290347, 290351,
290359, 290369, 290383, 290393, 290399, 290419, 290429, 290441, 290443, 290447, 290471, 290473, 290489,
290497, 290509, 290527, 290531, 290533, 290539, 290557, 290593, 290597, 290611, 290617, 290621, 290623,
290627, 290657, 290659, 290663, 290669, 290671, 290677, 290701, 290707, 290711, 290737, 290761, 290767,
290791, 290803, 290821, 290827, 290837, 290839, 290861, 290869, 290879, 290897, 290923, 290959, 290963,
290971, 290987, 290993, 290999, 291007, 291013, 291037, 291041, 291043, 291077, 291089, 291101, 291103,
291107, 291113, 291143, 291167, 291169, 291173, 291191, 291199, 291209, 291217, 291253, 291257, 291271,
291287, 291293, 291299, 291331, 291337, 291349, 291359, 291367, 291371, 291373, 291377, 291419, 291437,
291439, 291443, 291457, 291481, 291491, 291503, 291509, 291521, 291539, 291547, 291559, 291563, 291569,
291619, 291647, 291649, 291661, 291677, 291689, 291691, 291701, 291721, 291727, 291743, 291751, 291779,
291791, 291817, 291829, 291833, 291853, 291857, 291869, 291877, 291887, 291899, 291901, 291923, 291971,
291979, 291983, 291997, 292021, 292027, 292037, 292057, 292069, 292079, 292081, 292091, 292093, 292133,
292141, 292147, 292157, 292181, 292183, 292223, 292231, 292241, 292249, 292267, 292283, 292301, 292309,
292319, 292343, 292351, 292363, 292367, 292381, 292393, 292427, 292441, 292459, 292469, 292471, 292477,
292483, 292489, 292493, 292517, 292531, 292541, 292549, 292561, 292573, 292577, 292601, 292627, 292631,
292661, 292667, 292673, 292679, 292693, 292703, 292709, 292711, 292717, 292727, 292753, 292759, 292777,
292793, 292801, 292807, 292819, 292837, 292841, 292849, 292867, 292879, 292909, 292921, 292933, 292969,
292973, 292979, 292993, 293021, 293071, 293081, 293087, 293093, 293099, 293107, 293123, 293129, 293147,
293149, 293173, 293177, 293179, 293201, 293207, 293213, 293221, 293257, 293261, 293263, 293269, 293311,
293329, 293339, 293351, 293357, 293399, 293413, 293431, 293441, 293453, 293459, 293467, 293473, 293483,
293507, 293543, 293599, 293603, 293617, 293621, 293633, 293639, 293651, 293659, 293677, 293681, 293701,
293717, 293723, 293729, 293749, 293767, 293773, 293791, 293803, 293827, 293831, 293861, 293863, 293893,
293899, 293941, 293957, 293983, 293989, 293999, 294001, 294013, 294023, 294029, 294043, 294053, 294059,
294067, 294103, 294127, 294131, 294149, 294157, 294167, 294169, 294179, 294181, 294199, 294211, 294223,
294227, 294241, 294247, 294251, 294269, 294277, 294289, 294293, 294311, 294313, 294317, 294319, 294337,
294341, 294347, 294353, 294383, 294391, 294397, 294403, 294431, 294439, 294461, 294467, 294479, 294499,
294509, 294523, 294529, 294551, 294563, 294629, 294641, 294647, 294649, 294659, 294673, 294703, 294731,
294751, 294757, 294761, 294773, 294781, 294787, 294793, 294799, 294803, 294809, 294821, 294829, 294859,
294869, 294887, 294893, 294911, 294919, 294923, 294947, 294949, 294953, 294979, 294989, 294991, 294997,
295007, 295033, 295037, 295039, 295049, 295073, 295079, 295081, 295111, 295123, 295129, 295153, 295187,
295199, 295201, 295219, 295237, 295247, 295259, 295271, 295277, 295283, 295291, 295313, 295319, 295333,
295357, 295363, 295387, 295411, 295417, 295429, 295433, 295439, 295441, 295459, 295513, 295517, 295541,
295553, 295567, 295571, 295591, 295601, 295663, 295693, 295699, 295703, 295727, 295751, 295759, 295769,
295777, 295787, 295819, 295831, 295837, 295843, 295847, 295853, 295861, 295871, 295873, 295877, 295879,

295901, 295903, 295909, 295937, 295943, 295949, 295951, 295961, 295973, 295993, 296011, 296017, 296027,
296041, 296047, 296071, 296083, 296099, 296117, 296129, 296137, 296159, 296183, 296201, 296213, 296221,
296237, 296243, 296249, 296251, 296269, 296273, 296279, 296287, 296299, 296347, 296353, 296363, 296369,
296377, 296437, 296441, 296473, 296477, 296479, 296489, 296503, 296507, 296509, 296519, 296551, 296557,
296561, 296563, 296579, 296581, 296587, 296591, 296627, 296651, 296663, 296669, 296683, 296687, 296693,
296713, 296719, 296729, 296731, 296741, 296749, 296753, 296767, 296771, 296773, 296797, 296801, 296819,
296827, 296831, 296833, 296843, 296909, 296911, 296921, 296929, 296941, 296969, 296971, 296981, 296983,
296987, 297019, 297023, 297049, 297061, 297067, 297079, 297083, 297097, 297113, 297133, 297151, 297161,
297169, 297191, 297233, 297247, 297251, 297257, 297263, 297289, 297317, 297359, 297371, 297377, 297391,
297397, 297403, 297421, 297439, 297457, 297467, 297469, 297481, 297487, 297503, 297509, 297523, 297533,
297581, 297589, 297601, 297607, 297613, 297617, 297623, 297629, 297641, 297659, 297683, 297691, 297707,
297719, 297727, 297757, 297779, 297793, 297797, 297809, 297811, 297833, 297841, 297853, 297881, 297889,
297893, 297907, 297911, 297931, 297953, 297967, 297971, 297989, 297991, 298013, 298021, 298031, 298043,
298049, 298063, 298087, 298093, 298099, 298153, 298157, 298159, 298169, 298171, 298187, 298201, 298211,
298213, 298223, 298237, 298247, 298261, 298283, 298303, 298307, 298327, 298339, 298343, 298349, 298369,
298373, 298399, 298409, 298411, 298427, 298451, 298477, 298483, 298513, 298559, 298579, 298583, 298589,
298601, 298607, 298621, 298631, 298651, 298667, 298679, 298681, 298687, 298691, 298693, 298709, 298723,
298733, 298757, 298759, 298777, 298799, 298801, 298817, 298819, 298841, 298847, 298853, 298861, 298897,
298937, 298943, 298993, 298999, 299011, 299017, 299027, 299029, 299053, 299059, 299063, 299087, 299099,
299107, 299113, 299137, 299147, 299171, 299179, 299191, 299197, 299213, 299239, 299261, 299281, 299287,
299311, 299317, 299329, 299333, 299357, 299359, 299363, 299371, 299389, 299393, 299401, 299417, 299419,
299447, 299471, 299473, 299477, 299479, 299501, 299513, 299521, 299527, 299539, 299567, 299569, 299603,
299617, 299623, 299653, 299671, 299681, 299683, 299699, 299701, 299711, 299723, 299731, 299743, 299749,
299771, 299777, 299807, 299843, 299857, 299861, 299881, 299891, 299903, 299909, 299933, 299941, 299951,
299969, 299977, 299983, 299993, 300007, 300017, 300023, 300043, 300073, 300089, 300109, 300119, 300137,
300149, 300151, 300163, 300187, 300191, 300193, 300221, 300229, 300233, 300239, 300247, 300277, 300299,
300301, 300317, 300319, 300323, 300331, 300343, 300347, 300367, 300397, 300413, 300427, 300431, 300439,
300463, 300481, 300491, 300493, 300497, 300499, 300511, 300557, 300569, 300581, 300583, 300589, 300593,
300623, 300631, 300647, 300649, 300661, 300667, 300673, 300683, 300691, 300719, 300721, 300733, 300739,
300743, 300749, 300757, 300761, 300779, 300787, 300799, 300809, 300821, 300823, 300851, 300857, 300869,
300877, 300889, 300893, 300929, 300931, 300953, 300961, 300967, 300973, 300977, 300997, 301013, 301027,
301039, 301051, 301057, 301073, 301079, 301123, 301127, 301141, 301153, 301159, 301177, 301181, 301183,
301211, 301219, 301237, 301241, 301243, 301247, 301267, 301303, 301319, 301331, 301333, 301349, 301361,
301363, 301381, 301403, 301409, 301423, 301429, 301447, 301459, 301463, 301471, 301487, 301489, 301493,
301501, 301531, 301577, 301579, 301583, 301591, 301601, 301619, 301627, 301643, 301649, 301657, 301669,
301673, 301681, 301703, 301711, 301747, 301751, 301753, 301759, 301789, 301793, 301813, 301831, 301841,
301843, 301867, 301877, 301897, 301901, 301907, 301913, 301927, 301933, 301943, 301949, 301979, 301991,
301993, 301997, 301999, 302009, 302053, 302111, 302123, 302143, 302167, 302171, 302173, 302189, 302191,
302213, 302221, 302227, 302261, 302273, 302279, 302287, 302297, 302299, 302317, 302329, 302399, 302411,
302417, 302429, 302443, 302459, 302483, 302507, 302513, 302551, 302563, 302567, 302573, 302579, 302581,
302587, 302593, 302597, 302609, 302629, 302647, 302663, 302681, 302711, 302723, 302747, 302759, 302767,
302779, 302791, 302801, 302831, 302833, 302837, 302843, 302851, 302857, 302873, 302891, 302903, 302909,
302921, 302927, 302941, 302959, 302969, 302971, 302977, 302983, 302989, 302999, 303007, 303011, 303013,
303019, 303029, 303049, 303053, 303073, 303089, 303091, 303097, 303119, 303139, 303143, 303151, 303157,
303187, 303217, 303257, 303271, 303283, 303287, 303293, 303299, 303307, 303313, 303323, 303337, 303341,
303361, 303367, 303371, 303377, 303379, 303389, 303409, 303421, 303431, 303463, 303469, 303473, 303491,
303493, 303497, 303529, 303539, 303547, 303551, 303553, 303571, 303581, 303587, 303593, 303613, 303617,
303619, 303643, 303647, 303649, 303679, 303683, 303689, 303691, 303703, 303713, 303727, 303731, 303749,
303767, 303781, 303803, 303817, 303827, 303839, 303859, 303871, 303889, 303907, 303917, 303931, 303937,

303959, 303983, 303997, 304009, 304013, 304021, 304033, 304039, 304049, 304063, 304067, 304069, 304081,
304091, 304099, 304127, 304151, 304153, 304163, 304169, 304193, 304211, 304217, 304223, 304253, 304259,
304279, 304301, 304303, 304331, 304349, 304357, 304363, 304373, 304391, 304393, 304411, 304417, 304429,
304433, 304439, 304457, 304459, 304477, 304481, 304489, 304501, 304511, 304517, 304523, 304537, 304541,
304553, 304559, 304561, 304597, 304609, 304631, 304643, 304651, 304663, 304687, 304709, 304723, 304729,
304739, 304751, 304757, 304763, 304771, 304781, 304789, 304807, 304813, 304831, 304847, 304849, 304867,
304879, 304883, 304897, 304901, 304903, 304907, 304933, 304937, 304943, 304949, 304961, 304979, 304981,
305017, 305021, 305023, 305029, 305033, 305047, 305069, 305093, 305101, 305111, 305113, 305119, 305131,
305143, 305147, 305209, 305219, 305231, 305237, 305243, 305267, 305281, 305297, 305329, 305339, 305351,
305353, 305363, 305369, 305377, 305401, 305407, 305411, 305413, 305419, 305423, 305441, 305449, 305471,
305477, 305479, 305483, 305489, 305497, 305521, 305533, 305551, 305563, 305581, 305593, 305597, 305603,
305611, 305621, 305633, 305639, 305663, 305717, 305719, 305741, 305743, 305749, 305759, 305761, 305771,
305783, 305803, 305821, 305839, 305849, 305857, 305861, 305867, 305873, 305917, 305927, 305933, 305947,
305971, 305999, 306011, 306023, 306029, 306041, 306049, 306083, 306091, 306121, 306133, 306139, 306149,
306157, 306167, 306169, 306191, 306193, 306209, 306239, 306247, 306253, 306259, 306263, 306301, 306329,
306331, 306347, 306349, 306359, 306367, 306377, 306389, 306407, 306419, 306421, 306431, 306437, 306457,
306463, 306473, 306479, 306491, 306503, 306511, 306517, 306529, 306533, 306541, 306563, 306577, 306587,
306589, 306643, 306653, 306661, 306689, 306701, 306703, 306707, 306727, 306739, 306749, 306763, 306781,
306809, 306821, 306827, 306829, 306847, 306853, 306857, 306871, 306877, 306883, 306893, 306899, 306913,
306919, 306941, 306947, 306949, 306953, 306991, 307009, 307019, 307031, 307033, 307067, 307079, 307091,
307093, 307103, 307121, 307129, 307147, 307163, 307169, 307171, 307187, 307189, 307201, 307243, 307253,
307259, 307261, 307267, 307273, 307277, 307283, 307289, 307301, 307337, 307339, 307361, 307367, 307381,
307397, 307399, 307409, 307423, 307451, 307471, 307481, 307511, 307523, 307529, 307537, 307543, 307577,
307583, 307589, 307609, 307627, 307631, 307633, 307639, 307651, 307669, 307687, 307691, 307693, 307711,
307733, 307759, 307817, 307823, 307831, 307843, 307859, 307871, 307873, 307891, 307903, 307919, 307939,
307969, 308003, 308017, 308027, 308041, 308051, 308081, 308093, 308101, 308107, 308117, 308129, 308137,
308141, 308149, 308153, 308213, 308219, 308249, 308263, 308291, 308293, 308303, 308309, 308311, 308317,
308323, 308327, 308333, 308359, 308383, 308411, 308423, 308437, 308447, 308467, 308489, 308491, 308501,
308507, 308509, 308519, 308521, 308527, 308537, 308551, 308569, 308573, 308587, 308597, 308621, 308639,
308641, 308663, 308681, 308701, 308713, 308723, 308761, 308773, 308801, 308809, 308813, 308827, 308849,
308851, 308857, 308887, 308899, 308923, 308927, 308929, 308933, 308939, 308951, 308989, 308999, 309007,
309011, 309013, 309019, 309031, 309037, 309059, 309079, 309083, 309091, 309107, 309109, 309121, 309131,
309137, 309157, 309167, 309173, 309193, 309223, 309241, 309251, 309259, 309269, 309271, 309277, 309289,
309293, 309311, 309313, 309317, 309359, 309367, 309371, 309391, 309403, 309433, 309437, 309457, 309461,
309469, 309479, 309481, 309493, 309503, 309521, 309523, 309539, 309541, 309559, 309571, 309577, 309583,
309599, 309623, 309629, 309637, 309667, 309671, 309677, 309707, 309713, 309731, 309737, 309769, 309779,
309781, 309797, 309811, 309823, 309851, 309853, 309857, 309877, 309899, 309929, 309931, 309937, 309977,
309989, 310019, 310021, 310027, 310043, 310049, 310081, 310087, 310091, 310111, 310117, 310127, 310129,
310169, 310181, 310187, 310223, 310229, 310231, 310237, 310243, 310273, 310283, 310291, 310313, 310333,
310357, 310361, 310363, 310379, 310397, 310423, 310433, 310439, 310447, 310459, 310463, 310481, 310489,
310501, 310507, 310511, 310547, 310553, 310559, 310567, 310571, 310577, 310591, 310627, 310643, 310663,
310693, 310697, 310711, 310721, 310727, 310729, 310733, 310741, 310747, 310771, 310781, 310789, 310801,
310819, 310823, 310829, 310831, 310861, 310867, 310883, 310889, 310901, 310927, 310931, 310949, 310969,
310987, 310997, 311009, 311021, 311027, 311033, 311041, 311099, 311111, 311123, 311137, 311153, 311173,
311177, 311183, 311189, 311197, 311203, 311237, 311279, 311291, 311293, 311299, 311303, 311323, 311329,
311341, 311347, 311359, 311371, 311393, 311407, 311419, 311447, 311453, 311473, 311533, 311537, 311539,
311551, 311557, 311561, 311567, 311569, 311603, 311609, 311653, 311659, 311677, 311681, 311683, 311687,
311711, 311713, 311737, 311743, 311747, 311749, 311791, 311803, 311807, 311821, 311827, 311867, 311869,
311881, 311897, 311951, 311957, 311963, 311981, 312007, 312023, 312029, 312031, 312043, 312047, 312071,

312073, 312083, 312089, 312101, 312107, 312121, 312161, 312197, 312199, 312203, 312209, 312211, 312217,
312229, 312233, 312241, 312251, 312253, 312269, 312281, 312283, 312289, 312311, 312313, 312331, 312343,
312349, 312353, 312371, 312383, 312397, 312401, 312407, 312413, 312427, 312451, 312469, 312509, 312517,
312527, 312551, 312553, 312563, 312581, 312583, 312589, 312601, 312617, 312619, 312623, 312643, 312673,
312677, 312679, 312701, 312703, 312709, 312727, 312737, 312743, 312757, 312773, 312779, 312799, 312839,
312841, 312857, 312863, 312887, 312899, 312929, 312931, 312937, 312941, 312943, 312967, 312971, 312979,
312989, 313003, 313009, 313031, 313037, 313081, 313087, 313109, 313127, 313129, 313133, 313147, 313151,
313153, 313163, 313207, 313211, 313219, 313241, 313249, 313267, 313273, 313289, 313297, 313301, 313307,
313321, 313331, 313333, 313343, 313351, 313373, 313381, 313387, 313399, 313409, 313471, 313477, 313507,
313517, 313543, 313549, 313553, 313561, 313567, 313571, 313583, 313589, 313597, 313603, 313613, 313619,
313637, 313639, 313661, 313669, 313679, 313699, 313711, 313717, 313721, 313727, 313739, 313741, 313763,
313777, 313783, 313829, 313849, 313853, 313879, 313883, 313889, 313897, 313909, 313921, 313931, 313933,
313949, 313961, 313969, 313979, 313981, 313987, 313991, 313993, 313997, 314003, 314021, 314059, 314063,
314077, 314107, 314113, 314117, 314129, 314137, 314159, 314161, 314173, 314189, 314213, 314219, 314227,
314233, 314239, 314243, 314257, 314261, 314263, 314267, 314299, 314329, 314339, 314351, 314357, 314359,
314399, 314401, 314407, 314423, 314441, 314453, 314467, 314491, 314497, 314513, 314527, 314543, 314549,
314569, 314581, 314591, 314597, 314599, 314603, 314623, 314627, 314641, 314651, 314693, 314707, 314711,
314719, 314723, 314747, 314761, 314771, 314777, 314779, 314807, 314813, 314827, 314851, 314879, 314903,
314917, 314927, 314933, 314953, 314957, 314983, 314989, 315011, 315013, 315037, 315047, 315059, 315067,
315083, 315097, 315103, 315109, 315127, 315179, 315181, 315193, 315199, 315223, 315247, 315251, 315257,
315269, 315281, 315313, 315349, 315361, 315373, 315377, 315389, 315407, 315409, 315421, 315437, 315449,
315451, 315461, 315467, 315481, 315493, 315517, 315521, 315527, 315529, 315547, 315551, 315559, 315569,
315589, 315593, 315599, 315613, 315617, 315631, 315643, 315671, 315677, 315691, 315697, 315701, 315703,
315739, 315743, 315751, 315779, 315803, 315811, 315829, 315851, 315857, 315881, 315883, 315893, 315899,
315907, 315937, 315949, 315961, 315967, 315977, 316003, 316031, 316033, 316037, 316051, 316067, 316073,
316087, 316097, 316109, 316133, 316139, 316153, 316177, 316189, 316193, 316201, 316213, 316219, 316223,
316241, 316243, 316259, 316271, 316291, 316297, 316301, 316321, 316339, 316343, 316363, 316373, 316391,
316403, 316423, 316429, 316439, 316453, 316469, 316471, 316493, 316499, 316501, 316507, 316531, 316567,
316571, 316577, 316583, 316621, 316633, 316637, 316649, 316661, 316663, 316681, 316691, 316697, 316699,
316703, 316717, 316753, 316759, 316769, 316777, 316783, 316793, 316801, 316817, 316819, 316847, 316853,
316859, 316861, 316879, 316891, 316903, 316907, 316919, 316937, 316951, 316957, 316961, 316991, 317003,
317011, 317021, 317029, 317047, 317063, 317071, 317077, 317087, 317089, 317123, 317159, 317171, 317179,
317189, 317197, 317209, 317227, 317257, 317263, 317267, 317269, 317279, 317321, 317323, 317327, 317333,
317351, 317353, 317363, 317371, 317399, 317411, 317419, 317431, 317437, 317453, 317459, 317483, 317489,
317491, 317503, 317539, 317557, 317563, 317587, 317591, 317593, 317599, 317609, 317617, 317621, 317651,
317663, 317671, 317693, 317701, 317711, 317717, 317729, 317731, 317741, 317743, 317771, 317773, 317777,
317783, 317789, 317797, 317827, 317831, 317839, 317857, 317887, 317903, 317921, 317923, 317957, 317959,
317963, 317969, 317971, 317983, 317987, 318001, 318007, 318023, 318077, 318103, 318107, 318127, 318137,
318161, 318173, 318179, 318181, 318191, 318203, 318209, 318211, 318229, 318233, 318247, 318259, 318271,
318281, 318287, 318289, 318299, 318301, 318313, 318319, 318323, 318337, 318347, 318349, 318377, 318403,
318407, 318419, 318431, 318443, 318457, 318467, 318473, 318503, 318523, 318557, 318559, 318569, 318581,
318589, 318601, 318629, 318641, 318653, 318671, 318677, 318679, 318683, 318691, 318701, 318713, 318737,
318743, 318749, 318751, 318781, 318793, 318809, 318811, 318817, 318823, 318833, 318841, 318863, 318881,
318883, 318889, 318907, 318911, 318917, 318919, 318949, 318979, 319001, 319027, 319031, 319037, 319049,
319057, 319061, 319069, 319093, 319097, 319117, 319127, 319129, 319133, 319147, 319159, 319169, 319183,
319201, 319211, 319223, 319237, 319259, 319279, 319289, 319313, 319321, 319327, 319339, 319343, 319351,
319357, 319387, 319391, 319399, 319411, 319427, 319433, 319439, 319441, 319453, 319469, 319477, 319483,
319489, 319499, 319511, 319519, 319541, 319547, 319567, 319577, 319589, 319591, 319601, 319607, 319639,
319673, 319679, 319681, 319687, 319691, 319699, 319727, 319729, 319733, 319747, 319757, 319763, 319811,

319817, 319819, 319829, 319831, 319849, 319883, 319897, 319901, 319919, 319927, 319931, 319937, 319967,
319973, 319981, 319993, 320009, 320011, 320027, 320039, 320041, 320053, 320057, 320063, 320081, 320083,
320101, 320107, 320113, 320119, 320141, 320143, 320149, 320153, 320179, 320209, 320213, 320219, 320237,
320239, 320267, 320269, 320273, 320291, 320293, 320303, 320317, 320329, 320339, 320377, 320387, 320389,
320401, 320417, 320431, 320449, 320471, 320477, 320483, 320513, 320521, 320533, 320539, 320561, 320563,
320591, 320609, 320611, 320627, 320647, 320657, 320659, 320669, 320687, 320693, 320699, 320713, 320741,
320759, 320767, 320791, 320821, 320833, 320839, 320843, 320851, 320861, 320867, 320899, 320911, 320923,
320927, 320939, 320941, 320953, 321007, 321017, 321031, 321037, 321047, 321053, 321073, 321077, 321091,
321109, 321143, 321163, 321169, 321187, 321193, 321199, 321203, 321221, 321227, 321239, 321247, 321289,
321301, 321311, 321313, 321319, 321323, 321329, 321331, 321341, 321359, 321367, 321371, 321383, 321397,
321403, 321413, 321427, 321443, 321449, 321467, 321469, 321509, 321547, 321553, 321569, 321571, 321577,
321593, 321611, 321617, 321619, 321631, 321647, 321661, 321679, 321707, 321709, 321721, 321733, 321743,
321751, 321757, 321779, 321799, 321817, 321821, 321823, 321829, 321833, 321847, 321851, 321889, 321901,
321911, 321947, 321949, 321961, 321983, 321991, 322001, 322009, 322013, 322037, 322039, 322051, 322057,
322067, 322073, 322079, 322093, 322097, 322109, 322111, 322139, 322169, 322171, 322193, 322213, 322229,
322237, 322243, 322247, 322249, 322261, 322271, 322319, 322327, 322339, 322349, 322351, 322397, 322403,
322409, 322417, 322429, 322433, 322459, 322463, 322501, 322513, 322519, 322523, 322537, 322549, 322559,
322571, 322573, 322583, 322589, 322591, 322607, 322613, 322627, 322631, 322633, 322649, 322669, 322709,
322727, 322747, 322757, 322769, 322771, 322781, 322783, 322807, 322849, 322859, 322871, 322877, 322891,
322901, 322919, 322921, 322939, 322951, 322963, 322969, 322997, 322999, 323003, 323009, 323027, 323053,
323077, 323083, 323087, 323093, 323101, 323123, 323131, 323137, 323149, 323201, 323207, 323233, 323243,
323249, 323251, 323273, 323333, 323339, 323341, 323359, 323369, 323371, 323377, 323381, 323383, 323413,
323419, 323441, 323443, 323467, 323471, 323473, 323507, 323509, 323537, 323549, 323567, 323579, 323581,
323591, 323597, 323599, 323623, 323641, 323647, 323651, 323699, 323707, 323711, 323717, 323759, 323767,
323789, 323797, 323801, 323803, 323819, 323837, 323879, 323899, 323903, 323923, 323927, 323933, 323951,
323957, 323987, 324011, 324031, 324053, 324067, 324073, 324089, 324097, 324101, 324113, 324119, 324131,
324143, 324151, 324161, 324179, 324199, 324209, 324211, 324217, 324223, 324239, 324251, 324293, 324299,
324301, 324319, 324329, 324341, 324361, 324391, 324397, 324403, 324419, 324427, 324431, 324437, 324439,
324449, 324451, 324469, 324473, 324491, 324497, 324503, 324517, 324523, 324529, 324557, 324587, 324589,
324593, 324617, 324619, 324637, 324641, 324647, 324661, 324673, 324689, 324697, 324707, 324733, 324743,
324757, 324763, 324773, 324781, 324791, 324799, 324809, 324811, 324839, 324847, 324869, 324871, 324889,
324893, 324901, 324931, 324941, 324949, 324953, 324977, 324979, 324983, 324991, 324997, 325001, 325009,
325019, 325021, 325027, 325043, 325051, 325063, 325079, 325081, 325093, 325133, 325153, 325163, 325181,
325187, 325189, 325201, 325217, 325219, 325229, 325231, 325249, 325271, 325301, 325307, 325309, 325319,
325333, 325343, 325349, 325379, 325411, 325421, 325439, 325447, 325453, 325459, 325463, 325477, 325487,
325513, 325517, 325537, 325541, 325543, 325571, 325597, 325607, 325627, 325631, 325643, 325667, 325673,
325681, 325691, 325693, 325697, 325709, 325723, 325729, 325747, 325751, 325753, 325769, 325777, 325781,
325783, 325807, 325813, 325849, 325861, 325877, 325883, 325889, 325891, 325901, 325921, 325939, 325943,
325951, 325957, 325987, 325993, 325999, 326023, 326057, 326063, 326083, 326087, 326099, 326101, 326113,
326119, 326141, 326143, 326147, 326149, 326153, 326159, 326171, 326189, 326203, 326219, 326251, 326257,
326309, 326323, 326351, 326353, 326369, 326437, 326441, 326449, 326467, 326479, 326497, 326503, 326537,
326539, 326549, 326561, 326563, 326567, 326581, 326593, 326597, 326609, 326611, 326617, 326633, 326657,
326659, 326663, 326681, 326687, 326693, 326701, 326707, 326737, 326741, 326773, 326779, 326831, 326863,
326867, 326869, 326873, 326881, 326903, 326923, 326939, 326941, 326947, 326951, 326983, 326993, 326999,
327001, 327007, 327011, 327017, 327023, 327059, 327071, 327079, 327127, 327133, 327163, 327179, 327193,
327203, 327209, 327211, 327247, 327251, 327263, 327277, 327289, 327307, 327311, 327317, 327319, 327331,
327337, 327343, 327347, 327401, 327407, 327409, 327419, 327421, 327433, 327443, 327463, 327469, 327473,
327479, 327491, 327493, 327499, 327511, 327517, 327529, 327553, 327557, 327559, 327571, 327581, 327583,
327599, 327619, 327629, 327647, 327661, 327667, 327673, 327689, 327707, 327721, 327737, 327739, 327757,

327779, 327797, 327799, 327809, 327823, 327827, 327829, 327839, 327851, 327853, 327869, 327871, 327881,
327889, 327917, 327923, 327941, 327953, 327967, 327979, 327983, 328007, 328037, 328043, 328051, 328061,
328063, 328067, 328093, 328103, 328109, 328121, 328127, 328129, 328171, 328177, 328213, 328243, 328249,
328271, 328277, 328283, 328291, 328303, 328327, 328331, 328333, 328343, 328357, 328373, 328379, 328381,
328397, 328411, 328421, 328429, 328439, 328481, 328511, 328513, 328519, 328543, 328579, 328589, 328591,
328619, 328621, 328633, 328637, 328639, 328651, 328667, 328687, 328709, 328721, 328753, 328777, 328781,
328787, 328789, 328813, 328829, 328837, 328847, 328849, 328883, 328891, 328897, 328901, 328919, 328921,
328931, 328961, 328981, 329009, 329027, 329053, 329059, 329081, 329083, 329089, 329101, 329111, 329123,
329143, 329167, 329177, 329191, 329201, 329207, 329209, 329233, 329243, 329257, 329267, 329269, 329281,
329293, 329297, 329299, 329309, 329317, 329321, 329333, 329347, 329387, 329393, 329401, 329419, 329431,
329471, 329473, 329489, 329503, 329519, 329533, 329551, 329557, 329587, 329591, 329597, 329603, 329617,
329627, 329629, 329639, 329657, 329663, 329671, 329677, 329683, 329687, 329711, 329717, 329723, 329729,
329761, 329773, 329779, 329789, 329801, 329803, 329863, 329867, 329873, 329891, 329899, 329941, 329947,
329951, 329957, 329969, 329977, 329993, 329999, 330017, 330019, 330037, 330041, 330047, 330053, 330061,
330067, 330097, 330103, 330131, 330133, 330139, 330149, 330167, 330199, 330203, 330217, 330227, 330229,
330233, 330241, 330247, 330271, 330287, 330289, 330311, 330313, 330329, 330331, 330347, 330359, 330383,
330389, 330409, 330413, 330427, 330431, 330433, 330439, 330469, 330509, 330557, 330563, 330569, 330587,
330607, 330611, 330623, 330641, 330643, 330653, 330661, 330679, 330683, 330689, 330697, 330703, 330719,
330721, 330731, 330749, 330767, 330787, 330791, 330793, 330821, 330823, 330839, 330853, 330857, 330859,
330877, 330887, 330899, 330907, 330917, 330943, 330983, 330997, 331013, 331027, 331031, 331043, 331063,
331081, 331099, 331127, 331141, 331147, 331153, 331159, 331171, 331183, 331207, 331213, 331217, 331231,
331241, 331249, 331259, 331277, 331283, 331301, 331307, 331319, 331333, 331337, 331339, 331349, 331367,
331369, 331391, 331399, 331423, 331447, 331451, 331489, 331501, 331511, 331519, 331523, 331537, 331543,
331547, 331549, 331553, 331577, 331579, 331589, 331603, 331609, 331613, 331651, 331663, 331691, 331693,
331697, 331711, 331739, 331753, 331769, 331777, 331781, 331801, 331819, 331841, 331843, 331871, 331883,
331889, 331897, 331907, 331909, 331921, 331937, 331943, 331957, 331967, 331973, 331997, 331999, 332009,
332011, 332039, 332053, 332069, 332081, 332099, 332113, 332117, 332147, 332159, 332161, 332179, 332183,
332191, 332201, 332203, 332207, 332219, 332221, 332251, 332263, 332273, 332287, 332303, 332309, 332317,
332393, 332399, 332411, 332417, 332441, 332447, 332461, 332467, 332471, 332473, 332477, 332489, 332509,
332513, 332561, 332567, 332569, 332573, 332611, 332617, 332623, 332641, 332687, 332699, 332711, 332729,
332743, 332749, 332767, 332779, 332791, 332803, 332837, 332851, 332873, 332881, 332887, 332903, 332921,
332933, 332947, 332951, 332987, 332989, 332993, 333019, 333023, 333029, 333031, 333041, 333049, 333071,
333097, 333101, 333103, 333107, 333131, 333139, 333161, 333187, 333197, 333209, 333227, 333233, 333253,
333269, 333271, 333283, 333287, 333299, 333323, 333331, 333337, 333341, 333349, 333367, 333383, 333397,
333419, 333427, 333433, 333439, 333449, 333451, 333457, 333479, 333491, 333493, 333497, 333503, 333517,
333533, 333539, 333563, 333581, 333589, 333623, 333631, 333647, 333667, 333673, 333679, 333691, 333701,
333713, 333719, 333721, 333737, 333757, 333769, 333779, 333787, 333791, 333793, 333803, 333821, 333857,
333871, 333911, 333923, 333929, 333941, 333959, 333973, 333989, 333997, 334021, 334031, 334043, 334049,
334057, 334069, 334093, 334099, 334127, 334133, 334157, 334171, 334177, 334183, 334189, 334199, 334231,
334247, 334261, 334289, 334297, 334319, 334331, 334333, 334349, 334363, 334379, 334387, 334393, 334403,
334421, 334423, 334427, 334429, 334447, 334487, 334493, 334507, 334511, 334513, 334541, 334547, 334549,
334561, 334603, 334619, 334637, 334643, 334651, 334661, 334667, 334681, 334693, 334699, 334717, 334721,
334727, 334751, 334753, 334759, 334771, 334777, 334783, 334787, 334793, 334843, 334861, 334877, 334889,
334891, 334897, 334931, 334963, 334973, 334987, 334991, 334993, 335009, 335021, 335029, 335033, 335047,
335051, 335057, 335077, 335081, 335089, 335107, 335113, 335117, 335123, 335131, 335149, 335161, 335171,
335173, 335207, 335213, 335221, 335249, 335261, 335273, 335281, 335299, 335323, 335341, 335347, 335381,
335383, 335411, 335417, 335429, 335449, 335453, 335459, 335473, 335477, 335507, 335519, 335527, 335539,
335557, 335567, 335579, 335591, 335609, 335633, 335641, 335653, 335663, 335669, 335681, 335689, 335693,
335719, 335729, 335743, 335747, 335771, 335807, 335809, 335813, 335821, 335833, 335843, 335857, 335879,

335893, 335897, 335917, 335941, 335953, 335957, 335999, 336029, 336031, 336041, 336059, 336079, 336101,
336103, 336109, 336113, 336121, 336143, 336151, 336157, 336163, 336181, 336199, 336211, 336221, 336223,
336227, 336239, 336247, 336251, 336253, 336263, 336307, 336317, 336353, 336361, 336373, 336397, 336403,
336419, 336437, 336463, 336491, 336499, 336503, 336521, 336527, 336529, 336533, 336551, 336563, 336571,
336577, 336587, 336593, 336599, 336613, 336631, 336643, 336649, 336653, 336667, 336671, 336683, 336689,
336703, 336727, 336757, 336761, 336767, 336769, 336773, 336793, 336799, 336803, 336823, 336827, 336829,
336857, 336863, 336871, 336887, 336899, 336901, 336911, 336929, 336961, 336977, 336983, 336989, 336997,
337013, 337021, 337031, 337039, 337049, 337069, 337081, 337091, 337097, 337121, 337153, 337189, 337201,
337213, 337217, 337219, 337223, 337261, 337277, 337279, 337283, 337291, 337301, 337313, 337327, 337339,
337343, 337349, 337361, 337367, 337369, 337397, 337411, 337427, 337453, 337457, 337487, 337489, 337511,
337517, 337529, 337537, 337541, 337543, 337583, 337607, 337609, 337627, 337633, 337639, 337651, 337661,
337669, 337681, 337691, 337697, 337721, 337741, 337751, 337759, 337781, 337793, 337817, 337837, 337853,
337859, 337861, 337867, 337871, 337873, 337891, 337901, 337903, 337907, 337919, 337949, 337957, 337969,
337973, 337999, 338017, 338027, 338033, 338119, 338137, 338141, 338153, 338159, 338161, 338167, 338171,
338183, 338197, 338203, 338207, 338213, 338231, 338237, 338251, 338263, 338267, 338269, 338279, 338287,
338293, 338297, 338309, 338321, 338323, 338339, 338341, 338347, 338369, 338383, 338389, 338407, 338411,
338413, 338423, 338431, 338449, 338461, 338473, 338477, 338497, 338531, 338543, 338563, 338567, 338573,
338579, 338581, 338609, 338659, 338669, 338683, 338687, 338707, 338717, 338731, 338747, 338753, 338761,
338773, 338777, 338791, 338803, 338839, 338851, 338857, 338867, 338893, 338909, 338927, 338959, 338993,
338999, 339023, 339049, 339067, 339071, 339091, 339103, 339107, 339121, 339127, 339137, 339139, 339151,
339161, 339173, 339187, 339211, 339223, 339239, 339247, 339257, 339263, 339289, 339307, 339323, 339331,
339341, 339373, 339389, 339413, 339433, 339467, 339491, 339517, 339527, 339539, 339557, 339583, 339589,
339601, 339613, 339617, 339631, 339637, 339649, 339653, 339659, 339671, 339673, 339679, 339707, 339727,
339749, 339751, 339761, 339769, 339799, 339811, 339817, 339821, 339827, 339839, 339841, 339863, 339887,
339907, 339943, 339959, 339991, 340007, 340027, 340031, 340037, 340049, 340057, 340061, 340063, 340073,
340079, 340103, 340111, 340117, 340121, 340127, 340129, 340169, 340183, 340201, 340211, 340237, 340261,
340267, 340283, 340297, 340321, 340337, 340339, 340369, 340381, 340387, 340393, 340397, 340409, 340429,
340447, 340451, 340453, 340477, 340481, 340519, 340541, 340559, 340573, 340577, 340579, 340583, 340591,
340601, 340619, 340633, 340643, 340649, 340657, 340661, 340687, 340693, 340709, 340723, 340757, 340777,
340787, 340789, 340793, 340801, 340811, 340819, 340849, 340859, 340877, 340889, 340897, 340903, 340909,
340913, 340919, 340927, 340931, 340933, 340937, 340939, 340957, 340979, 340999, 341017, 341027, 341041,
341057, 341059, 341063, 341083, 341087, 341123, 341141, 341171, 341179, 341191, 341203, 341219, 341227,
341233, 341269, 341273, 341281, 341287, 341293, 341303, 341311, 341321, 341323, 341333, 341339, 341347,
341357, 341423, 341443, 341447, 341459, 341461, 341477, 341491, 341501, 341507, 341521, 341543, 341557,
341569, 341587, 341597, 341603, 341617, 341623, 341629, 341641, 341647, 341659, 341681, 341687, 341701,
341729, 341743, 341749, 341771, 341773, 341777, 341813, 341821, 341827, 341839, 341851, 341863, 341879,
341911, 341927, 341947, 341951, 341953, 341959, 341963, 341983, 341993, 342037, 342047, 342049, 342059,
342061, 342071, 342073, 342077, 342101, 342107, 342131, 342143, 342179, 342187, 342191, 342197, 342203,
342211, 342233, 342239, 342241, 342257, 342281, 342283, 342299, 342319, 342337, 342341, 342343, 342347,
342359, 342371, 342373, 342379, 342389, 342413, 342421, 342449, 342451, 342467, 342469, 342481, 342497,
342521, 342527, 342547, 342553, 342569, 342593, 342599, 342607, 342647, 342653, 342659, 342673, 342679,
342691, 342697, 342733, 342757, 342761, 342791, 342799, 342803, 342821, 342833, 342841, 342847, 342863,
342869, 342871, 342889, 342899, 342929, 342949, 342971, 342989, 343019, 343037, 343051, 343061, 343073,
343081, 343087, 343127, 343141, 343153, 343163, 343169, 343177, 343193, 343199, 343219, 343237, 343243,
343253, 343261, 343267, 343289, 343303, 343307, 343309, 343313, 343327, 343333, 343337, 343373, 343379,
343381, 343391, 343393, 343411, 343423, 343433, 343481, 343489, 343517, 343529, 343531, 343543, 343547,
343559, 343561, 343579, 343583, 343589, 343591, 343601, 343627, 343631, 343639, 343649, 343661, 343667,
343687, 343709, 343727, 343769, 343771, 343787, 343799, 343801, 343813, 343817, 343823, 343829, 343831,
343891, 343897, 343901, 343913, 343933, 343939, 343943, 343951, 343963, 343997, 344017, 344021, 344039,

344053, 344083, 344111, 344117, 344153, 344161, 344167, 344171, 344173, 344177, 344189, 344207, 344209,
344213, 344221, 344231, 344237, 344243, 344249, 344251, 344257, 344263, 344269, 344273, 344291, 344293,
344321, 344327, 344347, 344353, 344363, 344371, 344417, 344423, 344429, 344453, 344479, 344483, 344497,
344543, 344567, 344587, 344599, 344611, 344621, 344629, 344639, 344653, 344671, 344681, 344683, 344693,
344719, 344749, 344753, 344759, 344791, 344797, 344801, 344807, 344819, 344821, 344843, 344857, 344863,
344873, 344887, 344893, 344909, 344917, 344921, 344941, 344957, 344959, 344963, 344969, 344987, 345001,
345011, 345017, 345019, 345041, 345047, 345067, 345089, 345109, 345133, 345139, 345143, 345181, 345193,
345221, 345227, 345229, 345259, 345263, 345271, 345307, 345311, 345329, 345379, 345413, 345431, 345451,
345461, 345463, 345473, 345479, 345487, 345511, 345517, 345533, 345547, 345551, 345571, 345577, 345581,
345599, 345601, 345607, 345637, 345643, 345647, 345659, 345673, 345679, 345689, 345701, 345707, 345727,
345731, 345733, 345739, 345749, 345757, 345769, 345773, 345791, 345803, 345811, 345817, 345823, 345853,
345869, 345881, 345887, 345889, 345907, 345923, 345937, 345953, 345979, 345997, 346013, 346039, 346043,
346051, 346079, 346091, 346097, 346111, 346117, 346133, 346139, 346141, 346147, 346169, 346187, 346201,
346207, 346217, 346223, 346259, 346261, 346277, 346303, 346309, 346321, 346331, 346337, 346349, 346361,
346369, 346373, 346391, 346393, 346397, 346399, 346417, 346421, 346429, 346433, 346439, 346441, 346447,
346453, 346469, 346501, 346529, 346543, 346547, 346553, 346559, 346561, 346589, 346601, 346607, 346627,
346639, 346649, 346651, 346657, 346667, 346669, 346699, 346711, 346721, 346739, 346751, 346763, 346793,
346831, 346849, 346867, 346873, 346877, 346891, 346903, 346933, 346939, 346943, 346961, 346963, 347003,
347033, 347041, 347051, 347057, 347059, 347063, 347069, 347071, 347099, 347129, 347131, 347141, 347143,
347161, 347167, 347173, 347177, 347183, 347197, 347201, 347209, 347227, 347233, 347239, 347251, 347257,
347287, 347297, 347299, 347317, 347329, 347341, 347359, 347401, 347411, 347437, 347443, 347489, 347509,
347513, 347519, 347533, 347539, 347561, 347563, 347579, 347587, 347591, 347609, 347621, 347629, 347651,
347671, 347707, 347717, 347729, 347731, 347747, 347759, 347771, 347773, 347779, 347801, 347813, 347821,
347849, 347873, 347887, 347891, 347899, 347929, 347933, 347951, 347957, 347959, 347969, 347981, 347983,
347987, 347989, 347993, 348001, 348011, 348017, 348037, 348043, 348053, 348077, 348083, 348097, 348149,
348163, 348181, 348191, 348209, 348217, 348221, 348239, 348241, 348247, 348253, 348259, 348269, 348287,
348307, 348323, 348353, 348367, 348389, 348401, 348407, 348419, 348421, 348431, 348433, 348437, 348443,
348451, 348457, 348461, 348463, 348487, 348527, 348547, 348553, 348559, 348563, 348571, 348583, 348587,
348617, 348629, 348637, 348643, 348661, 348671, 348709, 348731, 348739, 348757, 348763, 348769, 348779,
348811, 348827, 348833, 348839, 348851, 348883, 348889, 348911, 348917, 348919, 348923, 348937, 348949,
348989, 348991, 349007, 349039, 349043, 349051, 349079, 349081, 349093, 349099, 349109, 349121, 349133,
349171, 349177, 349183, 349187, 349199, 349207, 349211, 349241, 349291, 349303, 349313, 349331, 349337,
349343, 349357, 349369, 349373, 349379, 349381, 349387, 349397, 349399, 349403, 349409, 349411, 349423,
349471, 349477, 349483, 349493, 349499, 349507, 349519, 349529, 349553, 349567, 349579, 349589, 349603,
349637, 349663, 349667, 349697, 349709, 349717, 349729, 349753, 349759, 349787, 349793, 349801, 349813,
349819, 349829, 349831, 349837, 349841, 349849, 349871, 349903, 349907, 349913, 349919, 349927, 349931,
349933, 349939, 349949, 349963, 349967, 349981, 350003, 350029, 350033, 350039, 350087, 350089, 350093,
350107, 350111, 350137, 350159, 350179, 350191, 350213, 350219, 350237, 350249, 350257, 350281, 350293,
350347, 350351, 350377, 350381, 350411, 350423, 350429, 350431, 350437, 350443, 350447, 350453, 350459,
350503, 350521, 350549, 350561, 350563, 350587, 350593, 350617, 350621, 350629, 350657, 350663, 350677,
350699, 350711, 350719, 350729, 350731, 350737, 350741, 350747, 350767, 350771, 350783, 350789, 350803,
350809, 350843, 350851, 350869, 350881, 350887, 350891, 350899, 350941, 350947, 350963, 350971, 350981,
350983, 350989, 351011, 351023, 351031, 351037, 351041, 351047, 351053, 351059, 351061, 351077, 351079,
351097, 351121, 351133, 351151, 351157, 351179, 351217, 351223, 351229, 351257, 351259, 351269, 351287,
351289, 351293, 351301, 351311, 351341, 351343, 351347, 351359, 351361, 351383, 351391, 351397, 351401,
351413, 351427, 351437, 351457, 351469, 351479, 351497, 351503, 351517, 351529, 351551, 351563, 351587,
351599, 351643, 351653, 351661, 351667, 351691, 351707, 351727, 351731, 351733, 351749, 351751, 351763,
351773, 351779, 351797, 351803, 351811, 351829, 351847, 351851, 351859, 351863, 351887, 351913, 351919,
351929, 351931, 351959, 351971, 351991, 352007, 352021, 352043, 352049, 352057, 352069, 352073, 352081,

352097, 352109, 352111, 352123, 352133, 352181, 352193, 352201, 352217, 352229, 352237, 352249, 352267,
352271, 352273, 352301, 352309, 352327, 352333, 352349, 352357, 352361, 352367, 352369, 352381, 352399,
352403, 352409, 352411, 352421, 352423, 352441, 352459, 352463, 352481, 352483, 352489, 352493, 352511,
352523, 352543, 352549, 352579, 352589, 352601, 352607, 352619, 352633, 352637, 352661, 352691, 352711,
352739, 352741, 352753, 352757, 352771, 352813, 352817, 352819, 352831, 352837, 352841, 352853, 352867,
352883, 352907, 352909, 352931, 352939, 352949, 352951, 352973, 352991, 353011, 353021, 353047, 353053,
353057, 353069, 353081, 353099, 353117, 353123, 353137, 353147, 353149, 353161, 353173, 353179, 353201,
353203, 353237, 353263, 353293, 353317, 353321, 353329, 353333, 353341, 353359, 353389, 353401, 353411,
353429, 353443, 353453, 353459, 353471, 353473, 353489, 353501, 353527, 353531, 353557, 353567, 353603,
353611, 353621, 353627, 353629, 353641, 353653, 353657, 353677, 353681, 353687, 353699, 353711, 353737,
353747, 353767, 353777, 353783, 353797, 353807, 353813, 353819, 353833, 353867, 353869, 353879, 353891,
353897, 353911, 353917, 353921, 353929, 353939, 353963, 354001, 354007, 354017, 354023, 354031, 354037,
354041, 354043, 354047, 354073, 354091, 354097, 354121, 354139, 354143, 354149, 354163, 354169, 354181,
354209, 354247, 354251, 354253, 354257, 354259, 354271, 354301, 354307, 354313, 354317, 354323, 354329,
354337, 354353, 354371, 354373, 354377, 354383, 354391, 354401, 354421, 354439, 354443, 354451, 354461,
354463, 354469, 354479, 354533, 354539, 354551, 354553, 354581, 354587, 354619, 354643, 354647, 354661,
354667, 354677, 354689, 354701, 354703, 354727, 354737, 354743, 354751, 354763, 354779, 354791, 354799,
354829, 354833, 354839, 354847, 354869, 354877, 354881, 354883, 354911, 354953, 354961, 354971, 354973,
354979, 354983, 354997, 355007, 355009, 355027, 355031, 355037, 355039, 355049, 355057, 355063, 355073,
355087, 355093, 355099, 355109, 355111, 355127, 355139, 355171, 355193, 355211, 355261, 355297, 355307,
355321, 355331, 355339, 355343, 355361, 355363, 355379, 355417, 355427, 355441, 355457, 355463, 355483,
355499, 355501, 355507, 355513, 355517, 355519, 355529, 355541, 355549, 355559, 355571, 355573, 355591,
355609, 355633, 355643, 355651, 355669, 355679, 355697, 355717, 355721, 355723, 355753, 355763, 355777,
355783, 355799, 355811, 355819, 355841, 355847, 355853, 355867, 355891, 355909, 355913, 355933, 355937,
355939, 355951, 355967, 355969, 356023, 356039, 356077, 356093, 356101, 356113, 356123, 356129, 356137,
356141, 356143, 356171, 356173, 356197, 356219, 356243, 356261, 356263, 356287, 356299, 356311, 356327,
356333, 356351, 356387, 356399, 356441, 356443, 356449, 356453, 356467, 356479, 356501, 356509, 356533,
356549, 356561, 356563, 356567, 356579, 356591, 356621, 356647, 356663, 356693, 356701, 356731, 356737,
356749, 356761, 356803, 356819, 356821, 356831, 356869, 356887, 356893, 356927, 356929, 356933, 356947,
356959, 356969, 356977, 356981, 356989, 356999, 357031, 357047, 357073, 357079, 357083, 357103, 357107,
357109, 357131, 357139, 357169, 357179, 357197, 357199, 357211, 357229, 357239, 357241, 357263, 357271,
357281, 357283, 357293, 357319, 357347, 357349, 357353, 357359, 357377, 357389, 357421, 357431, 357437,
357473, 357503, 357509, 357517, 357551, 357559, 357563, 357569, 357571, 357583, 357587, 357593, 357611,
357613, 357619, 357649, 357653, 357659, 357661, 357667, 357671, 357677, 357683, 357689, 357703, 357727,
357733, 357737, 357739, 357767, 357779, 357781, 357787, 357793, 357809, 357817, 357823, 357829, 357839,
357859, 357883, 357913, 357967, 357977, 357983, 357989, 357997, 358031, 358051, 358069, 358073, 358079,
358103, 358109, 358153, 358157, 358159, 358181, 358201, 358213, 358219, 358223, 358229, 358243, 358273,
358277, 358279, 358289, 358291, 358297, 358301, 358313, 358327, 358331, 358349, 358373, 358417, 358427,
358429, 358441, 358447, 358459, 358471, 358483, 358487, 358499, 358531, 358541, 358571, 358573, 358591,
358597, 358601, 358607, 358613, 358637, 358667, 358669, 358681, 358691, 358697, 358703, 358711, 358723,
358727, 358733, 358747, 358753, 358769, 358783, 358793, 358811, 358829, 358847, 358859, 358861, 358867,
358877, 358879, 358901, 358903, 358907, 358909, 358931, 358951, 358973, 358979, 358987, 358993, 358999,
359003, 359017, 359027, 359041, 359063, 359069, 359101, 359111, 359129, 359137, 359143, 359147, 359153,
359167, 359171, 359207, 359209, 359231, 359243, 359263, 359267, 359279, 359291, 359297, 359299, 359311,
359323, 359327, 359353, 359357, 359377, 359389, 359407, 359417, 359419, 359441, 359449, 359477, 359479,
359483, 359501, 359509, 359539, 359549, 359561, 359563, 359581, 359587, 359599, 359621, 359633, 359641,
359657, 359663, 359701, 359713, 359719, 359731, 359747, 359753, 359761, 359767, 359783, 359837, 359851,
359869, 359897, 359911, 359929, 359981, 359987, 360007, 360023, 360037, 360049, 360053, 360071, 360089,
360091, 360163, 360167, 360169, 360181, 360187, 360193, 360197, 360223, 360229, 360233, 360257, 360271,

360277, 360287, 360289, 360293, 360307, 360317, 360323, 360337, 360391, 360407, 360421, 360439, 360457,
360461, 360497, 360509, 360511, 360541, 360551, 360589, 360593, 360611, 360637, 360649, 360653, 360749,
360769, 360779, 360781, 360803, 360817, 360821, 360823, 360827, 360851, 360853, 360863, 360869, 360901,
360907, 360947, 360949, 360953, 360959, 360973, 360977, 360979, 360989, 361001, 361003, 361013, 361033,
361069, 361091, 361093, 361111, 361159, 361183, 361211, 361213, 361217, 361219, 361223, 361237, 361241,
361271, 361279, 361313, 361321, 361327, 361337, 361349, 361351, 361357, 361363, 361373, 361409, 361411,
361421, 361433, 361441, 361447, 361451, 361463, 361469, 361481, 361499, 361507, 361511, 361523, 361531,
361541, 361549, 361561, 361577, 361637, 361643, 361649, 361651, 361663, 361679, 361687, 361723, 361727,
361747, 361763, 361769, 361787, 361789, 361793, 361799, 361807, 361843, 361871, 361873, 361877, 361901,
361903, 361909, 361919, 361927, 361943, 361961, 361967, 361973, 361979, 361993, 362003, 362027, 362051,
362053, 362059, 362069, 362081, 362093, 362099, 362107, 362137, 362143, 362147, 362161, 362177, 362191,
362203, 362213, 362221, 362233, 362237, 362281, 362291, 362293, 362303, 362309, 362333, 362339, 362347,
362353, 362357, 362363, 362371, 362377, 362381, 362393, 362407, 362419, 362429, 362431, 362443, 362449,
362459, 362473, 362521, 362561, 362569, 362581, 362599, 362629, 362633, 362657, 362693, 362707, 362717,
362723, 362741, 362743, 362749, 362753, 362759, 362801, 362851, 362863, 362867, 362897, 362903, 362911,
362927, 362941, 362951, 362953, 362969, 362977, 362983, 362987, 363017, 363019, 363037, 363043, 363047,
363059, 363061, 363067, 363119, 363149, 363151, 363157, 363161, 363173, 363179, 363199, 363211, 363217,
363257, 363269, 363271, 363277, 363313, 363317, 363329, 363343, 363359, 363361, 363367, 363371, 363373,
363379, 363397, 363401, 363403, 363431, 363437, 363439, 363463, 363481, 363491, 363497, 363523, 363529,
363533, 363541, 363551, 363557, 363563, 363569, 363577, 363581, 363589, 363611, 363619, 363659, 363677,
363683, 363691, 363719, 363731, 363751, 363757, 363761, 363767, 363773, 363799, 363809, 363829, 363833,
363841, 363871, 363887, 363889, 363901, 363911, 363917, 363941, 363947, 363949, 363959, 363967, 363977,
363989, 364027, 364031, 364069, 364073, 364079, 364103, 364127, 364129, 364141, 364171, 364183, 364187,
364193, 364213, 364223, 364241, 364267, 364271, 364289, 364291, 364303, 364313, 364321, 364333, 364337,
364349, 364373, 364379, 364393, 364411, 364417, 364423, 364433, 364447, 364451, 364459, 364471, 364499,
364513, 364523, 364537, 364541, 364543, 364571, 364583, 364601, 364607, 364621, 364627, 364643, 364657,
364669, 364687, 364691, 364699, 364717, 364739, 364747, 364751, 364753, 364759, 364801, 364829, 364853,
364873, 364879, 364883, 364891, 364909, 364919, 364921, 364937, 364943, 364961, 364979, 364993, 364997,
365003, 365017, 365021, 365039, 365063, 365069, 365089, 365107, 365119, 365129, 365137, 365147, 365159,
365173, 365179, 365201, 365213, 365231, 365249, 365251, 365257, 365291, 365293, 365297, 365303, 365327,
365333, 365357, 365369, 365377, 365411, 365413, 365419, 365423, 365441, 365461, 365467, 365471, 365473,
365479, 365489, 365507, 365509, 365513, 365527, 365531, 365537, 365557, 365567, 365569, 365587, 365591,
365611, 365627, 365639, 365641, 365669, 365683, 365689, 365699, 365747, 365749, 365759, 365773, 365779,
365791, 365797, 365809, 365837, 365839, 365851, 365903, 365929, 365933, 365941, 365969, 365983, 366001,
366013, 366019, 366029, 366031, 366053, 366077, 366097, 366103, 366127, 366133, 366139, 366161, 366167,
366169, 366173, 366181, 366193, 366199, 366211, 366217, 366221, 366227, 366239, 366259, 366269, 366277,
366287, 366293, 366307, 366313, 366329, 366341, 366343, 366347, 366383, 366397, 366409, 366419, 366433,
366437, 366439, 366461, 366463, 366467, 366479, 366497, 366511, 366517, 366521, 366547, 366593, 366599,
366607, 366631, 366677, 366683, 366697, 366701, 366703, 366713, 366721, 366727, 366733, 366787, 366791,
366811, 366829, 366841, 366851, 366853, 366859, 366869, 366881, 366889, 366901, 366907, 366917, 366923,
366941, 366953, 366967, 366973, 366983, 366997, 367001, 367007, 367019, 367021, 367027, 367033, 367049,
367069, 367097, 367121, 367123, 367127, 367139, 367163, 367181, 367189, 367201, 367207, 367219, 367229,
367231, 367243, 367259, 367261, 367273, 367277, 367307, 367309, 367313, 367321, 367357, 367369, 367391,
367397, 367427, 367453, 367457, 367469, 367501, 367519, 367531, 367541, 367547, 367559, 367561, 367573,
367597, 367603, 367613, 367621, 367637, 367649, 367651, 367663, 367673, 367687, 367699, 367711, 367721,
367733, 367739, 367751, 367771, 367777, 367781, 367789, 367819, 367823, 367831, 367841, 367849, 367853,
367867, 367879, 367883, 367889, 367909, 367949, 367957, 368021, 368029, 368047, 368059, 368077, 368083,
368089, 368099, 368107, 368111, 368117, 368129, 368141, 368149, 368153, 368171, 368189, 368197, 368227,
368231, 368233, 368243, 368273, 368279, 368287, 368293, 368323, 368327, 368359, 368363, 368369, 368399,

368411, 368443, 368447, 368453, 368471, 368491, 368507, 368513, 368521, 368531, 368539, 368551, 368579,
368593, 368597, 368609, 368633, 368647, 368651, 368653, 368689, 368717, 368729, 368737, 368743, 368773,
368783, 368789, 368791, 368801, 368803, 368833, 368857, 368873, 368881, 368899, 368911, 368939, 368947,
368957, 369007, 369013, 369023, 369029, 369067, 369071, 369077, 369079, 369097, 369119, 369133, 369137,
369143, 369169, 369181, 369191, 369197, 369211, 369247, 369253, 369263, 369269, 369283, 369293, 369301,
369319, 369331, 369353, 369361, 369407, 369409, 369419, 369469, 369487, 369491, 369539, 369553, 369557,
369581, 369637, 369647, 369659, 369661, 369673, 369703, 369709, 369731, 369739, 369751, 369791, 369793,
369821, 369827, 369829, 369833, 369841, 369851, 369877, 369893, 369913, 369917, 369947, 369959, 369961,
369979, 369983, 369991, 369997, 370003, 370009, 370021, 370033, 370057, 370061, 370067, 370081, 370091,
370103, 370121, 370133, 370147, 370159, 370169, 370193, 370199, 370207, 370213, 370217, 370241, 370247,
370261, 370373, 370387, 370399, 370411, 370421, 370423, 370427, 370439, 370441, 370451, 370463, 370471,
370477, 370483, 370493, 370511, 370529, 370537, 370547, 370561, 370571, 370597, 370603, 370609, 370613,
370619, 370631, 370661, 370663, 370673, 370679, 370687, 370693, 370723, 370759, 370793, 370801, 370813,
370837, 370871, 370873, 370879, 370883, 370891, 370897, 370919, 370949, 371027, 371029, 371057, 371069,
371071, 371083, 371087, 371099, 371131, 371141, 371143, 371153, 371177, 371179, 371191, 371213, 371227,
371233, 371237, 371249, 371251, 371257, 371281, 371291, 371299, 371303, 371311, 371321, 371333, 371339,
371341, 371353, 371359, 371383, 371387, 371389, 371417, 371447, 371453, 371471, 371479, 371491, 371509,
371513, 371549, 371561, 371573, 371587, 371617, 371627, 371633, 371639, 371663, 371669, 371699, 371719,
371737, 371779, 371797, 371831, 371837, 371843, 371851, 371857, 371869, 371873, 371897, 371927, 371929,
371939, 371941, 371951, 371957, 371971, 371981, 371999, 372013, 372023, 372037, 372049, 372059, 372061,
372067, 372107, 372121, 372131, 372137, 372149, 372167, 372173, 372179, 372223, 372241, 372263, 372269,
372271, 372277, 372289, 372293, 372299, 372311, 372313, 372353, 372367, 372371, 372377, 372397, 372401,
372409, 372413, 372443, 372451, 372461, 372473, 372481, 372497, 372511, 372523, 372539, 372607, 372611,
372613, 372629, 372637, 372653, 372661, 372667, 372677, 372689, 372707, 372709, 372719, 372733, 372739,
372751, 372763, 372769, 372773, 372797, 372803, 372809, 372817, 372829, 372833, 372839, 372847, 372859,
372871, 372877, 372881, 372901, 372917, 372941, 372943, 372971, 372973, 372979, 373003, 373007, 373019,
373049, 373063, 373073, 373091, 373127, 373151, 373157, 373171, 373181, 373183, 373187, 373193, 373199,
373207, 373211, 373213, 373229, 373231, 373273, 373291, 373297, 373301, 373327, 373339, 373343, 373349,
373357, 373361, 373363, 373379, 373393, 373447, 373453, 373459, 373463, 373487, 373489, 373501, 373517,
373553, 373561, 373567, 373613, 373621, 373631, 373649, 373657, 373661, 373669, 373693, 373717, 373721,
373753, 373757, 373777, 373783, 373823, 373837, 373859, 373861, 373903, 373909, 373937, 373943, 373951,
373963, 373969, 373981, 373987, 373999, 374009, 374029, 374039, 374041, 374047, 374063, 374069, 374083,
374089, 374093, 374111, 374117, 374123, 374137, 374149, 374159, 374173, 374177, 374189, 374203, 374219,
374239, 374287, 374291, 374293, 374299, 374317, 374321, 374333, 374347, 374351, 374359, 374389, 374399,
374441, 374443, 374447, 374461, 374483, 374501, 374531, 374537, 374557, 374587, 374603, 374639, 374641,
374653, 374669, 374677, 374681, 374683, 374687, 374701, 374713, 374719, 374729, 374741, 374753, 374761,
374771, 374783, 374789, 374797, 374807, 374819, 374837, 374839, 374849, 374879, 374887, 374893, 374903,
374909, 374929, 374939, 374953, 374977, 374981, 374987, 374989, 374993, 375017, 375019, 375029, 375043,
375049, 375059, 375083, 375091, 375097, 375101, 375103, 375113, 375119, 375121, 375127, 375149, 375157,
375163, 375169, 375203, 375209, 375223, 375227, 375233, 375247, 375251, 375253, 375257, 375259, 375281,
375283, 375311, 375341, 375359, 375367, 375371, 375373, 375391, 375407, 375413, 375443, 375449, 375451,
375457, 375467, 375481, 375509, 375511, 375523, 375527, 375533, 375553, 375559, 375563, 375569, 375593,
375607, 375623, 375631, 375643, 375647, 375667, 375673, 375703, 375707, 375709, 375743, 375757, 375761,
375773, 375779, 375787, 375799, 375833, 375841, 375857, 375899, 375901, 375923, 375931, 375967, 375971,
375979, 375983, 375997, 376001, 376003, 376009, 376021, 376039, 376049, 376063, 376081, 376097, 376099,
376127, 376133, 376147, 376153, 376171, 376183, 376199, 376231, 376237, 376241, 376283, 376291, 376297,
376307, 376351, 376373, 376393, 376399, 376417, 376463, 376469, 376471, 376477, 376483, 376501, 376511,
376529, 376531, 376547, 376573, 376577, 376583, 376589, 376603, 376609, 376627, 376631, 376633, 376639,
376657, 376679, 376687, 376699, 376709, 376721, 376729, 376757, 376759, 376769, 376787, 376793, 376801,

376807, 376811, 376819, 376823, 376837, 376841, 376847, 376853, 376889, 376891, 376897, 376921, 376927,
376931, 376933, 376949, 376963, 376969, 377011, 377021, 377051, 377059, 377071, 377099, 377123, 377129,
377137, 377147, 377171, 377173, 377183, 377197, 377219, 377231, 377257, 377263, 377287, 377291, 377297,
377327, 377329, 377339, 377347, 377353, 377369, 377371, 377387, 377393, 377459, 377471, 377477, 377491,
377513, 377521, 377527, 377537, 377543, 377557, 377561, 377563, 377581, 377593, 377599, 377617, 377623,
377633, 377653, 377681, 377687, 377711, 377717, 377737, 377749, 377761, 377771, 377779, 377789, 377801,
377809, 377827, 377831, 377843, 377851, 377873, 377887, 377911, 377963, 377981, 377999, 378011, 378019,
378023, 378041, 378071, 378083, 378089, 378101, 378127, 378137, 378149, 378151, 378163, 378167, 378179,
378193, 378223, 378229, 378239, 378241, 378253, 378269, 378277, 378283, 378289, 378317, 378353, 378361,
378379, 378401, 378407, 378439, 378449, 378463, 378467, 378493, 378503, 378509, 378523, 378533, 378551,
378559, 378569, 378571, 378583, 378593, 378601, 378619, 378629, 378661, 378667, 378671, 378683, 378691,
378713, 378733, 378739, 378757, 378761, 378779, 378793, 378809, 378817, 378821, 378823, 378869, 378883,
378893, 378901, 378919, 378929, 378941, 378949, 378953, 378967, 378977, 378997, 379007, 379009, 379013,
379033, 379039, 379073, 379081, 379087, 379097, 379103, 379123, 379133, 379147, 379157, 379163, 379177,
379187, 379189, 379199, 379207, 379273, 379277, 379283, 379289, 379307, 379319, 379333, 379343, 379369,
379387, 379391, 379397, 379399, 379417, 379433, 379439, 379441, 379451, 379459, 379499, 379501, 379513,
379531, 379541, 379549, 379571, 379573, 379579, 379597, 379607, 379633, 379649, 379663, 379667, 379679,
379681, 379693, 379699, 379703, 379721, 379723, 379727, 379751, 379777, 379787, 379811, 379817, 379837,
379849, 379853, 379859, 379877, 379889, 379903, 379909, 379913, 379927, 379931, 379963, 379979, 379993,
379997, 379999, 380041, 380047, 380059, 380071, 380117, 380129, 380131, 380141, 380147, 380179, 380189,
380197, 380201, 380203, 380207, 380231, 380251, 380267, 380269, 380287, 380291, 380299, 380309, 380311,
380327, 380329, 380333, 380363, 380377, 380383, 380417, 380423, 380441, 380447, 380453, 380459, 380461,
380483, 380503, 380533, 380557, 380563, 380591, 380621, 380623, 380629, 380641, 380651, 380657, 380707,
380713, 380729, 380753, 380777, 380797, 380803, 380819, 380837, 380839, 380843, 380867, 380869, 380879,
380881, 380909, 380917, 380929, 380951, 380957, 380971, 380977, 380983, 381001, 381011, 381019, 381037,
381047, 381061, 381071, 381077, 381097, 381103, 381167, 381169, 381181, 381209, 381221, 381223, 381233,
381239, 381253, 381287, 381289, 381301, 381319, 381323, 381343, 381347, 381371, 381373, 381377, 381383,
381389, 381401, 381413, 381419, 381439, 381443, 381461, 381467, 381481, 381487, 381509, 381523, 381527,
381529, 381533, 381541, 381559, 381569, 381607, 381629, 381631, 381637, 381659, 381673, 381697, 381707,
381713, 381737, 381739, 381749, 381757, 381761, 381791, 381793, 381817, 381841, 381853, 381859, 381911,
381917, 381937, 381943, 381949, 381977, 381989, 381991, 382001, 382003, 382021, 382037, 382061, 382069,
382073, 382087, 382103, 382117, 382163, 382171, 382189, 382229, 382231, 382241, 382253, 382267, 382271,
382303, 382331, 382351, 382357, 382363, 382373, 382391, 382427, 382429, 382457, 382463, 382493, 382507,
382511, 382519, 382541, 382549, 382553, 382567, 382579, 382583, 382589, 382601, 382621, 382631, 382643,
382649, 382661, 382663, 382693, 382703, 382709, 382727, 382729, 382747, 382751, 382763, 382769, 382777,
382801, 382807, 382813, 382843, 382847, 382861, 382867, 382871, 382873, 382883, 382919, 382933, 382939,
382961, 382979, 382999, 383011, 383023, 383029, 383041, 383051, 383069, 383077, 383081, 383083, 383099,
383101, 383107, 383113, 383143, 383147, 383153, 383171, 383179, 383219, 383221, 383261, 383267, 383281,
383291, 383297, 383303, 383321, 383347, 383371, 383393, 383399, 383417, 383419, 383429, 383459, 383483,
383489, 383519, 383521, 383527, 383533, 383549, 383557, 383573, 383587, 383609, 383611, 383623, 383627,
383633, 383651, 383657, 383659, 383681, 383683, 383693, 383723, 383729, 383753, 383759, 383767, 383777,
383791, 383797, 383807, 383813, 383821, 383833, 383837, 383839, 383869, 383891, 383909, 383917, 383923,
383941, 383951, 383963, 383969, 383983, 383987, 384001, 384017, 384029, 384049, 384061, 384067, 384079,
384089, 384107, 384113, 384133, 384143, 384151, 384157, 384173, 384187, 384193, 384203, 384227, 384247,
384253, 384257, 384259, 384277, 384287, 384289, 384299, 384301, 384317, 384331, 384343, 384359, 384367,
384383, 384403, 384407, 384437, 384469, 384473, 384479, 384481, 384487, 384497, 384509, 384533, 384547,
384577, 384581, 384589, 384599, 384611, 384619, 384623, 384641, 384673, 384691, 384697, 384701, 384719,
384733, 384737, 384751, 384757, 384773, 384779, 384817, 384821, 384827, 384841, 384847, 384851, 384889,
384907, 384913, 384919, 384941, 384961, 384973, 385001, 385013, 385027, 385039, 385057, 385069, 385079,

385081, 385087, 385109, 385127, 385129, 385139, 385141, 385153, 385159, 385171, 385193, 385199, 385223,
385249, 385261, 385267, 385279, 385289, 385291, 385321, 385327, 385331, 385351, 385379, 385391, 385393,
385397, 385403, 385417, 385433, 385471, 385481, 385493, 385501, 385519, 385531, 385537, 385559, 385571,
385573, 385579, 385589, 385591, 385597, 385607, 385621, 385631, 385639, 385657, 385661, 385663, 385709,
385739, 385741, 385771, 385783, 385793, 385811, 385817, 385831, 385837, 385843, 385859, 385877, 385897,
385901, 385907, 385927, 385939, 385943, 385967, 385991, 385997, 386017, 386039, 386041, 386047, 386051,
386083, 386093, 386117, 386119, 386129, 386131, 386143, 386149, 386153, 386159, 386161, 386173, 386219,
386227, 386233, 386237, 386249, 386263, 386279, 386297, 386299, 386303, 386329, 386333, 386339, 386363,
386369, 386371, 386381, 386383, 386401, 386411, 386413, 386429, 386431, 386437, 386471, 386489, 386501,
386521, 386537, 386543, 386549, 386569, 386587, 386609, 386611, 386621, 386629, 386641, 386647, 386651,
386677, 386689, 386693, 386713, 386719, 386723, 386731, 386747, 386777, 386809, 386839, 386851, 386887,
386891, 386921, 386927, 386963, 386977, 386987, 386989, 386993, 387007, 387017, 387031, 387047, 387071,
387077, 387083, 387089, 387109, 387137, 387151, 387161, 387169, 387173, 387187, 387197, 387199, 387203,
387227, 387253, 387263, 387269, 387281, 387307, 387313, 387329, 387341, 387371, 387397, 387403, 387433,
387437, 387449, 387463, 387493, 387503, 387509, 387529, 387551, 387577, 387587, 387613, 387623, 387631,
387641, 387659, 387677, 387679, 387683, 387707, 387721, 387727, 387743, 387749, 387763, 387781, 387791,
387799, 387839, 387853, 387857, 387911, 387913, 387917, 387953, 387967, 387971, 387973, 387977, 388009,
388051, 388057, 388067, 388081, 388099, 388109, 388111, 388117, 388133, 388159, 388163, 388169, 388177,
388181, 388183, 388187, 388211, 388231, 388237, 388253, 388259, 388273, 388277, 388301, 388313, 388319,
388351, 388363, 388369, 388373, 388391, 388403, 388459, 388471, 388477, 388481, 388483, 388489, 388499,
388519, 388529, 388541, 388567, 388573, 388621, 388651, 388657, 388673, 388691, 388693, 388697, 388699,
388711, 388727, 388757, 388777, 388781, 388789, 388793, 388813, 388823, 388837, 388859, 388879, 388891,
388897, 388901, 388903, 388931, 388933, 388937, 388961, 388963, 388991, 389003, 389023, 389027, 389029,
389041, 389047, 389057, 389083, 389089, 389099, 389111, 389117, 389141, 389149, 389161, 389167, 389171,
389173, 389189, 389219, 389227, 389231, 389269, 389273, 389287, 389297, 389299, 389303, 389357, 389369,
389381, 389399, 389401, 389437, 389447, 389461, 389479, 389483, 389507, 389513, 389527, 389531, 389533,
389539, 389561, 389563, 389567, 389569, 389579, 389591, 389621, 389629, 389651, 389659, 389663, 389687,
389699, 389713, 389723, 389743, 389749, 389761, 389773, 389783, 389791, 389797, 389819, 389839, 389849,
389867, 389891, 389897, 389903, 389911, 389923, 389927, 389941, 389947, 389953, 389957, 389971, 389981,
389989, 389999, 390001, 390043, 390067, 390077, 390083, 390097, 390101, 390107, 390109, 390113, 390119,
390151, 390157, 390161, 390191, 390193, 390199, 390209, 390211, 390223, 390263, 390281, 390289, 390307,
390323, 390343, 390347, 390353, 390359, 390367, 390373, 390389, 390391, 390407, 390413, 390419, 390421,
390433, 390437, 390449, 390463, 390479, 390487, 390491, 390493, 390499, 390503, 390527, 390539, 390553,
390581, 390647, 390653, 390671, 390673, 390703, 390707, 390721, 390727, 390737, 390739, 390743, 390751,
390763, 390781, 390791, 390809, 390821, 390829, 390851, 390869, 390877, 390883, 390889, 390893, 390953,
390959, 390961, 390967, 390989, 390991, 391009, 391019, 391021, 391031, 391049, 391057, 391063, 391067,
391073, 391103, 391117, 391133, 391151, 391159, 391163, 391177, 391199, 391217, 391219, 391231, 391247,
391249, 391273, 391283, 391291, 391301, 391331, 391337, 391351, 391367, 391373, 391379, 391387, 391393,
391397, 391399, 391403, 391441, 391451, 391453, 391487, 391519, 391537, 391553, 391579, 391613, 391619,
391627, 391631, 391639, 391661, 391679, 391691, 391693, 391711, 391717, 391733, 391739, 391751, 391753,
391757, 391789, 391801, 391817, 391823, 391847, 391861, 391873, 391879, 391889, 391891, 391903, 391907,
391921, 391939, 391961, 391967, 391987, 391999, 392011, 392033, 392053, 392059, 392069, 392087, 392099,
392101, 392111, 392113, 392131, 392143, 392149, 392153, 392159, 392177, 392201, 392209, 392213, 392221,
392233, 392239, 392251, 392261, 392263, 392267, 392269, 392279, 392281, 392297, 392299, 392321, 392333,
392339, 392347, 392351, 392363, 392383, 392389, 392423, 392437, 392443, 392467, 392473, 392477, 392489,
392503, 392519, 392531, 392543, 392549, 392569, 392593, 392599, 392611, 392629, 392647, 392663, 392669,
392699, 392723, 392737, 392741, 392759, 392761, 392767, 392803, 392807, 392809, 392827, 392831, 392837,
392849, 392851, 392857, 392879, 392893, 392911, 392923, 392927, 392929, 392957, 392963, 392969, 392981,
392983, 393007, 393013, 393017, 393031, 393059, 393073, 393077, 393079, 393083, 393097, 393103, 393109,

393121, 393137, 393143, 393157, 393161, 393181, 393187, 393191, 393203, 393209, 393241, 393247, 393257,
393271, 393287, 393299, 393301, 393311, 393331, 393361, 393373, 393377, 393383, 393401, 393403, 393413,
393451, 393473, 393479, 393487, 393517, 393521, 393539, 393541, 393551, 393557, 393571, 393577, 393581,
393583, 393587, 393593, 393611, 393629, 393637, 393649, 393667, 393671, 393677, 393683, 393697, 393709,
393713, 393721, 393727, 393739, 393749, 393761, 393779, 393797, 393847, 393853, 393857, 393859, 393863,
393871, 393901, 393919, 393929, 393931, 393947, 393961, 393977, 393989, 393997, 394007, 394019, 394039,
394049, 394063, 394073, 394099, 394123, 394129, 394153, 394157, 394169, 394187, 394201, 394211, 394223,
394241, 394249, 394259, 394271, 394291, 394319, 394327, 394357, 394363, 394367, 394369, 394393, 394409,
394411, 394453, 394481, 394489, 394501, 394507, 394523, 394529, 394549, 394571, 394577, 394579, 394601,
394619, 394631, 394633, 394637, 394643, 394673, 394699, 394717, 394721, 394727, 394729, 394733, 394739,
394747, 394759, 394787, 394811, 394813, 394817, 394819, 394829, 394837, 394861, 394879, 394897, 394931,
394943, 394963, 394967, 394969, 394981, 394987, 394993, 395023, 395027, 395039, 395047, 395069, 395089,
395093, 395107, 395111, 395113, 395119, 395137, 395141, 395147, 395159, 395173, 395189, 395191, 395201,
395231, 395243, 395251, 395261, 395273, 395287, 395293, 395303, 395309, 395321, 395323, 395377, 395383,
395407, 395429, 395431, 395443, 395449, 395453, 395459, 395491, 395509, 395513, 395533, 395537, 395543,
395581, 395597, 395611, 395621, 395627, 395657, 395671, 395677, 395687, 395701, 395719, 395737, 395741,
395749, 395767, 395803, 395849, 395851, 395873, 395887, 395891, 395897, 395909, 395921, 395953, 395959,
395971, 396001, 396029, 396031, 396041, 396043, 396061, 396079, 396091, 396103, 396107, 396119, 396157,
396173, 396181, 396197, 396199, 396203, 396217, 396239, 396247, 396259, 396269, 396293, 396299, 396301,
396311, 396323, 396349, 396353, 396373, 396377, 396379, 396413, 396427, 396437, 396443, 396449, 396479,
396509, 396523, 396527, 396533, 396541, 396547, 396563, 396577, 396581, 396601, 396619, 396623, 396629,
396631, 396637, 396647, 396667, 396679, 396703, 396709, 396713, 396719, 396733, 396833, 396871, 396881,
396883, 396887, 396919, 396931, 396937, 396943, 396947, 396953, 396971, 396983, 396997, 397013, 397027,
397037, 397051, 397057, 397063, 397073, 397093, 397099, 397127, 397151, 397153, 397181, 397183, 397211,
397217, 397223, 397237, 397253, 397259, 397283, 397289, 397297, 397301, 397303, 397337, 397351, 397357,
397361, 397373, 397379, 397427, 397429, 397433, 397459, 397469, 397489, 397493, 397517, 397519, 397541,
397543, 397547, 397549, 397567, 397589, 397591, 397597, 397633, 397643, 397673, 397687, 397697, 397721,
397723, 397729, 397751, 397753, 397757, 397759, 397763, 397799, 397807, 397811, 397829, 397849, 397867,
397897, 397907, 397921, 397939, 397951, 397963, 397973, 397981, 398011, 398023, 398029, 398033, 398039,
398053, 398059, 398063, 398077, 398087, 398113, 398117, 398119, 398129, 398143, 398149, 398171, 398207,
398213, 398219, 398227, 398249, 398261, 398267, 398273, 398287, 398303, 398311, 398323, 398339, 398341,
398347, 398353, 398357, 398369, 398393, 398407, 398417, 398423, 398441, 398459, 398467, 398471, 398473,
398477, 398491, 398509, 398539, 398543, 398549, 398557, 398569, 398581, 398591, 398609, 398611, 398621,
398627, 398669, 398681, 398683, 398693, 398711, 398729, 398731, 398759, 398771, 398813, 398819, 398821,
398833, 398857, 398863, 398887, 398903, 398917, 398921, 398933, 398941, 398969, 398977, 398989, 399023,
399031, 399043, 399059, 399067, 399071, 399079, 399097, 399101, 399107, 399131, 399137, 399149, 399151,
399163, 399173, 399181, 399197, 399221, 399227, 399239, 399241, 399263, 399271, 399277, 399281, 399283,
399353, 399379, 399389, 399391, 399401, 399403, 399409, 399433, 399439, 399473, 399481, 399491, 399493,
399499, 399523, 399527, 399541, 399557, 399571, 399577, 399583, 399587, 399601, 399613, 399617, 399643,
399647, 399667, 399677, 399689, 399691, 399719, 399727, 399731, 399739, 399757, 399761, 399769, 399781,
399787, 399793, 399851, 399853, 399871, 399887, 399899, 399911, 399913, 399937, 399941, 399953, 399979,
399983, 399989, 400009, 400031, 400033, 400051, 400067, 400069, 400087, 400093, 400109, 400123, 400151,
400157, 400187, 400199, 400207, 400217, 400237, 400243, 400247, 400249, 400261, 400277, 400291, 400297,
400307, 400313, 400321, 400331, 400339, 400381, 400391, 400409, 400417, 400429, 400441, 400457, 400471,
400481, 400523, 400559, 400579, 400597, 400601, 400607, 400619, 400643, 400651, 400657, 400679, 400681,
400703, 400711, 400721, 400723, 400739, 400753, 400759, 400823, 400837, 400849, 400853, 400859, 400871,
400903, 400927, 400931, 400943, 400949, 400963, 400997, 401017, 401029, 401039, 401053, 401057, 401069,
401077, 401087, 401101, 401113, 401119, 401161, 401173, 401179, 401201, 401209, 401231, 401237, 401243,
401279, 401287, 401309, 401311, 401321, 401329, 401341, 401347, 401371, 401381, 401393, 401407, 401411,

401417, 401473, 401477, 401507, 401519, 401537, 401539, 401551, 401567, 401587, 401593, 401627, 401629,
401651, 401669, 401671, 401689, 401707, 401711, 401743, 401771, 401773, 401809, 401813, 401827, 401839,
401861, 401867, 401887, 401903, 401909, 401917, 401939, 401953, 401957, 401959, 401981, 401987, 401993,
402023, 402029, 402037, 402043, 402049, 402053, 402071, 402089, 402091, 402107, 402131, 402133, 402137,
402139, 402197, 402221, 402223, 402239, 402253, 402263, 402277, 402299, 402307, 402313, 402329, 402331,
402341, 402343, 402359, 402361, 402371, 402379, 402383, 402403, 402419, 402443, 402487, 402503, 402511,
402517, 402527, 402529, 402541, 402551, 402559, 402581, 402583, 402587, 402593, 402601, 402613, 402631,
402691, 402697, 402739, 402751, 402757, 402761, 402763, 402767, 402769, 402797, 402803, 402817, 402823,
402847, 402851, 402859, 402863, 402869, 402881, 402923, 402943, 402947, 402949, 402991, 403001, 403003,
403037, 403043, 403049, 403057, 403061, 403063, 403079, 403097, 403103, 403133, 403141, 403159, 403163,
403181, 403219, 403241, 403243, 403253, 403261, 403267, 403289, 403301, 403309, 403327, 403331, 403339,
403363, 403369, 403387, 403391, 403433, 403439, 403483, 403499, 403511, 403537, 403547, 403549, 403553,
403567, 403577, 403591, 403603, 403607, 403621, 403649, 403661, 403679, 403681, 403687, 403703, 403717,
403721, 403729, 403757, 403783, 403787, 403817, 403829, 403831, 403849, 403861, 403867, 403877, 403889,
403901, 403933, 403951, 403957, 403969, 403979, 403981, 403993, 404009, 404011, 404017, 404021, 404029,
404051, 404081, 404099, 404113, 404119, 404123, 404161, 404167, 404177, 404189, 404191, 404197, 404213,
404221, 404249, 404251, 404267, 404269, 404273, 404291, 404309, 404321, 404323, 404357, 404381, 404387,
404389, 404399, 404419, 404423, 404429, 404431, 404449, 404461, 404483, 404489, 404497, 404507, 404513,
404527, 404531, 404533, 404539, 404557, 404597, 404671, 404693, 404699, 404713, 404773, 404779, 404783,
404819, 404827, 404837, 404843, 404849, 404851, 404941, 404951, 404959, 404969, 404977, 404981, 404983,
405001, 405011, 405029, 405037, 405047, 405049, 405071, 405073, 405089, 405091, 405143, 405157, 405179,
405199, 405211, 405221, 405227, 405239, 405241, 405247, 405253, 405269, 405277, 405287, 405299, 405323,
405341, 405343, 405347, 405373, 405401, 405407, 405413, 405437, 405439, 405473, 405487, 405491, 405497,
405499, 405521, 405527, 405529, 405541, 405553, 405577, 405599, 405607, 405611, 405641, 405659, 405667,
405677, 405679, 405683, 405689, 405701, 405703, 405709, 405719, 405731, 405749, 405763, 405767, 405781,
405799, 405817, 405827, 405829, 405857, 405863, 405869, 405871, 405893, 405901, 405917, 405947, 405949,
405959, 405967, 405989, 405991, 405997, 406013, 406027, 406037, 406067, 406073, 406093, 406117, 406123,
406169, 406171, 406177, 406183, 406207, 406247, 406253, 406267, 406271, 406309, 406313, 406327, 406331,
406339, 406349, 406361, 406381, 406397, 406403, 406423, 406447, 406481, 406499, 406501, 406507, 406513,
406517, 406531, 406547, 406559, 406561, 406573, 406577, 406579, 406583, 406591, 406631, 406633, 406649,
406661, 406673, 406697, 406699, 406717, 406729, 406739, 406789, 406807, 406811, 406817, 406837, 406859,
406873, 406883, 406907, 406951, 406969, 406981, 406993, 407023, 407047, 407059, 407083, 407119, 407137,
407149, 407153, 407177, 407179, 407191, 407203, 407207, 407219, 407221, 407233, 407249, 407257, 407263,
407273, 407287, 407291, 407299, 407311, 407317, 407321, 407347, 407357, 407359, 407369, 407377, 407383,
407401, 407437, 407471, 407483, 407489, 407501, 407503, 407509, 407521, 407527, 407567, 407573, 407579,
407587, 407599, 407621, 407633, 407639, 407651, 407657, 407669, 407699, 407707, 407713, 407717, 407723,
407741, 407747, 407783, 407789, 407791, 407801, 407807, 407821, 407833, 407843, 407857, 407861, 407879,
407893, 407899, 407917, 407923, 407947, 407959, 407969, 407971, 407977, 407993, 408011, 408019, 408041,
408049, 408071, 408077, 408091, 408127, 408131, 408137, 408169, 408173, 408197, 408203, 408209, 408211,
408217, 408223, 408229, 408241, 408251, 408263, 408271, 408283, 408311, 408337, 408341, 408347, 408361,
408379, 408389, 408403, 408413, 408427, 408431, 408433, 408437, 408461, 408469, 408479, 408491, 408497,
408533, 408539, 408553, 408563, 408607, 408623, 408631, 408637, 408643, 408659, 408677, 408689, 408691,
408701, 408703, 408713, 408719, 408743, 408763, 408769, 408773, 408787, 408803, 408809, 408817, 408841,
408857, 408869, 408911, 408913, 408923, 408943, 408953, 408959, 408971, 408979, 408997, 409007, 409021,
409027, 409033, 409043, 409063, 409069, 409081, 409099, 409121, 409153, 409163, 409177, 409187, 409217,
409237, 409259, 409261, 409267, 409271, 409289, 409291, 409327, 409333, 409337, 409349, 409351, 409369,
409379, 409391, 409397, 409429, 409433, 409441, 409463, 409471, 409477, 409483, 409499, 409517, 409523,
409529, 409543, 409573, 409579, 409589, 409597, 409609, 409639, 409657, 409691, 409693, 409709, 409711,
409723, 409729, 409733, 409753, 409769, 409777, 409781, 409813, 409817, 409823, 409831, 409841, 409861,

409867, 409879, 409889, 409891, 409897, 409901, 409909, 409933, 409943, 409951, 409961, 409967, 409987,
409993, 409999, 410009, 410029, 410063, 410087, 410093, 410117, 410119, 410141, 410143, 410149, 410171,
410173, 410203, 410231, 410233, 410239, 410243, 410257, 410279, 410281, 410299, 410317, 410323, 410339,
410341, 410353, 410359, 410383, 410387, 410393, 410401, 410411, 410413, 410453, 410461, 410477, 410489,
410491, 410497, 410507, 410513, 410519, 410551, 410561, 410587, 410617, 410621, 410623, 410629, 410651,
410659, 410671, 410687, 410701, 410717, 410731, 410741, 410747, 410749, 410759, 410783, 410789, 410801,
410807, 410819, 410833, 410857, 410899, 410903, 410929, 410953, 410983, 410999, 411001, 411007, 411011,
411013, 411031, 411041, 411049, 411067, 411071, 411083, 411101, 411113, 411119, 411127, 411143, 411157,
411167, 411193, 411197, 411211, 411233, 411241, 411251, 411253, 411259, 411287, 411311, 411337, 411347,
411361, 411371, 411379, 411409, 411421, 411443, 411449, 411469, 411473, 411479, 411491, 411503, 411527,
411529, 411557, 411563, 411569, 411577, 411583, 411589, 411611, 411613, 411617, 411637, 411641, 411667,
411679, 411683, 411703, 411707, 411709, 411721, 411727, 411737, 411739, 411743, 411751, 411779, 411799,
411809, 411821, 411823, 411833, 411841, 411883, 411919, 411923, 411937, 411941, 411947, 411967, 411991,
412001, 412007, 412019, 412031, 412033, 412037, 412039, 412051, 412067, 412073, 412081, 412099, 412109,
412123, 412127, 412133, 412147, 412157, 412171, 412187, 412189, 412193, 412201, 412211, 412213, 412219,
412249, 412253, 412273, 412277, 412289, 412303, 412333, 412339, 412343, 412387, 412397, 412411, 412457,
412463, 412481, 412487, 412493, 412537, 412561, 412567, 412571, 412589, 412591, 412603, 412609, 412619,
412627, 412637, 412639, 412651, 412663, 412667, 412717, 412739, 412771, 412793, 412807, 412831, 412849,
412859, 412891, 412901, 412903, 412939, 412943, 412949, 412967, 412987, 413009, 413027, 413033, 413053,
413069, 413071, 413081, 413087, 413089, 413093, 413111, 413113, 413129, 413141, 413143, 413159, 413167,
413183, 413197, 413201, 413207, 413233, 413243, 413251, 413263, 413267, 413293, 413299, 413353, 413411,
413417, 413429, 413443, 413461, 413477, 413521, 413527, 413533, 413537, 413551, 413557, 413579, 413587,
413597, 413629, 413653, 413681, 413683, 413689, 413711, 413713, 413719, 413737, 413753, 413759, 413779,
413783, 413807, 413827, 413849, 413863, 413867, 413869, 413879, 413887, 413911, 413923, 413951, 413981,
414013, 414017, 414019, 414031, 414049, 414053, 414061, 414077, 414083, 414097, 414101, 414107, 414109,
414131, 414157, 414179, 414199, 414203, 414209, 414217, 414221, 414241, 414259, 414269, 414277, 414283,
414311, 414313, 414329, 414331, 414347, 414361, 414367, 414383, 414389, 414397, 414413, 414431, 414433,
414451, 414457, 414461, 414467, 414487, 414503, 414521, 414539, 414553, 414559, 414571, 414577, 414607,
414611, 414629, 414641, 414643, 414653, 414677, 414679, 414683, 414691, 414697, 414703, 414707, 414709,
414721, 414731, 414737, 414763, 414767, 414769, 414773, 414779, 414793, 414803, 414809, 414833, 414857,
414871, 414889, 414893, 414899, 414913, 414923, 414929, 414949, 414959, 414971, 414977, 414991, 415013,
415031, 415039, 415061, 415069, 415073, 415087, 415097, 415109, 415111, 415133, 415141, 415147, 415153,
415159, 415171, 415187, 415189, 415201, 415213, 415231, 415253, 415271, 415273, 415319, 415343, 415379,
415381, 415391, 415409, 415427, 415447, 415469, 415477, 415489, 415507, 415517, 415523, 415543, 415553,
415559, 415567, 415577, 415603, 415607, 415609, 415627, 415631, 415643, 415651, 415661, 415669, 415673,
415687, 415691, 415697, 415717, 415721, 415729, 415759, 415783, 415787, 415799, 415801, 415819, 415823,
415861, 415873, 415879, 415901, 415931, 415937, 415949, 415951, 415957, 415963, 415969, 415979, 415993,
415999, 416011, 416023, 416071, 416077, 416089, 416107, 416147, 416149, 416153, 416159, 416167, 416201,
416219, 416239, 416243, 416249, 416257, 416263, 416281, 416291, 416333, 416359, 416387, 416389, 416393,
416399, 416401, 416407, 416413, 416417, 416419, 416441, 416443, 416459, 416473, 416477, 416491, 416497,
416501, 416503, 416513, 416531, 416543, 416573, 416579, 416593, 416621, 416623, 416629, 416659, 416677,
416693, 416719, 416761, 416797, 416821, 416833, 416839, 416849, 416851, 416873, 416881, 416887, 416947,
416957, 416963, 416989, 417007, 417017, 417019, 417023, 417037, 417089, 417097, 417113, 417119, 417127,
417133, 417161, 417169, 417173, 417181, 417187, 417191, 417203, 417217, 417227, 417239, 417251, 417271,
417283, 417293, 417311, 417317, 417331, 417337, 417371, 417377, 417379, 417383, 417419, 417437, 417451,
417457, 417479, 417491, 417493, 417509, 417511, 417523, 417541, 417553, 417559, 417577, 417581, 417583,
417617, 417623, 417631, 417643, 417649, 417671, 417691, 417719, 417721, 417727, 417731, 417733, 417737,
417751, 417763, 417773, 417793, 417811, 417821, 417839, 417863, 417869, 417881, 417883, 417899, 417931,
417941, 417947, 417953, 417959, 417961, 417983, 417997, 418007, 418009, 418027, 418031, 418043, 418051,

418069, 418073, 418079, 418087, 418109, 418129, 418157, 418169, 418177, 418181, 418189, 418199, 418207,
418219, 418259, 418273, 418279, 418289, 418303, 418321, 418331, 418337, 418339, 418343, 418349, 418351,
418357, 418373, 418381, 418391, 418423, 418427, 418447, 418459, 418471, 418493, 418511, 418553, 418559,
418597, 418601, 418603, 418631, 418633, 418637, 418657, 418667, 418699, 418709, 418721, 418739, 418751,
418763, 418771, 418783, 418787, 418793, 418799, 418811, 418813, 418819, 418837, 418843, 418849, 418861,
418867, 418871, 418883, 418889, 418909, 418921, 418927, 418933, 418939, 418961, 418981, 418987, 418993,
418997, 419047, 419051, 419053, 419057, 419059, 419087, 419141, 419147, 419161, 419171, 419183, 419189,
419191, 419201, 419231, 419249, 419261, 419281, 419291, 419297, 419303, 419317, 419329, 419351, 419383,
419401, 419417, 419423, 419429, 419443, 419449, 419459, 419467, 419473, 419477, 419483, 419491, 419513,
419527, 419537, 419557, 419561, 419563, 419567, 419579, 419591, 419597, 419599, 419603, 419609, 419623,
419651, 419687, 419693, 419701, 419711, 419743, 419753, 419777, 419789, 419791, 419801, 419803, 419821,
419827, 419831, 419873, 419893, 419921, 419927, 419929, 419933, 419953, 419959, 419999, 420001, 420029,
420037, 420041, 420047, 420073, 420097, 420103, 420149, 420163, 420191, 420193, 420221, 420241, 420253,
420263, 420269, 420271, 420293, 420307, 420313, 420317, 420319, 420323, 420331, 420341, 420349, 420353,
420361, 420367, 420383, 420397, 420419, 420421, 420439, 420457, 420467, 420479, 420481, 420499, 420503,
420521, 420551, 420557, 420569, 420571, 420593, 420599, 420613, 420671, 420677, 420683, 420691, 420731,
420737, 420743, 420757, 420769, 420779, 420781, 420799, 420803, 420809, 420811, 420851, 420853, 420857,
420859, 420899, 420919, 420929, 420941, 420967, 420977, 420997, 421009, 421019, 421033, 421037, 421049,
421079, 421081, 421093, 421103, 421121, 421123, 421133, 421147, 421159, 421163, 421177, 421181, 421189,
421207, 421241, 421273, 421279, 421303, 421313, 421331, 421339, 421349, 421361, 421381, 421397, 421409,
421417, 421423, 421433, 421453, 421459, 421469, 421471, 421483, 421493, 421501, 421517, 421559, 421607,
421609, 421621, 421633, 421639, 421643, 421657, 421661, 421691, 421697, 421699, 421703, 421709, 421711,
421717, 421727, 421739, 421741, 421783, 421801, 421807, 421831, 421847, 421891, 421907, 421913, 421943,
421973, 421987, 421997, 422029, 422041, 422057, 422063, 422069, 422077, 422083, 422087, 422089, 422099,
422101, 422111, 422113, 422129, 422137, 422141, 422183, 422203, 422209, 422231, 422239, 422243, 422249,
422267, 422287, 422291, 422309, 422311, 422321, 422339, 422353, 422363, 422369, 422377, 422393, 422407,
422431, 422453, 422459, 422479, 422537, 422549, 422551, 422557, 422563, 422567, 422573, 422581, 422621,
422627, 422657, 422689, 422701, 422707, 422711, 422749, 422753, 422759, 422761, 422789, 422797, 422803,
422827, 422857, 422861, 422867, 422869, 422879, 422881, 422893, 422897, 422899, 422911, 422923, 422927,
422969, 422987, 423001, 423013, 423019, 423043, 423053, 423061, 423067, 423083, 423091, 423097, 423103,
423109, 423121, 423127, 423133, 423173, 423179, 423191, 423209, 423221, 423229, 423233, 423251, 423257,
423259, 423277, 423281, 423287, 423289, 423299, 423307, 423323, 423341, 423347, 423389, 423403, 423413,
423427, 423431, 423439, 423457, 423461, 423463, 423469, 423481, 423497, 423503, 423509, 423541, 423547,
423557, 423559, 423581, 423587, 423601, 423617, 423649, 423667, 423697, 423707, 423713, 423727, 423749,
423751, 423763, 423769, 423779, 423781, 423791, 423803, 423823, 423847, 423853, 423859, 423869, 423883,
423887, 423931, 423949, 423961, 423977, 423989, 423991, 424001, 424003, 424007, 424019, 424027, 424037,
424079, 424091, 424093, 424103, 424117, 424121, 424129, 424139, 424147, 424157, 424163, 424169, 424187,
424199, 424223, 424231, 424243, 424247, 424261, 424267, 424271, 424273, 424313, 424331, 424339, 424343,
424351, 424397, 424423, 424429, 424433, 424451, 424471, 424481, 424493, 424519, 424537, 424547, 424549,
424559, 424573, 424577, 424597, 424601, 424639, 424661, 424667, 424679, 424687, 424693, 424709, 424727,
424729, 424757, 424769, 424771, 424777, 424811, 424817, 424819, 424829, 424841, 424843, 424849, 424861,
424867, 424889, 424891, 424903, 424909, 424913, 424939, 424961, 424967, 424997, 425003, 425027, 425039,
425057, 425059, 425071, 425083, 425101, 425107, 425123, 425147, 425149, 425189, 425197, 425207, 425233,
425237, 425251, 425273, 425279, 425281, 425291, 425297, 425309, 425317, 425329, 425333, 425363, 425377,
425387, 425393, 425417, 425419, 425423, 425441, 425443, 425471, 425473, 425489, 425501, 425519, 425521,
425533, 425549, 425563, 425591, 425603, 425609, 425641, 425653, 425681, 425701, 425713, 425779, 425783,
425791, 425801, 425813, 425819, 425837, 425839, 425851, 425857, 425861, 425869, 425879, 425899, 425903,
425911, 425939, 425959, 425977, 425987, 425989, 426007, 426011, 426061, 426073, 426077, 426089, 426091,
426103, 426131, 426161, 426163, 426193, 426197, 426211, 426229, 426233, 426253, 426287, 426301, 426311,

426319, 426331, 426353, 426383, 426389, 426401, 426407, 426421, 426427, 426469, 426487, 426527, 426541,
426551, 426553, 426563, 426583, 426611, 426631, 426637, 426641, 426661, 426691, 426697, 426707, 426709,
426731, 426737, 426739, 426743, 426757, 426761, 426763, 426773, 426779, 426787, 426799, 426841, 426859,
426863, 426871, 426889, 426893, 426913, 426917, 426919, 426931, 426941, 426971, 426973, 426997, 427001,
427013, 427039, 427043, 427067, 427069, 427073, 427079, 427081, 427103, 427117, 427151, 427169, 427181,
427213, 427237, 427241, 427243, 427247, 427249, 427279, 427283, 427307, 427309, 427327, 427333, 427351,
427369, 427379, 427381, 427403, 427417, 427421, 427423, 427429, 427433, 427439, 427447, 427451, 427457,
427477, 427513, 427517, 427523, 427529, 427541, 427579, 427591, 427597, 427619, 427621, 427681, 427711,
427717, 427723, 427727, 427733, 427751, 427781, 427787, 427789, 427813, 427849, 427859, 427877, 427879,
427883, 427913, 427919, 427939, 427949, 427951, 427957, 427967, 427969, 427991, 427993, 427997, 428003,
428023, 428027, 428033, 428039, 428041, 428047, 428083, 428093, 428137, 428143, 428147, 428149, 428161,
428167, 428173, 428177, 428221, 428227, 428231, 428249, 428251, 428273, 428297, 428299, 428303, 428339,
428353, 428369, 428401, 428411, 428429, 428471, 428473, 428489, 428503, 428509, 428531, 428539, 428551,
428557, 428563, 428567, 428569, 428579, 428629, 428633, 428639, 428657, 428663, 428671, 428677, 428683,
428693, 428731, 428741, 428759, 428777, 428797, 428801, 428807, 428809, 428833, 428843, 428851, 428863,
428873, 428899, 428951, 428957, 428977, 429007, 429017, 429043, 429083, 429101, 429109, 429119, 429127,
429137, 429139, 429161, 429181, 429197, 429211, 429217, 429223, 429227, 429241, 429259, 429271, 429277,
429281, 429283, 429329, 429347, 429349, 429361, 429367, 429389, 429397, 429409, 429413, 429427, 429431,
429449, 429463, 429467, 429469, 429487, 429497, 429503, 429509, 429511, 429521, 429529, 429547, 429551,
429563, 429581, 429587, 429589, 429599, 429631, 429643, 429659, 429661, 429673, 429677, 429679, 429683,
429701, 429719, 429727, 429731, 429733, 429773, 429791, 429797, 429817, 429823, 429827, 429851, 429853,
429881, 429887, 429889, 429899, 429901, 429907, 429911, 429917, 429929, 429931, 429937, 429943, 429953,
429971, 429973, 429991, 430007, 430009, 430013, 430019, 430057, 430061, 430081, 430091, 430093, 430121,
430139, 430147, 430193, 430259, 430267, 430277, 430279, 430289, 430303, 430319, 430333, 430343, 430357,
430393, 430411, 430427, 430433, 430453, 430487, 430499, 430511, 430513, 430517, 430543, 430553, 430571,
430579, 430589, 430601, 430603, 430649, 430663, 430691, 430697, 430699, 430709, 430723, 430739, 430741,
430747, 430751, 430753, 430769, 430783, 430789, 430799, 430811, 430819, 430823, 430841, 430847, 430861,
430873, 430879, 430883, 430891, 430897, 430907, 430909, 430921, 430949, 430957, 430979, 430981, 430987,
430999, 431017, 431021, 431029, 431047, 431051, 431063, 431077, 431083, 431099, 431107, 431141, 431147,
431153, 431173, 431191, 431203, 431213, 431219, 431237, 431251, 431257, 431267, 431269, 431287, 431297,
431311, 431329, 431339, 431363, 431369, 431377, 431381, 431399, 431423, 431429, 431441, 431447, 431449,
431479, 431513, 431521, 431533, 431567, 431581, 431597, 431603, 431611, 431617, 431621, 431657, 431659,
431663, 431671, 431693, 431707, 431729, 431731, 431759, 431777, 431797, 431801, 431803, 431807, 431831,
431833, 431857, 431863, 431867, 431869, 431881, 431887, 431891, 431903, 431911, 431929, 431933, 431947,
431983, 431993, 432001, 432007, 432023, 432031, 432037, 432043, 432053, 432059, 432067, 432073, 432097,
432121, 432137, 432139, 432143, 432149, 432161, 432163, 432167, 432199, 432203, 432227, 432241, 432251,
432277, 432281, 432287, 432301, 432317, 432323, 432337, 432343, 432349, 432359, 432373, 432389, 432391,
432401, 432413, 432433, 432437, 432449, 432457, 432479, 432491, 432499, 432503, 432511, 432527, 432539,
432557, 432559, 432569, 432577, 432587, 432589, 432613, 432631, 432637, 432659, 432661, 432713, 432721,
432727, 432737, 432743, 432749, 432781, 432793, 432797, 432799, 432833, 432847, 432857, 432869, 432893,
432907, 432923, 432931, 432959, 432961, 432979, 432983, 432989, 433003, 433033, 433049, 433051, 433061,
433073, 433079, 433087, 433093, 433099, 433117, 433123, 433141, 433151, 433187, 433193, 433201, 433207,
433229, 433241, 433249, 433253, 433259, 433261, 433267, 433271, 433291, 433309, 433319, 433337, 433351,
433357, 433361, 433369, 433373, 433393, 433399, 433421, 433429, 433439, 433453, 433469, 433471, 433501,
433507, 433513, 433549, 433571, 433577, 433607, 433627, 433633, 433639, 433651, 433661, 433663, 433673,
433679, 433681, 433703, 433723, 433729, 433747, 433759, 433777, 433781, 433787, 433813, 433817, 433847,
433859, 433861, 433877, 433883, 433889, 433931, 433943, 433963, 433967, 433981, 434009, 434011, 434029,
434039, 434081, 434087, 434107, 434111, 434113, 434117, 434141, 434167, 434179, 434191, 434201, 434209,
434221, 434237, 434243, 434249, 434261, 434267, 434293, 434297, 434303, 434311, 434323, 434347, 434353,

434363, 434377, 434383, 434387, 434389, 434407, 434411, 434431, 434437, 434459, 434461, 434471, 434479,
434501, 434509, 434521, 434561, 434563, 434573, 434593, 434597, 434611, 434647, 434659, 434683, 434689,
434699, 434717, 434719, 434743, 434761, 434783, 434803, 434807, 434813, 434821, 434827, 434831, 434839,
434849, 434857, 434867, 434873, 434881, 434909, 434921, 434923, 434927, 434933, 434939, 434947, 434957,
434963, 434977, 434981, 434989, 435037, 435041, 435059, 435103, 435107, 435109, 435131, 435139, 435143,
435151, 435161, 435179, 435181, 435187, 435191, 435221, 435223, 435247, 435257, 435263, 435277, 435283,
435287, 435307, 435317, 435343, 435349, 435359, 435371, 435397, 435401, 435403, 435419, 435427, 435437,
435439, 435451, 435481, 435503, 435529, 435541, 435553, 435559, 435563, 435569, 435571, 435577, 435583,
435593, 435619, 435623, 435637, 435641, 435647, 435649, 435653, 435661, 435679, 435709, 435731, 435733,
435739, 435751, 435763, 435769, 435779, 435817, 435839, 435847, 435857, 435859, 435881, 435889, 435893,
435907, 435913, 435923, 435947, 435949, 435973, 435983, 435997, 436003, 436013, 436027, 436061, 436081,
436087, 436091, 436097, 436127, 436147, 436151, 436157, 436171, 436181, 436217, 436231, 436253, 436273,
436279, 436283, 436291, 436307, 436309, 436313, 436343, 436357, 436399, 436417, 436427, 436439, 436459,
436463, 436477, 436481, 436483, 436507, 436523, 436529, 436531, 436547, 436549, 436571, 436591, 436607,
436621, 436627, 436649, 436651, 436673, 436687, 436693, 436717, 436727, 436729, 436739, 436741, 436757,
436801, 436811, 436819, 436831, 436841, 436853, 436871, 436889, 436913, 436957, 436963, 436967, 436973,
436979, 436993, 436999, 437011, 437033, 437071, 437077, 437083, 437093, 437111, 437113, 437137, 437141,
437149, 437153, 437159, 437191, 437201, 437219, 437237, 437243, 437263, 437273, 437279, 437287, 437293,
437321, 437351, 437357, 437363, 437387, 437389, 437401, 437413, 437467, 437471, 437473, 437497, 437501,
437509, 437519, 437527, 437533, 437539, 437543, 437557, 437587, 437629, 437641, 437651, 437653, 437677,
437681, 437687, 437693, 437719, 437729, 437743, 437753, 437771, 437809, 437819, 437837, 437849, 437861,
437867, 437881, 437909, 437923, 437947, 437953, 437959, 437977, 438001, 438017, 438029, 438047, 438049,
438091, 438131, 438133, 438143, 438169, 438203, 438211, 438223, 438233, 438241, 438253, 438259, 438271,
438281, 438287, 438301, 438313, 438329, 438341, 438377, 438391, 438401, 438409, 438419, 438439, 438443,
438467, 438479, 438499, 438517, 438521, 438523, 438527, 438533, 438551, 438569, 438589, 438601, 438611,
438623, 438631, 438637, 438661, 438667, 438671, 438701, 438707, 438721, 438733, 438761, 438769, 438793,
438827, 438829, 438833, 438847, 438853, 438869, 438877, 438887, 438899, 438913, 438937, 438941, 438953,
438961, 438967, 438979, 438983, 438989, 439007, 439009, 439063, 439081, 439123, 439133, 439141, 439157,
439163, 439171, 439183, 439199, 439217, 439253, 439273, 439279, 439289, 439303, 439339, 439349, 439357,
439367, 439381, 439409, 439421, 439427, 439429, 439441, 439459, 439463, 439471, 439493, 439511, 439519,
439541, 439559, 439567, 439573, 439577, 439583, 439601, 439613, 439631, 439639, 439661, 439667, 439687,
439693, 439697, 439709, 439723, 439729, 439753, 439759, 439763, 439771, 439781, 439787, 439799, 439811,
439823, 439849, 439853, 439861, 439867, 439883, 439891, 439903, 439919, 439949, 439961, 439969, 439973,
439981, 439991, 440009, 440023, 440039, 440047, 440087, 440093, 440101, 440131, 440159, 440171, 440177,
440179, 440183, 440203, 440207, 440221, 440227, 440239, 440261, 440269, 440281, 440303, 440311, 440329,
440333, 440339, 440347, 440371, 440383, 440389, 440393, 440399, 440431, 440441, 440443, 440471, 440497,
440501, 440507, 440509, 440527, 440537, 440543, 440549, 440551, 440567, 440569, 440579, 440581, 440641,
440651, 440653, 440669, 440677, 440681, 440683, 440711, 440717, 440723, 440731, 440753, 440761, 440773,
440807, 440809, 440821, 440831, 440849, 440863, 440893, 440903, 440911, 440939, 440941, 440953, 440959,
440983, 440987, 440989, 441011, 441029, 441041, 441043, 441053, 441073, 441079, 441101, 441107, 441109,
441113, 441121, 441127, 441157, 441169, 441179, 441187, 441191, 441193, 441229, 441247, 441251, 441257,
441263, 441281, 441307, 441319, 441349, 441359, 441361, 441403, 441421, 441443, 441449, 441461, 441479,
441499, 441517, 441523, 441527, 441547, 441557, 441563, 441569, 441587, 441607, 441613, 441619, 441631,
441647, 441667, 441697, 441703, 441713, 441737, 441751, 441787, 441797, 441799, 441811, 441827, 441829,
441839, 441841, 441877, 441887, 441907, 441913, 441923, 441937, 441953, 441971, 442003, 442007, 442009,
442019, 442027, 442031, 442033, 442061, 442069, 442097, 442109, 442121, 442139, 442147, 442151, 442157,
442171, 442177, 442181, 442193, 442201, 442207, 442217, 442229, 442237, 442243, 442271, 442283, 442291,
442319, 442327, 442333, 442363, 442367, 442397, 442399, 442439, 442447, 442457, 442469, 442487, 442489,
442499, 442501, 442517, 442531, 442537, 442571, 442573, 442577, 442579, 442601, 442609, 442619, 442633,

442691, 442699, 442703, 442721, 442733, 442747, 442753, 442763, 442769, 442777, 442781, 442789, 442807,
442817, 442823, 442829, 442831, 442837, 442843, 442861, 442879, 442903, 442919, 442961, 442963, 442973,
442979, 442987, 442991, 442997, 443011, 443017, 443039, 443041, 443057, 443059, 443063, 443077, 443089,
443117, 443123, 443129, 443147, 443153, 443159, 443161, 443167, 443171, 443189, 443203, 443221, 443227,
443231, 443237, 443243, 443249, 443263, 443273, 443281, 443291, 443293, 443341, 443347, 443353, 443363,
443369, 443389, 443407, 443413, 443419, 443423, 443431, 443437, 443453, 443467, 443489, 443501, 443533,
443543, 443551, 443561, 443563, 443567, 443587, 443591, 443603, 443609, 443629, 443659, 443687, 443689,
443701, 443711, 443731, 443749, 443753, 443759, 443761, 443771, 443777, 443791, 443837, 443851, 443867,
443869, 443873, 443879, 443881, 443893, 443899, 443909, 443917, 443939, 443941, 443953, 443983, 443987,
443999, 444001, 444007, 444029, 444043, 444047, 444079, 444089, 444109, 444113, 444121, 444127, 444131,
444151, 444167, 444173, 444179, 444181, 444187, 444209, 444253, 444271, 444281, 444287, 444289, 444293,
444307, 444341, 444343, 444347, 444349, 444401, 444403, 444421, 444443, 444449, 444461, 444463, 444469,
444473, 444487, 444517, 444523, 444527, 444529, 444539, 444547, 444553, 444557, 444569, 444589, 444607,
444623, 444637, 444641, 444649, 444671, 444677, 444701, 444713, 444739, 444767, 444791, 444793, 444803,
444811, 444817, 444833, 444841, 444859, 444863, 444869, 444877, 444883, 444887, 444893, 444901, 444929,
444937, 444953, 444967, 444971, 444979, 445001, 445019, 445021, 445031, 445033, 445069, 445087, 445091,
445097, 445103, 445141, 445157, 445169, 445183, 445187, 445199, 445229, 445261, 445271, 445279, 445283,
445297, 445307, 445321, 445339, 445363, 445427, 445433, 445447, 445453, 445463, 445477, 445499, 445507,
445537, 445541, 445567, 445573, 445583, 445589, 445597, 445619, 445631, 445633, 445649, 445657, 445691,
445699, 445703, 445741, 445747, 445769, 445771, 445789, 445799, 445807, 445829, 445847, 445853, 445871,
445877, 445883, 445891, 445931, 445937, 445943, 445967, 445969, 446003, 446009, 446041, 446053, 446081,
446087, 446111, 446123, 446129, 446141, 446179, 446189, 446191, 446197, 446221, 446227, 446231, 446261,
446263, 446273, 446279, 446293, 446309, 446323, 446333, 446353, 446363, 446387, 446389, 446399, 446401,
446417, 446441, 446447, 446461, 446473, 446477, 446503, 446533, 446549, 446561, 446569, 446597, 446603,
446609, 446647, 446657, 446713, 446717, 446731, 446753, 446759, 446767, 446773, 446819, 446827, 446839,
446863, 446881, 446893, 446909, 446911, 446921, 446933, 446951, 446969, 446983, 447001, 447011,
447019, 447053, 447067, 447079, 447101, 447107, 447119, 447133, 447137, 447173, 447179, 447193, 447197,
447211, 447217, 447221, 447233, 447247, 447257, 447259, 447263, 447311, 447319, 447323, 447331, 447353,
447401, 447409, 447427, 447439, 447443, 447449, 447451, 447463, 447467, 447481, 447509, 447521, 447527,
447541, 447569, 447571, 447611, 447617, 447637, 447641, 447677, 447683, 447701, 447703, 447743, 447749,
447757, 447779, 447791, 447793, 447817, 447823, 447827, 447829, 447841, 447859, 447877, 447883, 447893,
447901, 447907, 447943, 447961, 447983, 447991, 448003, 448013, 448027, 448031, 448057, 448067, 448073,
448093, 448111, 448121, 448139, 448141, 448157, 448159, 448169, 448177, 448187, 448193, 448199, 448207,
448241, 448249, 448303, 448309, 448313, 448321, 448351, 448363, 448367, 448373, 448379, 448387, 448397,
448421, 448451, 448519, 448531, 448561, 448597, 448607, 448627, 448631, 448633, 448667, 448687, 448697,
448703, 448727, 448733, 448741, 448769, 448793, 448801, 448807, 448829, 448843, 448853, 448859, 448867,
448871, 448873, 448879, 448883, 448907, 448927, 448939, 448969, 448993, 448997, 448999, 449003, 449011,
449051, 449077, 449083, 449093, 449107, 449117, 449129, 449131, 449149, 449153, 449161, 449171, 449173,
449201, 449203, 449209, 449227, 449243, 449249, 449261, 449263, 449269, 449287, 449299, 449303, 449311,
449321, 449333, 449347, 449353, 449363, 449381, 449399, 449411, 449417, 449419, 449437, 449441, 449459,
449473, 449543, 449549, 449557, 449563, 449567, 449569, 449591, 449609, 449621, 449629, 449653, 449663,
449671, 449677, 449681, 449689, 449693, 449699, 449741, 449759, 449767, 449773, 449783, 449797, 449807,
449821, 449833, 449851, 449879, 449921, 449929, 449941, 449951, 449959, 449963, 449971, 449987, 449989,
450001, 450011, 450019, 450029, 450067, 450071, 450077, 450083, 450101, 450103, 450113, 450127, 450137,
450161, 450169, 450193, 450199, 450209, 450217, 450223, 450227, 450239, 450257, 450259, 450277, 450287,
450293, 450299, 450301, 450311, 450343, 450349, 450361, 450367, 450377, 450383, 450391, 450403, 450413,
450421, 450431, 450451, 450473, 450479, 450481, 450487, 450493, 450503, 450529, 450533, 450557, 450563,
450581, 450587, 450599, 450601, 450617, 450641, 450643, 450649, 450677, 450691, 450707, 450719, 450727,
450761, 450767, 450787, 450797, 450799, 450803, 450809, 450811, 450817, 450829, 450839, 450841, 450847,

450859, 450881, 450883, 450887, 450893, 450899, 450913, 450917, 450929, 450943, 450949, 450971, 450991,
450997, 451013, 451039, 451051, 451057, 451069, 451093, 451097, 451103, 451109, 451159, 451177, 451181,
451183, 451201, 451207, 451249, 451277, 451279, 451301, 451303, 451309, 451313, 451331, 451337, 451343,
451361, 451387, 451397, 451411, 451439, 451441, 451481, 451499, 451519, 451523, 451541, 451547, 451553,
451579, 451601, 451609, 451621, 451637, 451657, 451663, 451667, 451669, 451679, 451681, 451691, 451699,
451709, 451723, 451747, 451753, 451771, 451783, 451793, 451799, 451823, 451831, 451837, 451859, 451873,
451879, 451897, 451901, 451903, 451909, 451921, 451933, 451937, 451939, 451961, 451967, 451987, 452009,
452017, 452027, 452033, 452041, 452077, 452083, 452087, 452131, 452159, 452161, 452171, 452191, 452201,
452213, 452227, 452233, 452239, 452269, 452279, 452293, 452297, 452329, 452363, 452377, 452393, 452401,
452443, 452453, 452497, 452519, 452521, 452531, 452533, 452537, 452539, 452549, 452579, 452587, 452597,
452611, 452629, 452633, 452671, 452687, 452689, 452701, 452731, 452759, 452773, 452797, 452807, 452813,
452821, 452831, 452857, 452869, 452873, 452923, 452953, 452957, 452983, 452989, 453023, 453029, 453053,
453073, 453107, 453119, 453133, 453137, 453143, 453157, 453161, 453181, 453197, 453199, 453209, 453217,
453227, 453239, 453247, 453269, 453289, 453293, 453301, 453311, 453317, 453329, 453347, 453367, 453371,
453377, 453379, 453421, 453451, 453461, 453527, 453553, 453559, 453569, 453571, 453599, 453601, 453617,
453631, 453637, 453641, 453643, 453659, 453667, 453671, 453683, 453703, 453707, 453709, 453737, 453757,
453797, 453799, 453823, 453833, 453847, 453851, 453877, 453889, 453907, 453913, 453923, 453931, 453949,
453961, 453977, 453983, 453991, 454009, 454021, 454031, 454033, 454039, 454061, 454063, 454079, 454109,
454141, 454151, 454159, 454183, 454199, 454211, 454213, 454219, 454229, 454231, 454247, 454253, 454277,
454297, 454303, 454313, 454331, 454351, 454357, 454361, 454379, 454387, 454409, 454417, 454451, 454453,
454483, 454501, 454507, 454513, 454541, 454543, 454547, 454577, 454579, 454603, 454609, 454627, 454637,
454673, 454679, 454709, 454711, 454721, 454723, 454759, 454763, 454777, 454799, 454823, 454843, 454847,
454849, 454859, 454889, 454891, 454907, 454919, 454921, 454931, 454943, 454967, 454969, 454973, 454991,
455003, 455011, 455033, 455047, 455053, 455093, 455099, 455123, 455149, 455159, 455167, 455171, 455177,
455201, 455219, 455227, 455233, 455237, 455261, 455263, 455269, 455291, 455309, 455317, 455321, 455333,
455339, 455341, 455353, 455381, 455393, 455401, 455407, 455419, 455431, 455437, 455443, 455461, 455471,
455473, 455479, 455489, 455491, 455513, 455527, 455531, 455537, 455557, 455573, 455579, 455597, 455599,
455603, 455627, 455647, 455659, 455681, 455683, 455687, 455701, 455711, 455717, 455737, 455761, 455783,
455789, 455809, 455827, 455831, 455849, 455863, 455881, 455899, 455921, 455933, 455941, 455953, 455969,
455977, 455989, 455993, 455999, 456007, 456013, 456023, 456037, 456047, 456061, 456091, 456107, 456109,
456119, 456149, 456151, 456167, 456193, 456223, 456233, 456241, 456283, 456293, 456329, 456349, 456353,
456367, 456377, 456403, 456409, 456427, 456439, 456451, 456457, 456461, 456499, 456503, 456517, 456523,
456529, 456539, 456553, 456557, 456559, 456571, 456581, 456587, 456607, 456611, 456613, 456623, 456641,
456647, 456649, 456653, 456679, 456683, 456697, 456727, 456737, 456763, 456767, 456769, 456791, 456809,
456811, 456821, 456871, 456877, 456881, 456899, 456901, 456923, 456949, 456959, 456979, 456991, 457001,
457003, 457013, 457021, 457043, 457049, 457057, 457087, 457091, 457097, 457099, 457117, 457139, 457151,
457153, 457183, 457189, 457201, 457213, 457229, 457241, 457253, 457267, 457271, 457277, 457279, 457307,
457319, 457333, 457339, 457363, 457367, 457381, 457393, 457397, 457399, 457403, 457411, 457421, 457433,
457459, 457469, 457507, 457511, 457517, 457547, 457553, 457559, 457571, 457607, 457609, 457621, 457643,
457651, 457661, 457669, 457673, 457679, 457687, 457697, 457711, 457739, 457757, 457789, 457799, 457813,
457817, 457829, 457837, 457871, 457889, 457903, 457913, 457943, 457979, 457981, 457987, 458009, 458027,
458039, 458047, 458053, 458057, 458063, 458069, 458119, 458123, 458173, 458179, 458189, 458191, 458197,
458207, 458219, 458239, 458309, 458317, 458323, 458327, 458333, 458357, 458363, 458377, 458399, 458401,
458407, 458449, 458477, 458483, 458501, 458531, 458533, 458543, 458567, 458569, 458573, 458593, 458599,
458611, 458621, 458629, 458639, 458651, 458663, 458669, 458683, 458701, 458719, 458729, 458747, 458789,
458791, 458797, 458807, 458819, 458849, 458863, 458879, 458891, 458897, 458917, 458921, 458929, 458947,
458957, 458959, 458963, 458971, 458977, 458981, 458987, 458993, 459007, 459013, 459023, 459029, 459031,
459037, 459047, 459089, 459091, 459113, 459127, 459167, 459169, 459181, 459209, 459223, 459229, 459233,
459257, 459271, 459293, 459301, 459313, 459317, 459337, 459341, 459343, 459353, 459373, 459377, 459383,

459397, 459421, 459427, 459443, 459463, 459467, 459469, 459479, 459509, 459521, 459523, 459593, 459607,
459611, 459619, 459623, 459631, 459647, 459649, 459671, 459677, 459691, 459703, 459749, 459763, 459791,
459803, 459817, 459829, 459841, 459847, 459883, 459913, 459923, 459929, 459937, 459961, 460013, 460039,
460051, 460063, 460073, 460079, 460081, 460087, 460091, 460099, 460111, 460127, 460147, 460157, 460171,
460181, 460189, 460211, 460217, 460231, 460247, 460267, 460289, 460297, 460301, 460337, 460349, 460373,
460379, 460387, 460393, 460403, 460409, 460417, 460451, 460463, 460477, 460531, 460543, 460561, 460571,
460589, 460609, 460619, 460627, 460633, 460637, 460643, 460657, 460673, 460697, 460709, 460711, 460721,
460771, 460777, 460787, 460793, 460813, 460829, 460841, 460843, 460871, 460891, 460903, 460907, 460913,
460919, 460937, 460949, 460951, 460969, 460973, 460979, 460981, 460987, 460991, 461009, 461011, 461017,
461051, 461053, 461059, 461093, 461101, 461119, 461143, 461147, 461171, 461183, 461191, 461207, 461233,
461239, 461257, 461269, 461273, 461297, 461299, 461309, 461317, 461323, 461327, 461333, 461359, 461381,
461393, 461407, 461411, 461413, 461437, 461441, 461443, 461467, 461479, 461507, 461521, 461561, 461569,
461581, 461599, 461603, 461609, 461627, 461639, 461653, 461677, 461687, 461689, 461693, 461707, 461717,
461801, 461803, 461819, 461843, 461861, 461887, 461891, 461917, 461921, 461933, 461957, 461971, 461977,
461983, 462013, 462041, 462067, 462073, 462079, 462097, 462103, 462109, 462113, 462131, 462149, 462181,
462191, 462199, 462221, 462239, 462263, 462271, 462307, 462311, 462331, 462337, 462361, 462373, 462377,
462401, 462409, 462419, 462421, 462437, 462443, 462467, 462481, 462491, 462493, 462499, 462529, 462541,
462547, 462557, 462569, 462571, 462577, 462589, 462607, 462629, 462641, 462643, 462653, 462659, 462667,
462673, 462677, 462697, 462713, 462719, 462727, 462733, 462739, 462773, 462827, 462841, 462851, 462863,
462871, 462881, 462887, 462899, 462901, 462911, 462937, 462947, 462953, 462983, 463003, 463031, 463033,
463093, 463103, 463157, 463181, 463189, 463207, 463213, 463219, 463231, 463237, 463247, 463249, 463261,
463283, 463291, 463297, 463303, 463313, 463319, 463321, 463339, 463343, 463363, 463387, 463399, 463433,
463447, 463451, 463453, 463457, 463459, 463483, 463501, 463511, 463513, 463523, 463531, 463537, 463549,
463579, 463613, 463627, 463633, 463643, 463649, 463663, 463679, 463693, 463711, 463717, 463741, 463747,
463753, 463763, 463781, 463787, 463807, 463823, 463829, 463831, 463849, 463861, 463867, 463873, 463889,
463891, 463907, 463919, 463921, 463949, 463963, 463973, 463987, 463993, 464003, 464011, 464021, 464033,
464047, 464069, 464081, 464089, 464119, 464129, 464131, 464137, 464141, 464143, 464171, 464173, 464197,
464201, 464213, 464237, 464251, 464257, 464263, 464279, 464281, 464291, 464309, 464311, 464327, 464351,
464371, 464381, 464383, 464413, 464419, 464437, 464447, 464459, 464467, 464479, 464483, 464521, 464537,
464539, 464549, 464557, 464561, 464587, 464591, 464603, 464617, 464621, 464647, 464663, 464687, 464699,
464741, 464747, 464749, 464753, 464767, 464771, 464773, 464777, 464801, 464803, 464809, 464813, 464819,
464843, 464857, 464879, 464897, 464909, 464917, 464923, 464927, 464939, 464941, 464951, 464953, 464963,
464983, 464993, 464999, 465007, 465011, 465013, 465019, 465041, 465061, 465067, 465071, 465077, 465079,
465089, 465107, 465119, 465133, 465151, 465161, 465163, 465167, 465169, 465173, 465187, 465209, 465211,
465259, 465271, 465277, 465281, 465293, 465299, 465317, 465319, 465331, 465337, 465373, 465379, 465383,
465407, 465419, 465433, 465463, 465469, 465523, 465529, 465541, 465551, 465581, 465587, 465611, 465631,
465643, 465649, 465659, 465679, 465701, 465721, 465739, 465743, 465761, 465781, 465797, 465799, 465809,
465821, 465833, 465841, 465887, 465893, 465901, 465917, 465929, 465931, 465947, 465977, 465989, 466009,
466019, 466027, 466033, 466043, 466061, 466069, 466073, 466079, 466087, 466091, 466121, 466139, 466153,
466171, 466181, 466183, 466201, 466243, 466247, 466261, 466267, 466273, 466283, 466303, 466321, 466331,
466339, 466357, 466369, 466373, 466409, 466423, 466441, 466451, 466483, 466517, 466537, 466547, 466553,
466561, 466567, 466573, 466579, 466603, 466619, 466637, 466649, 466651, 466673, 466717, 466723, 466729,
466733, 466747, 466751, 466777, 466787, 466801, 466819, 466853, 466859, 466897, 466909, 466913, 466919,
466951, 466957, 466997, 467003, 467009, 467017, 467021, 467063, 467081, 467083, 467101, 467119, 467123,
467141, 467147, 467171, 467183, 467197, 467209, 467213, 467237, 467239, 467261, 467293, 467297, 467317,
467329, 467333, 467353, 467371, 467399, 467417, 467431, 467437, 467447, 467471, 467473, 467477, 467479,
467491, 467497, 467503, 467507, 467527, 467531, 467543, 467549, 467557, 467587, 467591, 467611, 467617,
467627, 467629, 467633, 467641, 467651, 467657, 467669, 467671, 467681, 467689, 467699, 467713, 467729,
467737, 467743, 467749, 467773, 467783, 467813, 467827, 467833, 467867, 467869, 467879, 467881, 467893,

467897, 467899, 467903, 467927, 467941, 467953, 467963, 467977, 468001, 468011, 468019, 468029, 468049,
468059, 468067, 468071, 468079, 468107, 468109, 468113, 468121, 468133, 468137, 468151, 468157, 468173,
468187, 468191, 468199, 468239, 468241, 468253, 468271, 468277, 468289, 468319, 468323, 468353, 468359,
468371, 468389, 468421, 468439, 468451, 468463, 468473, 468491, 468493, 468499, 468509, 468527, 468551,
468557, 468577, 468581, 468593, 468599, 468613, 468619, 468623, 468641, 468647, 468653, 468661, 468667,
468683, 468691, 468697, 468703, 468709, 468719, 468737, 468739, 468761, 468773, 468781, 468803, 468817,
468821, 468841, 468851, 468859, 468869, 468883, 468887, 468889, 468893, 468899, 468913, 468953, 468967,
468973, 468983, 469009, 469031, 469037, 469069, 469099, 469121, 469127, 469141, 469153, 469169, 469193,
469207, 469219, 469229, 469237, 469241, 469253, 469267, 469279, 469283, 469303, 469321, 469331, 469351,
469363, 469367, 469369, 469379, 469397, 469411, 469429, 469439, 469457, 469487, 469501, 469529, 469541,
469543, 469561, 469583, 469589, 469613, 469627, 469631, 469649, 469657, 469673, 469687, 469691, 469717,
469723, 469747, 469753, 469757, 469769, 469787, 469793, 469801, 469811, 469823, 469841, 469849, 469877,
469879, 469891, 469907, 469919, 469939, 469957, 469969, 469979, 469993, 470021, 470039, 470059, 470077,
470081, 470083, 470087, 470089, 470131, 470149, 470153, 470161, 470167, 470179, 470201, 470207, 470209,
470213, 470219, 470227, 470243, 470251, 470263, 470279, 470297, 470299, 470303, 470317, 470333, 470347,
470359, 470389, 470399, 470411, 470413, 470417, 470429, 470443, 470447, 470453, 470461, 470471, 470473,
470489, 470501, 470513, 470521, 470531, 470539, 470551, 470579, 470593, 470597, 470599, 470609, 470621,
470627, 470647, 470651, 470653, 470663, 470669, 470689, 470711, 470719, 470731, 470749, 470779, 470783,
470791, 470819, 470831, 470837, 470863, 470867, 470881, 470887, 470891, 470903, 470927, 470933, 470941,
470947, 470957, 470959, 470993, 470999, 471007, 471041, 471061, 471073, 471089, 471091, 471101, 471137,
471139, 471161, 471173, 471179, 471187, 471193, 471209, 471217, 471241, 471253, 471259, 471277, 471281,
471283, 471299, 471301, 471313, 471353, 471389, 471391, 471403, 471407, 471439, 471451, 471467, 471481,
471487, 471503, 471509, 471521, 471533, 471539, 471553, 471571, 471589, 471593, 471607, 471617, 471619,
471641, 471649, 471659, 471671, 471673, 471677, 471683, 471697, 471703, 471719, 471721, 471749, 471769,
471781, 471791, 471803, 471817, 471841, 471847, 471853, 471871, 471893, 471901, 471907, 471923, 471929,
471931, 471943, 471949, 471959, 471997, 472019, 472027, 472051, 472057, 472063, 472067, 472103, 472111,
472123, 472127, 472133, 472139, 472151, 472159, 472163, 472189, 472193, 472247, 472249, 472253, 472261,
472273, 472289, 472301, 472309, 472319, 472331, 472333, 472349, 472369, 472391, 472393, 472399, 472411,
472421, 472457, 472469, 472477, 472523, 472541, 472543, 472559, 472561, 472573, 472597, 472631, 472639,
472643, 472669, 472687, 472691, 472697, 472709, 472711, 472721, 472741, 472751, 472763, 472793, 472799,
472817, 472831, 472837, 472847, 472859, 472883, 472907, 472909, 472921, 472937, 472939, 472963, 472993,
473009, 473021, 473027, 473089, 473101, 473117, 473141, 473147, 473159, 473167, 473173, 473191, 473197,
473201, 473203, 473219, 473227, 473257, 473279, 473287, 473293, 473311, 473321, 473327, 473351, 473353,
473377, 473381, 473383, 473411, 473419, 473441, 473443, 473453, 473471, 473477, 473479, 473497, 473503,
473507, 473513, 473519, 473527, 473531, 473533, 473549, 473579, 473597, 473611, 473617, 473633, 473647,
473659, 473719, 473723, 473729, 473741, 473743, 473761, 473789, 473833, 473839, 473857, 473861, 473867,
473887, 473899, 473911, 473923, 473927, 473929, 473939, 473951, 473953, 473971, 473981, 473987, 473999,
474017, 474029, 474037, 474043, 474049, 474059, 474073, 474077, 474101, 474119, 474127, 474137, 474143,
474151, 474163, 474169, 474197, 474211, 474223, 474241, 474263, 474289, 474307, 474311, 474319, 474337,
474343, 474347, 474359, 474379, 474389, 474391, 474413, 474433, 474437, 474443, 474479, 474491, 474497,
474499, 474503, 474533, 474541, 474547, 474557, 474569, 474571, 474581, 474583, 474619, 474629, 474647,
474659, 474667, 474671, 474707, 474709, 474737, 474751, 474757, 474769, 474779, 474787, 474809, 474811,
474839, 474847, 474857, 474899, 474907, 474911, 474917, 474923, 474931, 474937, 474941, 474949, 474959,
474977, 474983, 475037, 475051, 475073, 475081, 475091, 475093, 475103, 475109, 475141, 475147, 475151,
475159, 475169, 475207, 475219, 475229, 475243, 475271, 475273, 475283, 475289, 475297, 475301, 475327,
475331, 475333, 475351, 475367, 475369, 475379, 475381, 475403, 475417, 475421, 475427, 475429, 475441,
475457, 475469, 475483, 475511, 475523, 475529, 475549, 475583, 475597, 475613, 475619, 475621, 475637,
475639, 475649, 475669, 475679, 475681, 475691, 475693, 475697, 475721, 475729, 475751, 475753, 475759,
475763, 475777, 475789, 475793, 475807, 475823, 475831, 475837, 475841, 475859, 475877, 475879, 475889,

475897, 475903, 475907, 475921, 475927, 475933, 475957, 475973, 475991, 475997, 476009, 476023, 476027,
476029, 476039, 476041, 476059, 476081, 476087, 476089, 476101, 476107, 476111, 476137, 476143, 476167,
476183, 476219, 476233, 476237, 476243, 476249, 476279, 476299, 476317, 476347, 476351, 476363, 476369,
476381, 476401, 476407, 476419, 476423, 476429, 476467, 476477, 476479, 476507, 476513, 476519, 476579,
476587, 476591, 476599, 476603, 476611, 476633, 476639, 476647, 476659, 476681, 476683, 476701, 476713,
476719, 476737, 476743, 476753, 476759, 476783, 476803, 476831, 476849, 476851, 476863, 476869, 476887,
476891, 476911, 476921, 476929, 476977, 476981, 476989, 477011, 477013, 477017, 477019, 477031, 477047,
477073, 477077, 477091, 477131, 477149, 477163, 477209, 477221, 477229, 477259, 477277, 477293, 477313,
477317, 477329, 477341, 477359, 477361, 477383, 477409, 477439, 477461, 477469, 477497, 477511, 477517,
477523, 477539, 477551, 477553, 477557, 477571, 477577, 477593, 477619, 477623, 477637, 477671, 477677,
477721, 477727, 477731, 477739, 477767, 477769, 477791, 477797, 477809, 477811, 477821, 477823, 477839,
477847, 477857, 477863, 477881, 477899, 477913, 477941, 477947, 477973, 477977, 477991, 478001, 478039,
478063, 478067, 478069, 478087, 478099, 478111, 478129, 478139, 478157, 478169, 478171, 478189, 478199,
478207, 478213, 478241, 478243, 478253, 478259, 478271, 478273, 478321, 478339, 478343, 478351, 478391,
478399, 478403, 478411, 478417, 478421, 478427, 478433, 478441, 478451, 478453, 478459, 478481, 478483,
478493, 478523, 478531, 478571, 478573, 478579, 478589, 478603, 478627, 478631, 478637, 478651, 478679,
478697, 478711, 478727, 478729, 478739, 478741, 478747, 478763, 478769, 478787, 478801, 478811, 478813,
478823, 478831, 478843, 478853, 478861, 478871, 478879, 478897, 478901, 478913, 478927, 478931, 478937,
478943, 478963, 478967, 478991, 478999, 479023, 479027, 479029, 479041, 479081, 479131, 479137, 479147,
479153, 479189, 479191, 479201, 479209, 479221, 479231, 479239, 479243, 479263, 479267, 479287, 479299,
479309, 479317, 479327, 479357, 479371, 479377, 479387, 479419, 479429, 479431, 479441, 479461, 479473,
479489, 479497, 479509, 479513, 479533, 479543, 479561, 479569, 479581, 479593, 479599, 479623, 479629,
479639, 479701, 479749, 479753, 479761, 479771, 479777, 479783, 479797, 479813, 479821, 479833, 479839,
479861, 479879, 479881, 479891, 479903, 479909, 479939, 479951, 479953, 479957, 479971, 480013, 480017,
480019, 480023, 480043, 480047, 480049, 480059, 480061, 480071, 480091, 480101, 480107, 480113, 480133,
480143, 480157, 480167, 480169, 480203, 480209, 480287, 480299, 480317, 480329, 480341, 480343, 480349,
480367, 480373, 480379, 480383, 480391, 480409, 480419, 480427, 480449, 480451, 480461, 480463, 480499,
480503, 480509, 480517, 480521, 480527, 480533, 480541, 480553, 480563, 480569, 480583, 480587, 480647,
480661, 480707, 480713, 480731, 480737, 480749, 480761, 480773, 480787, 480803, 480827, 480839, 480853,
480881, 480911, 480919, 480929, 480937, 480941, 480959, 480967, 480979, 480989, 481001, 481003, 481009,
481021, 481043, 481051, 481067, 481073, 481087, 481093, 481097, 481109, 481123, 481133, 481141, 481147,
481153, 481157, 481171, 481177, 481181, 481199, 481207, 481211, 481231, 481249, 481297, 481301, 481303,
481307, 481343, 481363, 481373, 481379, 481387, 481409, 481417, 481433, 481447, 481469, 481489, 481501,
481513, 481531, 481549, 481571, 481577, 481589, 481619, 481633, 481639, 481651, 481667, 481673, 481681,
481693, 481697, 481699, 481721, 481751, 481753, 481769, 481787, 481801, 481807, 481813, 481837, 481843,
481847, 481849, 481861, 481867, 481879, 481883, 481909, 481939, 481963, 481997, 482017, 482021, 482029,
482033, 482039, 482051, 482071, 482093, 482099, 482101, 482117, 482123, 482179, 482189, 482203, 482213,
482227, 482231, 482233, 482243, 482263, 482281, 482309, 482323, 482347, 482351, 482359, 482371, 482387,
482393, 482399, 482401, 482407, 482413, 482423, 482437, 482441, 482483, 482501, 482507, 482509, 482513,
482519, 482527, 482539, 482569, 482593, 482597, 482621, 482627, 482633, 482641, 482659, 482663, 482683,
482687, 482689, 482707, 482711, 482717, 482719, 482731, 482743, 482753, 482759, 482767, 482773, 482789,
482803, 482819, 482827, 482837, 482861, 482863, 482873, 482897, 482899, 482917, 482941, 482947, 482957,
482971, 483017, 483031, 483061, 483071, 483097, 483127, 483139, 483163, 483167, 483179, 483209, 483211,
483221, 483229, 483233, 483239, 483247, 483251, 483281, 483289, 483317, 483323, 483337, 483347, 483367,
483377, 483389, 483397, 483407, 483409, 483433, 483443, 483467, 483481, 483491, 483499, 483503, 483523,
483541, 483551, 483557, 483563, 483577, 483611, 483619, 483629, 483643, 483649, 483671, 483697, 483709,
483719, 483727, 483733, 483751, 483757, 483761, 483767, 483773, 483787, 483809, 483811, 483827, 483829,
483839, 483853, 483863, 483869, 483883, 483907, 483929, 483937, 483953, 483971, 483991, 484019, 484027,
484037, 484061, 484067, 484079, 484091, 484111, 484117, 484123, 484129, 484151, 484153, 484171, 484181,

484193, 484201, 484207, 484229, 484243, 484259, 484283, 484301, 484303, 484327, 484339, 484361, 484369,
484373, 484397, 484411, 484417, 484439, 484447, 484457, 484459, 484487, 484489, 484493, 484531, 484543,
484577, 484597, 484607, 484609, 484613, 484621, 484639, 484643, 484691, 484703, 484727, 484733, 484751,
484763, 484769, 484777, 484787, 484829, 484853, 484867, 484927, 484951, 484987, 484999, 485021, 485029,
485041, 485053, 485059, 485063, 485081, 485101, 485113, 485123, 485131, 485137, 485161, 485167, 485171,
485201, 485207, 485209, 485263, 485311, 485347, 485351, 485363, 485371, 485383, 485389, 485411, 485417,
485423, 485437, 485447, 485479, 485497, 485509, 485519, 485543, 485567, 485587, 485593, 485603, 485609,
485647, 485657, 485671, 485689, 485701, 485717, 485729, 485731, 485753, 485777, 485819, 485827, 485831,
485833, 485893, 485899, 485909, 485923, 485941, 485959, 485977, 485993, 486023, 486037, 486041, 486043,
486053, 486061, 486071, 486091, 486103, 486119, 486133, 486139, 486163, 486179, 486181, 486193, 486203,
486221, 486223, 486247, 486281, 486293, 486307, 486313, 486323, 486329, 486331, 486341, 486349, 486377,
486379, 486389, 486391, 486397, 486407, 486433, 486443, 486449, 486481, 486491, 486503, 486509, 486511,
486527, 486539, 486559, 486569, 486583, 486589, 486601, 486617, 486637, 486641, 486643, 486653, 486667,
486671, 486677, 486679, 486683, 486697, 486713, 486721, 486757, 486767, 486769, 486781, 486797, 486817,
486821, 486833, 486839, 486869, 486907, 486923, 486929, 486943, 486947, 486949, 486971, 486977, 486991,
487007, 487013, 487021, 487049, 487051, 487057, 487073, 487079, 487093, 487099, 487111, 487133, 487177,
487183, 487187, 487211, 487213, 487219, 487247, 487261, 487283, 487303, 487307, 487313, 487349, 487363,
487381, 487387, 487391, 487397, 487423, 487427, 487429, 487447, 487457, 487463, 487469, 487471, 487477,
487481, 487489, 487507, 487561, 487589, 487601, 487603, 487607, 487637, 487649, 487651, 487657, 487681,
487691, 487703, 487709, 487717, 487727, 487733, 487741, 487757, 487769, 487783, 487789, 487793, 487811,
487819, 487829, 487831, 487843, 487873, 487889, 487891, 487897, 487933, 487943, 487973, 487979, 487997,
488003, 488009, 488011, 488021, 488051, 488057, 488069, 488119, 488143, 488149, 488153, 488161, 488171,
488197, 488203, 488207, 488209, 488227, 488231, 488233, 488239, 488249, 488261, 488263, 488287, 488303,
488309, 488311, 488317, 488321, 488329, 488333, 488339, 488347, 488353, 488381, 488399, 488401, 488407,
488417, 488419, 488441, 488459, 488473, 488503, 488513, 488539, 488567, 488573, 488603, 488611, 488617,
488627, 488633, 488639, 488641, 488651, 488687, 488689, 488701, 488711, 488717, 488723, 488729, 488743,
488749, 488759, 488779, 488791, 488797, 488821, 488827, 488833, 488861, 488879, 488893, 488897, 488909,
488921, 488947, 488959, 488981, 488993, 489001, 489011, 489019, 489043, 489053, 489061, 489101, 489109,
489113, 489127, 489133, 489157, 489161, 489179, 489191, 489197, 489217, 489239, 489241, 489257, 489263,
489283, 489299, 489329, 489337, 489343, 489361, 489367, 489389, 489407, 489409, 489427, 489431, 489439,
489449, 489457, 489479, 489487, 489493, 489529, 489539, 489551, 489553, 489557, 489571, 489613, 489631,
489653, 489659, 489673, 489677, 489679, 489689, 489691, 489733, 489743, 489761, 489791, 489793, 489799,
489803, 489817, 489823, 489833, 489847, 489851, 489869, 489871, 489887, 489901, 489911, 489913, 489941,
489943, 489959, 489961, 489977, 489989, 490001, 490003, 490019, 490031, 490033, 490057, 490097, 490103,
490111, 490117, 490121, 490151, 490159, 490169, 490183, 490201, 490207, 490223, 490241, 490247, 490249,
490267, 490271, 490277, 490283, 490309, 490313, 490339, 490367, 490393, 490417, 490421, 490453, 490459,
490463, 490481, 490493, 490499, 490519, 490537, 490541, 490543, 490549, 490559, 490571, 490573, 490577,
490579, 490591, 490619, 490627, 490631, 490643, 490661, 490663, 490697, 490733, 490741, 490769, 490771,
490783, 490829, 490837, 490849, 490859, 490877, 490891, 490913, 490921, 490927, 490937, 490949, 490951,
490957, 490967, 490969, 490991, 490993, 491003, 491039, 491041, 491059, 491081, 491083, 491129, 491137,
491149, 491159, 491167, 491171, 491201, 491213, 491219, 491251, 491261, 491273, 491279, 491297, 491299,
491327, 491329, 491333, 491339, 491341, 491353, 491357, 491371, 491377, 491417, 491423, 491429, 491461,
491483, 491489, 491497, 491501, 491503, 491527, 491531, 491537, 491539, 491581, 491591, 491593, 491611,
491627, 491633, 491639, 491651, 491653, 491669, 491677, 491707, 491719, 491731, 491737, 491747, 491773,
491783, 491789, 491797, 491819, 491833, 491837, 491851, 491857, 491867, 491873, 491899, 491923, 491951,
491969, 491977, 491983, 492007, 492013, 492017, 492029, 492047, 492053, 492059, 492061, 492067, 492077,
492083, 492103, 492113, 492227, 492251, 492253, 492257, 492281, 492293, 492299, 492319, 492377, 492389,
492397, 492403, 492409, 492413, 492421, 492431, 492463, 492467, 492487, 492491, 492511, 492523, 492551,
492563, 492587, 492601, 492617, 492619, 492629, 492631, 492641, 492647, 492659, 492671, 492673, 492707,

492719, 492721, 492731, 492757, 492761, 492763, 492769, 492781, 492799, 492839, 492853, 492871, 492883,
492893, 492901, 492911, 492967, 492979, 493001, 493013, 493021, 493027, 493043, 493049, 493067, 493093,
493109, 493111, 493121, 493123, 493127, 493133, 493139, 493147, 493159, 493169, 493177, 493193, 493201,
493211, 493217, 493219, 493231, 493243, 493249, 493277, 493279, 493291, 493301, 493313, 493333, 493351,
493369, 493393, 493397, 493399, 493403, 493433, 493447, 493457, 493463, 493481, 493523, 493531, 493541,
493567, 493573, 493579, 493583, 493607, 493621, 493627, 493643, 493657, 493693, 493709, 493711, 493721,
493729, 493733, 493747, 493777, 493793, 493807, 493811, 493813, 493817, 493853, 493859, 493873, 493877,
493897, 493919, 493931, 493937, 493939, 493967, 493973, 493979, 493993, 494023, 494029, 494041, 494051,
494069, 494077, 494083, 494093, 494101, 494107, 494129, 494141, 494147, 494167, 494191, 494213, 494237,
494251, 494257, 494267, 494269, 494281, 494287, 494317, 494327, 494341, 494353, 494359, 494369, 494381,
494383, 494387, 494407, 494413, 494441, 494443, 494471, 494497, 494519, 494521, 494539, 494561, 494563,
494567, 494587, 494591, 494609, 494617, 494621, 494639, 494647, 494651, 494671, 494677, 494687, 494693,
494699, 494713, 494719, 494723, 494731, 494737, 494743, 494749, 494759, 494761, 494783, 494789, 494803,
494843, 494849, 494873, 494899, 494903, 494917, 494927, 494933, 494939, 494959, 494987, 495017, 495037,
495041, 495043, 495067, 495071, 495109, 495113, 495119, 495133, 495139, 495149, 495151, 495161, 495181,
495199, 495211, 495221, 495241, 495269, 495277, 495289, 495301, 495307, 495323, 495337, 495343, 495347,
495359, 495361, 495371, 495377, 495389, 495401, 495413, 495421, 495433, 495437, 495449, 495457, 495461,
495491, 495511, 495527, 495557, 495559, 495563, 495569, 495571, 495587, 495589, 495611, 495613, 495617,
495619, 495629, 495637, 495647, 495667, 495679, 495701, 495707, 495713, 495749, 495751, 495757, 495769,
495773, 495787, 495791, 495797, 495799, 495821, 495827, 495829, 495851, 495877, 495893, 495899, 495923,
495931, 495947, 495953, 495959, 495967, 495973, 495983, 496007, 496019, 496039, 496051, 496063, 496073,
496079, 496123, 496127, 496163, 496187, 496193, 496211, 496229, 496231, 496259, 496283, 496289, 496291,
496297, 496303, 496313, 496333, 496339, 496343, 496381, 496399, 496427, 496439, 496453, 496459, 496471,
496477, 496481, 496487, 496493, 496499, 496511, 496549, 496579, 496583, 496609, 496631, 496669, 496681,
496687, 496703, 496711, 496733, 496747, 496763, 496789, 496813, 496817, 496841, 496849, 496871, 496877,
496889, 496891, 496897, 496901, 496913, 496919, 496949, 496957, 496963, 496997, 496999, 497011, 497017, 497041,
497047, 497051, 497069, 497093, 497111, 497113, 497117, 497137, 497141, 497153, 497171, 497177, 497197,
497239, 497257, 497261, 497269, 497279, 497281, 497291, 497297, 497303, 497309, 497323, 497339, 497351,
497389, 497411, 497417, 497423, 497449, 497461, 497473, 497479, 497491, 497501, 497507, 497509, 497521,
497537, 497551, 497557, 497561, 497579, 497587, 497597, 497603, 497633, 497659, 497663, 497671, 497677,
497689, 497701, 497711, 497719, 497729, 497737, 497741, 497771, 497773, 497801, 497813, 497831, 497839,
497851, 497867, 497869, 497873, 497899, 497929, 497957, 497963, 497969, 497977, 497989, 497993, 497999,
498013, 498053, 498061, 498073, 498089, 498101, 498103, 498119, 498143, 498163, 498167, 498181, 498209,
498227, 498257, 498259, 498271, 498301, 498331, 498343, 498361, 498367, 498391, 498397, 498401, 498403,
498409, 498439, 498461, 498467, 498469, 498493, 498497, 498521, 498523, 498527, 498551, 498557, 498577,
498583, 498599, 498611, 498613, 498643, 498647, 498653, 498679, 498689, 498691, 498733, 498739, 498749,
498761, 498767, 498779, 498781, 498787, 498791, 498803, 498833, 498857, 498859, 498881, 498907, 498923,
498931, 498937, 498947, 498961, 498973, 498977, 498989, 499021, 499027, 499033, 499039, 499063, 499067,
499099, 499117, 499127, 499129, 499133, 499139, 499141, 499151, 499157, 499159, 499181, 499183, 499189,
499211, 499229, 499253, 499267, 499277, 499283, 499309, 499321, 499327, 499349, 499361, 499363, 499391,
499397, 499403, 499423, 499439, 499459, 499481, 499483, 499493, 499507, 499519, 499523, 499549, 499559,
499571, 499591, 499601, 499607, 499621, 499633, 499637, 499649, 499661, 499663, 499669, 499673, 499679,
499687, 499691, 499693, 499711, 499717, 499729, 499739, 499747, 499781, 499787, 499801, 499819, 499853,
499879, 499883, 499897, 499903, 499927, 499943, 499957, 499969, 499973, 499979, 500000, 500009, 500029, 500041,
500057, 500069, 500083, 500107, 500111, 500113, 500119, 500153, 500167, 500173, 500177, 500179, 500197,
500209, 500231, 500233, 500237, 500239, 500249, 500257, 500287, 500299, 500317, 500321, 500333, 500341,
500363, 500369, 500389, 500393, 500413, 500417, 500431, 500443, 500459, 500471, 500473, 500483, 500501,
500509, 500519, 500527, 500567, 500579, 500587, 500603, 500629, 500671, 500677, 500693, 500699, 500713,
500719, 500723, 500729, 500741, 500777, 500791, 500807, 500809, 500831, 500839, 500861, 500873, 500881,

500887, 500891, 500909, 500911, 500921, 500923, 500933, 500947, 500953, 500957, 500977, 501001, 501013,
501019, 501029, 501031, 501037, 501043, 501077, 501089, 501103, 501121, 501131, 501133, 501139, 501157,
501173, 501187, 501191, 501197, 501203, 501209, 501217, 501223, 501229, 501233, 501257, 501271, 501287,
501299, 501317, 501341, 501343, 501367, 501383, 501401, 501409, 501419, 501427, 501451, 501463, 501493,
501503, 501511, 501563, 501577, 501593, 501601, 501617, 501623, 501637, 501659, 501691, 501701, 501703,
501707, 501719, 501731, 501769, 501779, 501803, 501817, 501821, 501827, 501829, 501841, 501863, 501889,
501911, 501931, 501947, 501953, 501967, 501971, 501997, 502001, 502013, 502039, 502043, 502057, 502063,
502079, 502081, 502087, 502093, 502121, 502133, 502141, 502171, 502181, 502217, 502237, 502247, 502259,
502261, 502277, 502301, 502321, 502339, 502393, 502409, 502421, 502429, 502441, 502451, 502487, 502499,
502501, 502507, 502517, 502543, 502549, 502553, 502591, 502597, 502613, 502631, 502633, 502643, 502651,
502669, 502687, 502699, 502703, 502717, 502729, 502769, 502771, 502781, 502787, 502807, 502819, 502829,
502841, 502847, 502861, 502883, 502919, 502921, 502937, 502961, 502973, 503003, 503017, 503039, 503053,
503077, 503123, 503131, 503137, 503147, 503159, 503197, 503207, 503213, 503227, 503231, 503233, 503249,
503267, 503287, 503297, 503303, 503317, 503339, 503351, 503359, 503369, 503381, 503383, 503389, 503407,
503413, 503423, 503431, 503441, 503453, 503483, 503501, 503543, 503549, 503551, 503563, 503593, 503599,
503609, 503611, 503621, 503623, 503647, 503653, 503663, 503707, 503717, 503743, 503753, 503771, 503777,
503779, 503791, 503803, 503819, 503821, 503827, 503851, 503857, 503869, 503879, 503911, 503927, 503929,
503939, 503947, 503959, 503963, 503969, 503983, 503989, 504001, 504011, 504017, 504047, 504061, 504073,
504103, 504121, 504139, 504143, 504149, 504151, 504157, 504181, 504187, 504197, 504209, 504221, 504247,
504269, 504289, 504299, 504307, 504311, 504323, 504337, 504349, 504353, 504359, 504377, 504379, 504389,
504403, 504457, 504461, 504473, 504479, 504521, 504523, 504527, 504547, 504563, 504593, 504599, 504607,
504617, 504619, 504631, 504661, 504667, 504671, 504677, 504683, 504727, 504767, 504787, 504797, 504799,
504817, 504821, 504851, 504853, 504857, 504871, 504877, 504893, 504901, 504929, 504937, 504943, 504947,
504953, 504967, 504983, 504989, 504991, 505027, 505031, 505033, 505049, 505051, 505061, 505067, 505073,
505091, 505097, 505111, 505117, 505123, 505129, 505139, 505157, 505159, 505181, 505187, 505201, 505213,
505231, 505237, 505277, 505279, 505283, 505301, 505313, 505319, 505321, 505327, 505339, 505357, 505367,
505369, 505399, 505409, 505411, 505429, 505447, 505459, 505469, 505481, 505493, 505501, 505511, 505513,
505523, 505537, 505559, 505573, 505601, 505607, 505613, 505619, 505633, 505639, 505643, 505657, 505663,
505669, 505691, 505693, 505709, 505711, 505727, 505759, 505763, 505777, 505781, 505811, 505819, 505823,
505867, 505871, 505877, 505907, 505919, 505927, 505949, 505961, 505969, 505979, 506047, 506071, 506083,
506101, 506113, 506119, 506131, 506147, 506171, 506173, 506183, 506201, 506213, 506251, 506263, 506269,
506281, 506291, 506327, 506329, 506333, 506339, 506347, 506351, 506357, 506381, 506393, 506417, 506423,
506449, 506459, 506461, 506479, 506491, 506501, 506507, 506531, 506533, 506537, 506551, 506563, 506573,
506591, 506593, 506599, 506609, 506629, 506647, 506663, 506683, 506687, 506689, 506699, 506729, 506731,
506743, 506773, 506783, 506791, 506797, 506809, 506837, 506843, 506861, 506873, 506887, 506893, 506899,
506903, 506911, 506929, 506941, 506963, 506983, 506993, 506999, 507029, 507049, 507071, 507077, 507079,
507103, 507109, 507113, 507119, 507137, 507139, 507149, 507151, 507163, 507193, 507197, 507217, 507289,
507301, 507313, 507317, 507329, 507347, 507349, 507359, 507361, 507371, 507383, 507401, 507421, 507431,
507461, 507491, 507497, 507499, 507503, 507523, 507557, 507571, 507589, 507593, 507599, 507607, 507631,
507641, 507667, 507673, 507691, 507697, 507713, 507719, 507743, 507757, 507779, 507781, 507797, 507803,
507809, 507821, 507827, 507839, 507883, 507901, 507907, 507917, 507919, 507937, 507953, 507961, 507971,
507979, 508009, 508019, 508021, 508033, 508037, 508073, 508087, 508091, 508097, 508103, 508129, 508159,
508171, 508187, 508213, 508223, 508229, 508237, 508243, 508259, 508271, 508273, 508297, 508301, 508327,
508331, 508349, 508363, 508367, 508373, 508393, 508433, 508439, 508451, 508471, 508477, 508489, 508499,
508513, 508517, 508531, 508549, 508559, 508567, 508577, 508579, 508583, 508619, 508621, 508637, 508643,
508661, 508693, 508709, 508727, 508771, 508789, 508799, 508811, 508817, 508841, 508847, 508867, 508901,
508903, 508909, 508913, 508919, 508931, 508943, 508951, 508957, 508961, 508969, 508973, 508987, 509023,
509027, 509053, 509063, 509071, 509087, 509101, 509123, 509137, 509147, 509149, 509203, 509221, 509227,
509239, 509263, 509281, 509287, 509293, 509297, 509317, 509329, 509359, 509363, 509389, 509393, 509413,

509417, 509429, 509441, 509449, 509477, 509513, 509521, 509543, 509549, 509557, 509563, 509569, 509573,
509581, 509591, 509603, 509623, 509633, 509647, 509653, 509659, 509681, 509687, 509689, 509693, 509699,
509723, 509731, 509737, 509741, 509767, 509783, 509797, 509801, 509833, 509837, 509843, 509863, 509867,
509879, 509909, 509911, 509921, 509939, 509947, 509959, 509963, 509989, 510007, 510031, 510047, 510049,
510061, 510067, 510073, 510077, 510079, 510089, 510101, 510121, 510127, 510137, 510157, 510179, 510199,
510203, 510217, 510227, 510233, 510241, 510247, 510253, 510271, 510287, 510299, 510311, 510319, 510331,
510361, 510379, 510383, 510401, 510403, 510449, 510451, 510457, 510463, 510481, 510529, 510551, 510553,
510569, 510581, 510583, 510589, 510611, 510613, 510617, 510619, 510677, 510683, 510691, 510707, 510709,
510751, 510767, 510773, 510793, 510803, 510817, 510823, 510827, 510847, 510889, 510907, 510919, 510931,
510941, 510943, 510989, 511001, 511013, 511019, 511033, 511039, 511057, 511061, 511087, 511109, 511111,
511123, 511151, 511153, 511163, 511169, 511171, 511177, 511193, 511201, 511211, 511213, 511223, 511237,
511243, 511261, 511279, 511289, 511297, 511327, 511333, 511337, 511351, 511361, 511387, 511391, 511409,
511417, 511439, 511447, 511453, 511457, 511463, 511477, 511487, 511507, 511519, 511523, 511541, 511549,
511559, 511573, 511579, 511583, 511591, 511603, 511627, 511631, 511633, 511669, 511691, 511703, 511711,
511723, 511757, 511787, 511793, 511801, 511811, 511831, 511843, 511859, 511867, 511873, 511891, 511897,
511909, 511933, 511939, 511961, 511963, 511991, 511997, 512009, 512011, 512021, 512047, 512059, 512093,
512101, 512137, 512147, 512167, 512207, 512249, 512251, 512269, 512287, 512311, 512321, 512333, 512353,
512389, 512419, 512429, 512443, 512467, 512497, 512503, 512507, 512521, 512531, 512537, 512543, 512569,
512573, 512579, 512581, 512591, 512593, 512597, 512609, 512621, 512641, 512657, 512663, 512671, 512683,
512711, 512713, 512717, 512741, 512747, 512761, 512767, 512779, 512797, 512803, 512819, 512821, 512843,
512849, 512891, 512899, 512903, 512917, 512921, 512927, 512929, 512959, 512977, 512989, 512999, 513001,
513013, 513017, 513031, 513041, 513047, 513053, 513059, 513067, 513083, 513101, 513103, 513109, 513131,
513137, 513157, 513167, 513169, 513173, 513203, 513239, 513257, 513269, 513277, 513283, 513307, 513311,
513313, 513319, 513341, 513347, 513353, 513367, 513371, 513397, 513407, 513419, 513427, 513431, 513439,
513473, 513479, 513481, 513509, 513511, 513529, 513533, 513593, 513631, 513641, 513649, 513673, 513679,
513683, 513691, 513697, 513719, 513727, 513731, 513739, 513749, 513761, 513767, 513769, 513781, 513829,
513839, 513841, 513871, 513881, 513899, 513917, 513923, 513937, 513943, 513977, 513991, 514001, 514009,
514013, 514021, 514049, 514051, 514057, 514061, 514079, 514081, 514093, 514103, 514117, 514123, 514127,
514147, 514177, 514187, 514201, 514219, 514229, 514243, 514247, 514249, 514271, 514277, 514289, 514309,
514313, 514333, 514343, 514357, 514361, 514379, 514399, 514417, 514429, 514433, 514453, 514499, 514513,
514519, 514523, 514529, 514531, 514543, 514561, 514571, 514621, 514637, 514639, 514643, 514649, 514651,
514669, 514681, 514711, 514733, 514739, 514741, 514747, 514751, 514757, 514769, 514783, 514793, 514819,
514823, 514831, 514841, 514847, 514853, 514859, 514867, 514873, 514889, 514903, 514933, 514939, 514949,
514967, 515041, 515087, 515089, 515111, 515143, 515149, 515153, 515173, 515191, 515227, 515231, 515233,
515237, 515279, 515293, 515311, 515323, 515351, 515357, 515369, 515371, 515377, 515381, 515401, 515429,
515477, 515507, 515519, 515539, 515563, 515579, 515587, 515597, 515611, 515621, 515639, 515651, 515653,
515663, 515677, 515681, 515687, 515693, 515701, 515737, 515741, 515761, 515771, 515773, 515777, 515783,
515803, 515813, 515839, 515843, 515857, 515861, 515873, 515887, 515917, 515923, 515929, 515941, 515951,
515969, 515993, 516017, 516023, 516049, 516053, 516077, 516091, 516127, 516151, 516157, 516161, 516163,
516169, 516179, 516193, 516199, 516209, 516223, 516227, 516233, 516247, 516251, 516253, 516277, 516283,
516293, 516319, 516323, 516349, 516359, 516361, 516371, 516377, 516391, 516407, 516421, 516431, 516433,
516437, 516449, 516457, 516469, 516493, 516499, 516517, 516521, 516539, 516541, 516563, 516587, 516589,
516599, 516611, 516617, 516619, 516623, 516643, 516653, 516673, 516679, 516689, 516701, 516709, 516713,
516721, 516727, 516757, 516793, 516811, 516821, 516829, 516839, 516847, 516871, 516877, 516883, 516907,
516911, 516931, 516947, 516949, 516959, 516973, 516977, 516979, 516991, 517003, 517043, 517061, 517067,
517073, 517079, 517081, 517087, 517091, 517129, 517151, 517169, 517177, 517183, 517189, 517207, 517211,
517217, 517229, 517241, 517243, 517249, 517261, 517267, 517277, 517289, 517303, 517337, 517343, 517367,
517373, 517381, 517393, 517399, 517403, 517411, 517417, 517457, 517459, 517469, 517471, 517481, 517487,
517499, 517501, 517507, 517511, 517513, 517547, 517549, 517553, 517571, 517577, 517589, 517597, 517603,

517609, 517613, 517619, 517637, 517639, 517711, 517717, 517721, 517729, 517733, 517739, 517747, 517817,
517823, 517831, 517861, 517873, 517877, 517901, 517919, 517927, 517931, 517949, 517967, 517981, 517991,
517999, 518017, 518047, 518057, 518059, 518083, 518099, 518101, 518113, 518123, 518129, 518131, 518137,
518153, 518159, 518171, 518179, 518191, 518207, 518209, 518233, 518237, 518239, 518249, 518261, 518291,
518299, 518311, 518327, 518341, 518387, 518389, 518411, 518417, 518429, 518431, 518447, 518467, 518471,
518473, 518509, 518521, 518533, 518543, 518579, 518587, 518597, 518611, 518621, 518657, 518689, 518699,
518717, 518729, 518737, 518741, 518743, 518747, 518759, 518761, 518767, 518779, 518801, 518803, 518807,
518809, 518813, 518831, 518863, 518867, 518893, 518911, 518933, 518953, 518981, 518983, 518989, 519011,
519031, 519037, 519067, 519083, 519089, 519091, 519097, 519107, 519119, 519121, 519131, 519151, 519161,
519193, 519217, 519227, 519229, 519247, 519257, 519269, 519283, 519287, 519301, 519307, 519349, 519353,
519359, 519371, 519373, 519383, 519391, 519413, 519427, 519433, 519457, 519487, 519499, 519509, 519521,
519523, 519527, 519539, 519551, 519553, 519577, 519581, 519587, 519611, 519619, 519643, 519647, 519667,
519683, 519691, 519703, 519713, 519733, 519737, 519769, 519787, 519793, 519797, 519803, 519817, 519863,
519881, 519889, 519907, 519917, 519919, 519923, 519931, 519943, 519947, 519971, 519989, 519997, 520019,
520021, 520031, 520043, 520063, 520067, 520073, 520103, 520111, 520123, 520129, 520151, 520193, 520213,
520241, 520279, 520291, 520297, 520307, 520309, 520313, 520339, 520349, 520357, 520361, 520363, 520369,
520379, 520381, 520393, 520409, 520411, 520423, 520427, 520433, 520447, 520451, 520529, 520547, 520549,
520567, 520571, 520589, 520607, 520609, 520621, 520631, 520633, 520649, 520679, 520691, 520699, 520703,
520717, 520721, 520747, 520759, 520763, 520787, 520813, 520837, 520841, 520853, 520867, 520889, 520913,
520921, 520943, 520957, 520963, 520967, 520969, 520981, 521009, 521021, 521023, 521039, 521041, 521047,
521051, 521063, 521107, 521119, 521137, 521153, 521161, 521167, 521173, 521177, 521179, 521201, 521231,
521243, 521251, 521267, 521281, 521299, 521309, 521317, 521329, 521357, 521359, 521363, 521369, 521377,
521393, 521399, 521401, 521429, 521447, 521471, 521483, 521491, 521497, 521503, 521519, 521527, 521533,
521537, 521539, 521551, 521557, 521567, 521581, 521603, 521641, 521657, 521659, 521669, 521671, 521693,
521707, 521723, 521743, 521749, 521753, 521767, 521777, 521789, 521791, 521809, 521813, 521819, 521831,
521861, 521869, 521879, 521881, 521887, 521897, 521903, 521923, 521929, 521981, 521993, 521999, 522017,
522037, 522047, 522059, 522061, 522073, 522079, 522083, 522113, 522127, 522157, 522161, 522167, 522191,
522199, 522211, 522227, 522229, 522233, 522239, 522251, 522259, 522281, 522283, 522289, 522317, 522323,
522337, 522371, 522373, 522383, 522391, 522409, 522413, 522439, 522449, 522469, 522479, 522497, 522517,
522521, 522523, 522541, 522553, 522569, 522601, 522623, 522637, 522659, 522661, 522673, 522677, 522679,
522689, 522703, 522707, 522719, 522737, 522749, 522757, 522761, 522763, 522787, 522811, 522827, 522829,
522839, 522853, 522857, 522871, 522881, 522883, 522887, 522919, 522943, 522947, 522959, 522961, 522989,
523007, 523021, 523031, 523049, 523093, 523097, 523109, 523129, 523169, 523177, 523207, 523213, 523219,
523261, 523297, 523307, 523333, 523349, 523351, 523357, 523387, 523403, 523417, 523427, 523433, 523459,
523463, 523487, 523489, 523493, 523511, 523519, 523541, 523543, 523553, 523571, 523573, 523577, 523597,
523603, 523631, 523637, 523639, 523657, 523667, 523669, 523673, 523681, 523717, 523729, 523741, 523759,
523763, 523771, 523777, 523793, 523801, 523829, 523847, 523867, 523877, 523903, 523907, 523927, 523937,
523949, 523969, 523987, 523997, 524047, 524053, 524057, 524063, 524071, 524081, 524087, 524099, 524113,
524119, 524123, 524149, 524171, 524189, 524197, 524201, 524203, 524219, 524221, 524231, 524243, 524257,
524261, 524269, 524287, 524309, 524341, 524347, 524351, 524353, 524369, 524387, 524389, 524411, 524413,
524429, 524453, 524497, 524507, 524509, 524519, 524521, 524591, 524593, 524599, 524633, 524669, 524681,
524683, 524701, 524707, 524731, 524743, 524789, 524801, 524803, 524827, 524831, 524857, 524863, 524869,
524873, 524893, 524899, 524921, 524933, 524939, 524941, 524947, 524957, 524959, 524963, 524969, 524971,
524981, 524983, 524999, 525001, 525013, 525017, 525029, 525043, 525101, 525127, 525137, 525143, 525157,
525163, 525167, 525191, 525193, 525199, 525209, 525221, 525241, 525247, 525253, 525257, 525299, 525313,
525353, 525359, 525361, 525373, 525377, 525379, 525391, 525397, 525409, 525431, 525433, 525439, 525457,
525461, 525467, 525491, 525493, 525517, 525529, 525533, 525541, 525571, 525583, 525593, 525599, 525607,
525641, 525649, 525671, 525677, 525697, 525709, 525713, 525719, 525727, 525731, 525739, 525769, 525773,
525781, 525809, 525817, 525839, 525869, 525871, 525887, 525893, 525913, 525923, 525937, 525947, 525949,

525953, 525961, 525979, 525983, 526027, 526037, 526049, 526051, 526063, 526067, 526069, 526073, 526087, 526117, 526121, 526139, 526157, 526159, 526189, 526193, 526199, 526213, 526223, 526231, 526249, 526271, 526283, 526289, 526291, 526297, 526307, 526367, 526373, 526381, 526387, 526391, 526397, 526423, 526429, 526441, 526453, 526459, 526483, 526499, 526501, 526511, 526531, 526543, 526571, 526573, 526583, 526601, 526619, 526627, 526633, 526637, 526649, 526651, 526657, 526667, 526679, 526681, 526703, 526709, 526717, 526733, 526739, 526741, 526759, 526763, 526777, 526781, 526829, 526831, 526837, 526853, 526859, 526871, 526909, 526913, 526931, 526937, 526943, 526951, 526957, 526963, 526993, 526997, 527053, 527057, 527063, 527069, 527071, 527081, 527099, 527123, 527129, 527143, 527159, 527161, 527173, 527179, 527203, 527207, 527209, 527237, 527251, 527273, 527281, 527291, 527327, 527333, 527347, 527353, 527377, 527381, 527393, 527399, 527407, 527411, 527419, 527441, 527447, 527453, 527489, 527507, 527533, 527557, 527563, 527581, 527591, 527599, 527603, 527623, 527627, 527633, 527671, 527699, 527701, 527729, 527741, 527749, 527753, 527789, 527803, 527809, 527819, 527843, 527851, 527869, 527881, 527897, 527909, 527921, 527929, 527941, 527981, 527983, 527987, 527993, 528001, 528013, 528041, 528043, 528053, 528091, 528097, 528107, 528127, 528131, 528137, 528163, 528167, 528191, 528197, 528217, 528223, 528247, 528263, 528289, 528299, 528313, 528317, 528329, 528373, 528383, 528391, 528401, 528403, 528413, 528419, 528433, 528469, 528487, 528491, 528509, 528511, 528527, 528559, 528611, 528623, 528629, 528631, 528659, 528667, 528673, 528679, 528691, 528707, 528709, 528719, 528763, 528779, 528791, 528799, 528811, 528821, 528823, 528833, 528863, 528887, 528881, 528883, 528911, 528929, 528947, 528967, 528971, 528973, 528991, 529003, 529007, 529027, 529033, 529037, 529043, 529049, 529051, 529097, 529103, 529117, 529121, 529127, 529129, 529153, 529157, 529181, 529183, 529213, 529229, 529237, 529241, 529259, 529271, 529273, 529301, 529307, 529313, 529327, 529343, 529349, 529357, 529381, 529393, 529411, 529421, 529423, 529471, 529489, 529513, 529517, 529519, 529531, 529547, 529577, 529579, 529603, 529619, 529637, 529649, 529657, 529673, 529681, 529687, 529691, 529693, 529709, 529723, 529741, 529747, 529751, 529807, 529811, 529813, 529819, 529829, 529847, 529871, 529927, 529933, 529939, 529957, 529961, 529973, 529979, 529981, 529987, 529999, 530017, 530021, 530027, 530041, 530051, 530063, 530087, 530093, 530129, 530137, 530143, 530177, 530183, 530197, 530203, 530209, 530227, 530237, 530249, 530251, 530261, 530267, 530279, 530293, 530297, 530303, 530329, 530339, 530353, 530359, 530389, 530393, 530401, 530429, 530443, 530447, 530501, 530507, 530513, 530527, 530531, 530533, 530539, 530549, 530567, 530597, 530599, 530603, 530609, 530641, 530653, 530659, 530669, 530693, 530701, 530711, 530713, 530731, 530741, 530743, 530753, 530767, 530773, 530797, 530807, 530833, 530837, 530843, 530851, 530857, 530861, 530869, 530897, 530911, 530947, 530969, 530977, 530983, 530989, 531017, 531023, 531043, 531071, 531079, 531101, 531103, 531121, 531133, 531143, 531163, 531169, 531173, 531197, 531203, 531229, 531239, 531253, 531263, 531281, 531287, 531299, 531331, 531337, 531343, 531347, 531353, 531359, 531383, 531457, 531481, 531497, 531521, 531547, 531551, 531569, 531571, 531581, 531589, 531611, 531613, 531623, 531631, 531637, 531667, 531673, 531689, 531701, 531731, 531793, 531799, 531821, 531823, 531827, 531833, 531841, 531847, 531857, 531863, 531871, 531877, 531901, 531911, 531919, 531977, 531983, 531989, 531997, 532001, 532009, 532027, 532033, 532061, 532069, 532093, 532099, 532141, 532153, 532159, 532163, 532183, 532187, 532193, 532199, 532241, 532249, 532261, 532267, 532277, 532283, 532307, 532313, 532327, 532331, 532333, 532349, 532373, 532379, 532391, 532403, 532417, 532421, 532439, 532447, 532451, 532453, 532489, 532501, 532523, 532529, 532531, 532537, 532547, 532561, 532601, 532603, 532607, 532619, 532621, 532633, 532639, 532663, 532669, 532687, 532691, 532709, 532733, 532739, 532751, 532757, 532771, 532781, 532783, 532789, 532801, 532811, 532823, 532849, 532853, 532867, 532907, 532919, 532949, 532951, 532981, 532993, 532999, 533003, 533009, 533011, 533033, 533051, 533053, 533063, 533077, 533089, 533111, 533129, 533149, 533167, 533177, 533189, 533191, 533213, 533219, 533227, 533237, 533249, 533257, 533261, 533263, 533297, 533303, 533317, 533321, 533323, 533353, 533363, 533371, 533389, 533399, 533413, 533447, 533453, 533459, 533509, 533543, 533549, 533573, 533581, 533593, 533633, 533641, 533671, 533693, 533711, 533713, 533719, 533723, 533737, 533747, 533777, 533801, 533809, 533821, 533831, 533837, 533857, 533879, 533887, 533893, 533909, 533921, 533927, 533959, 533963, 533969, 533971, 533989, 533993, 533999, 534007, 534013, 534019, 534029, 534043, 534047, 534049, 534059, 534073, 534077, 534091, 534101, 534113, 534137, 534167, 534173, 534199, 534203, 534211, 534229, 534241, 534253, 534283, 534301, 534307, 534311, 534323, 534329,

534341, 534367, 534371, 534403, 534407, 534431, 534439, 534473, 534491, 534511, 534529, 534553, 534571,
534577, 534581, 534601, 534607, 534617, 534629, 534631, 534637, 534647, 534649, 534659, 534661, 534671,
534697, 534707, 534739, 534799, 534811, 534827, 534839, 534841, 534851, 534857, 534883, 534889, 534913,
534923, 534931, 534943, 534949, 534971, 535013, 535019, 535033, 535037, 535061, 535099, 535103, 535123,
535133, 535151, 535159, 535169, 535181, 535193, 535207, 535219, 535229, 535237, 535243, 535273, 535303,
535319, 535333, 535349, 535351, 535361, 535387, 535391, 535399, 535481, 535487, 535489, 535499, 535511,
535523, 535529, 535547, 535571, 535573, 535589, 535607, 535609, 535627, 535637, 535663, 535669, 535673,
535679, 535697, 535709, 535727, 535741, 535751, 535757, 535771, 535783, 535793, 535811, 535849, 535859,
535861, 535879, 535919, 535937, 535939, 535943, 535957, 535967, 535973, 535991, 535999, 536017, 536023,
536051, 536057, 536059, 536069, 536087, 536099, 536101, 536111, 536141, 536147, 536149, 536189, 536191,
536203, 536213, 536219, 536227, 536233, 536243, 536267, 536273, 536279, 536281, 536287, 536293, 536311,
536323, 536353, 536357, 536377, 536399, 536407, 536423, 536441, 536443, 536447, 536449, 536453, 536461,
536467, 536479, 536491, 536509, 536513, 536531, 536533, 536561, 536563, 536593, 536609, 536621, 536633,
536651, 536671, 536677, 536687, 536699, 536717, 536719, 536729, 536743, 536749, 536771, 536773, 536777,
536779, 536791, 536801, 536803, 536839, 536849, 536857, 536867, 536869, 536891, 536909, 536917, 536923,
536929, 536933, 536947, 536953, 536971, 536989, 536999, 537001, 537007, 537011, 537023, 537029, 537037,
537041, 537067, 537071, 537079, 537091, 537127, 537133, 537143, 537157, 537169, 537181, 537191, 537197,
537221, 537233, 537241, 537269, 537281, 537287, 537307, 537331, 537343, 537347, 537373, 537379, 537401,
537403, 537413, 537497, 537527, 537547, 537569, 537583, 537587, 537599, 537611, 537637, 537661, 537673,
537679, 537703, 537709, 537739, 537743, 537749, 537769, 537773, 537781, 537787, 537793, 537811, 537841,
537847, 537853, 537877, 537883, 537899, 537913, 537919, 537941, 537991, 538001, 538019, 538049, 538051,
538073, 538079, 538093, 538117, 538121, 538123, 538127, 538147, 538151, 538157, 538159, 538163, 538199,
538201, 538247, 538249, 538259, 538267, 538283, 538297, 538301, 538303, 538309, 538331, 538333, 538357,
538367, 538397, 538399, 538411, 538423, 538457, 538471, 538481, 538487, 538511, 538513, 538519, 538523,
538529, 538553, 538561, 538567, 538579, 538589, 538597, 538621, 538649, 538651, 538697, 538709, 538711,
538721, 538723, 538739, 538751, 538763, 538771, 538777, 538789, 538799, 538801, 538817, 538823, 538829,
538841, 538871, 538877, 538921, 538927, 538931, 538939, 538943, 538987, 539003, 539009, 539039, 539047,
539089, 539093, 539101, 539107, 539111, 539113, 539129, 539141, 539153, 539159, 539167, 539171, 539207,
539219, 539233, 539237, 539261, 539267, 539269, 539293, 539303, 539309, 539311, 539321, 539323, 539339,
539347, 539351, 539389, 539401, 539447, 539449, 539479, 539501, 539503, 539507, 539509, 539533, 539573,
539621, 539629, 539633, 539639, 539641, 539653, 539663, 539677, 539687, 539711, 539713, 539723, 539729,
539743, 539761, 539783, 539797, 539837, 539839, 539843, 539849, 539863, 539881, 539897, 539899, 539921,
539947, 539993, 540041, 540061, 540079, 540101, 540119, 540121, 540139, 540149, 540157, 540167, 540173,
540179, 540181, 540187, 540203, 540217, 540233, 540251, 540269, 540271, 540283, 540301, 540307, 540343,
540347, 540349, 540367, 540373, 540377, 540383, 540389, 540391, 540433, 540437, 540461, 540469, 540509,
540511, 540517, 540539, 540541, 540557, 540559, 540577, 540587, 540599, 540611, 540613, 540619, 540629,
540677, 540679, 540689, 540691, 540697, 540703, 540713, 540751, 540769, 540773, 540779, 540781, 540803,
540809, 540823, 540851, 540863, 540871, 540877, 540901, 540907, 540961, 540989, 541001, 541007, 541027,
541049, 541061, 541087, 541097, 541129, 541133, 541141, 541153, 541181, 541193, 541201, 541217, 541231,
541237, 541249, 541267, 541271, 541283, 541301, 541309, 541339, 541349, 541361, 541363, 541369, 541381,
541391, 541417, 541439, 541447, 541469, 541483, 541507, 541511, 541523, 541529, 541531, 541537, 541543,
541547, 541549, 541571, 541577, 541579, 541589, 541613, 541631, 541657, 541661, 541669, 541693, 541699,
541711, 541721, 541727, 541759, 541763, 541771, 541777, 541781, 541799, 541817, 541831, 541837, 541859,
541889, 541901, 541927, 541951, 541967, 541987, 541991, 541993, 542021, 542023, 542027, 542053,
542063, 542071, 542081, 542083, 542093, 542111, 542117, 542119, 542123, 542131, 542141, 542149, 542153,
542167, 542183, 542189, 542197, 542207, 542219, 542237, 542251, 542261, 542263, 542281, 542293, 542299,
542323, 542371, 542401, 542441, 542447, 542461, 542467, 542483, 542489, 542497, 542519, 542533, 542537,
542539, 542551, 542557, 542567, 542579, 542587, 542599, 542603, 542683, 542687, 542693, 542713, 542719,
542723, 542747, 542761, 542771, 542783, 542791, 542797, 542821, 542831, 542837, 542873, 542891, 542911,

542921, 542923, 542933, 542939, 542947, 542951, 542981, 542987, 542999, 543017, 543019, 543029, 543061,
543097, 543113, 543131, 543139, 543143, 543149, 543157, 543161, 543163, 543187, 543203, 543217, 543223,
543227, 543233, 543241, 543253, 543259, 543281, 543287, 543289, 543299, 543307, 543311, 543313, 543341,
543349, 543353, 543359, 543379, 543383, 543407, 543427, 543463, 543497, 543503, 543509, 543539, 543551,
543553, 543593, 543601, 543607, 543611, 543617, 543637, 543659, 543661, 543671, 543679, 543689, 543703,
543707, 543713, 543769, 543773, 543787, 543791, 543793, 543797, 543811, 543827, 543841, 543853, 543857,
543859, 543871, 543877, 543883, 543887, 543889, 543901, 543911, 543929, 543967, 543971, 543997, 544001,
544007, 544009, 544013, 544021, 544031, 544097, 544099, 544109, 544123, 544129, 544133, 544139, 544171,
544177, 544183, 544199, 544223, 544259, 544273, 544277, 544279, 544367, 544373, 544399, 544403, 544429,
544451, 544471, 544477, 544487, 544501, 544513, 544517, 544543, 544549, 544601, 544613, 544627, 544631,
544651, 544667, 544699, 544717, 544721, 544723, 544727, 544757, 544759, 544771, 544781, 544793, 544807,
544813, 544837, 544861, 544877, 544879, 544883, 544889, 544897, 544903, 544919, 544927, 544937, 544961,
544963, 544979, 545023, 545029, 545033, 545057, 545063, 545087, 545089, 545093, 545117, 545131, 545141,
545143, 545161, 545189, 545203, 545213, 545231, 545239, 545257, 545267, 545291, 545329, 545371, 545387,
545429, 545437, 545443, 545449, 545473, 545477, 545483, 545497, 545521, 545527, 545533, 545543, 545549,
545551, 545579, 545599, 545609, 545617, 545621, 545641, 545647, 545651, 545663, 545711, 545723, 545731,
545747, 545749, 545759, 545773, 545789, 545791, 545827, 545843, 545863, 545873, 545893, 545899, 545911,
545917, 545929, 545933, 545939, 545947, 545959, 546001, 546017, 546019, 546031, 546047, 546053, 546067,
546071, 546097, 546101, 546103, 546109, 546137, 546149, 546151, 546173, 546179, 546197, 546211, 546233,
546239, 546241, 546253, 546263, 546283, 546289, 546317, 546323, 546341, 546349, 546353, 546361, 546367,
546373, 546391, 546461, 546467, 546479, 546509, 546523, 546547, 546569, 546583, 546587, 546599, 546613,
546617, 546619, 546631, 546643, 546661, 546671, 546677, 546683, 546691, 546709, 546719, 546731, 546739,
546781, 546841, 546859, 546863, 546869, 546881, 546893, 546919, 546937, 546943, 546947, 546961, 546967,
546977, 547007, 547021, 547037, 547061, 547087, 547093, 547097, 547103, 547121, 547133, 547139, 547171,
547223, 547229, 547237, 547241, 547249, 547271, 547273, 547291, 547301, 547321, 547357, 547361, 547363,
547369, 547373, 547387, 547397, 547399, 547411, 547441, 547453, 547471, 547483, 547487, 547493, 547499,
547501, 547513, 547529, 547537, 547559, 547567, 547577, 547583, 547601, 547609, 547619, 547627, 547639,
547643, 547661, 547663, 547681, 547709, 547727, 547741, 547747, 547753, 547763, 547769, 547787, 547817,
547819, 547823, 547831, 547849, 547853, 547871, 547889, 547901, 547909, 547951, 547957, 547999, 548003,
548039, 548059, 548069, 548083, 548089, 548099, 548117, 548123, 548143, 548153, 548189, 548201, 548213,
548221, 548227, 548239, 548243, 548263, 548291, 548309, 548323, 548347, 548351, 548363, 548371, 548393,
548399, 548407, 548417, 548423, 548441, 548453, 548459, 548461, 548489, 548501, 548503, 548519, 548521,
548533, 548543, 548557, 548567, 548579, 548591, 548623, 548629, 548657, 548671, 548677, 548687, 548693,
548707, 548719, 548749, 548753, 548761, 548771, 548783, 548791, 548827, 548831, 548833, 548837, 548843,
548851, 548861, 548869, 548893, 548897, 548903, 548909, 548927, 548953, 548957, 548963, 549001, 549011,
549013, 549019, 549023, 549037, 549071, 549089, 549091, 549097, 549121, 549139, 549149, 549161, 549163,
549167, 549169, 549193, 549203, 549221, 549229, 549247, 549257, 549259, 549281, 549313, 549319, 549323,
549331, 549379, 549391, 549403, 549421, 549431, 549443, 549449, 549481, 549503, 549509, 549511, 549517,
549533, 549547, 549551, 549553, 549569, 549587, 549589, 549607, 549623, 549641, 549643, 549649, 549667,
549683, 549691, 549701, 549707, 549713, 549719, 549733, 549737, 549739, 549749, 549751, 549767, 549817,
549833, 549839, 549863, 549877, 549883, 549911, 549937, 549943, 549949, 549977, 549979, 550007, 550009,
550027, 550049, 550061, 550063, 550073, 550111, 550117, 550127, 550129, 550139, 550163, 550169, 550177,
550181, 550189, 550211, 550213, 550241, 550267, 550279, 550283, 550289, 550309, 550337, 550351, 550369,
550379, 550427, 550439, 550441, 550447, 550457, 550469, 550471, 550489, 550513, 550519, 550531, 550541,
550553, 550577, 550607, 550609, 550621, 550631, 550637, 550651, 550657, 550661, 550663, 550679, 550691,
550703, 550717, 550721, 550733, 550757, 550763, 550789, 550801, 550811, 550813, 550831, 550841, 550843,
550859, 550861, 550903, 550909, 550937, 550939, 550951, 550961, 550969, 550973, 550993, 550997, 551003,
551017, 551027, 551039, 551059, 551063, 551069, 551093, 551099, 551107, 551113, 551129, 551143, 551179,
551197, 551207, 551219, 551231, 551233, 551269, 551281, 551297, 551311, 551321, 551339, 551347, 551363,

551381, 551387, 551407, 551423, 551443, 551461, 551483, 551489, 551503, 551519, 551539, 551543, 551549,
551557, 551569, 551581, 551587, 551597, 551651, 551653, 551659, 551671, 551689, 551693, 551713, 551717,
551723, 551729, 551731, 551743, 551753, 551767, 551773, 551801, 551809, 551813, 551843, 551849, 551861,
551909, 551911, 551917, 551927, 551933, 551951, 551959, 551963, 551981, 552001, 552011, 552029, 552031,
552047, 552053, 552059, 552089, 552091, 552103, 552107, 552113, 552127, 552137, 552179, 552193, 552217,
552239, 552241, 552259, 552263, 552271, 552283, 552301, 552317, 552341, 552353, 552379, 552397, 552401,
552403, 552469, 552473, 552481, 552491, 552493, 552511, 552523, 552527, 552553, 552581, 552583, 552589,
552611, 552649, 552659, 552677, 552703, 552707, 552709, 552731, 552749, 552751, 552757, 552787, 552791,
552793, 552809, 552821, 552833, 552841, 552847, 552859, 552883, 552887, 552899, 552913, 552917, 552971,
552983, 552991, 553013, 553037, 553043, 553051, 553057, 553067, 553073, 553093, 553097, 553099, 553103,
553123, 553139, 553141, 553153, 553171, 553181, 553193, 553207, 553211, 553229, 553249, 553253, 553277,
553279, 553309, 553351, 553363, 553369, 553411, 553417, 553433, 553439, 553447, 553457, 553463, 553471,
553481, 553507, 553513, 553517, 553529, 553543, 553549, 553561, 553573, 553583, 553589, 553591, 553601,
553607, 553627, 553643, 553649, 553667, 553681, 553687, 553699, 553703, 553727, 553733, 553747, 553757,
553759, 553769, 553789, 553811, 553837, 553849, 553867, 553873, 553897, 553901, 553919, 553921, 553933,
553961, 553963, 553981, 553991, 554003, 554011, 554017, 554051, 554077, 554087, 554089, 554117, 554123,
554129, 554137, 554167, 554171, 554179, 554189, 554207, 554209, 554233, 554237, 554263, 554269, 554293,
554299, 554303, 554317, 554347, 554377, 554383, 554417, 554419, 554431, 554447, 554453, 554467, 554503,
554527, 554531, 554569, 554573, 554597, 554611, 554627, 554633, 554639, 554641, 554663, 554669, 554677,
554699, 554707, 554711, 554731, 554747, 554753, 554759, 554767, 554779, 554789, 554791, 554797, 554803,
554821, 554833, 554837, 554839, 554843, 554849, 554887, 554891, 554893, 554899, 554923, 554927, 554951,
554959, 554969, 554977, 555029, 555041, 555043, 555053, 555073, 555077, 555083, 555091, 555097, 555109,
555119, 555143, 555167, 555209, 555221, 555251, 555253, 555257, 555277, 555287, 555293, 555301, 555307,
555337, 555349, 555361, 555383, 555391, 555419, 555421, 555439, 555461, 555487, 555491, 555521, 555523,
555557, 555589, 555593, 555637, 555661, 555671, 555683, 555691, 555697, 555707, 555739, 555743,
555761, 555767, 555823, 555827, 555829, 555853, 555857, 555871, 555931, 555941, 555953, 555967, 556007,
556021, 556027, 556037, 556043, 556051, 556067, 556069, 556093, 556103, 556123, 556159, 556177, 556181,
556211, 556219, 556229, 556243, 556253, 556261, 556267, 556271, 556273, 556279, 556289, 556313, 556321,
556327, 556331, 556343, 556351, 556373, 556399, 556403, 556441, 556459, 556477, 556483, 556487, 556513,
556519, 556537, 556559, 556573, 556579, 556583, 556601, 556607, 556609, 556613, 556627, 556639, 556651,
556679, 556687, 556691, 556693, 556697, 556709, 556723, 556727, 556741, 556753, 556763, 556769, 556781,
556789, 556793, 556799, 556811, 556817, 556819, 556823, 556841, 556849, 556859, 556861, 556867, 556883,
556891, 556931, 556939, 556943, 556957, 556967, 556981, 556987, 556999, 557017, 557021, 557027, 557033,
557041, 557057, 557059, 557069, 557087, 557093, 557153, 557159, 557197, 557201, 557261, 557269, 557273,
557281, 557303, 557309, 557321, 557329, 557339, 557369, 557371, 557377, 557423, 557443, 557449, 557461,
557483, 557489, 557519, 557521, 557533, 557537, 557551, 557567, 557573, 557591, 557611, 557633, 557639,
557663, 557671, 557693, 557717, 557729, 557731, 557741, 557743, 557747, 557759, 557761, 557779, 557789,
557801, 557803, 557831, 557857, 557861, 557863, 557891, 557899, 557903, 557927, 557981, 557987, 558007,
558017, 558029, 558053, 558067, 558083, 558091, 558109, 558113, 558121, 558139, 558149, 558167, 558179,
558197, 558203, 558209, 558223, 558241, 558251, 558253, 558287, 558289, 558307, 558319, 558343, 558401,
558413, 558421, 558427, 558431, 558457, 558469, 558473, 558479, 558491, 558497, 558499, 558521, 558529,
558533, 558539, 558541, 558563, 558583, 558587, 558599, 558611, 558629, 558643, 558661, 558683, 558703,
558721, 558731, 558757, 558769, 558781, 558787, 558791, 558793, 558827, 558829, 558863, 558869, 558881,
558893, 558913, 558931, 558937, 558947, 558973, 558979, 558997, 559001, 559049, 559051, 559067, 559081,
559093, 559099, 559123, 559133, 559157, 559177, 559183, 559201, 559211, 559213, 559217, 559219, 559231,
559243, 559259, 559277, 559297, 559313, 559319, 559343, 559357, 559367, 559369, 559397, 559421, 559451,
559459, 559469, 559483, 559511, 559513, 559523, 559529, 559541, 559547, 559549, 559561, 559571, 559577,
559583, 559591, 559597, 559631, 559633, 559639, 559649, 559667, 559673, 559679, 559687, 559703, 559709,
559739, 559747, 559777, 559781, 559799, 559807, 559813, 559831, 559841, 559849, 559859, 559877, 559883,

559901, 559907, 559913, 559939, 559967, 559973, 559991, 560017, 560023, 560029, 560039, 560047, 560081,
560083, 560089, 560093, 560107, 560113, 560117, 560123, 560137, 560149, 560159, 560171, 560173, 560179,
560191, 560207, 560213, 560221, 560227, 560233, 560237, 560239, 560243, 560249, 560281, 560293, 560297,
560299, 560311, 560317, 560341, 560353, 560393, 560411, 560437, 560447, 560459, 560471, 560477, 560479,
560489, 560491, 560501, 560503, 560531, 560543, 560551, 560561, 560597, 560617, 560621, 560639, 560641,
560653, 560669, 560683, 560689, 560701, 560719, 560737, 560753, 560761, 560767, 560771, 560783, 560797,
560803, 560827, 560837, 560863, 560869, 560873, 560887, 560891, 560893, 560897, 560929, 560939, 560941,
560969, 560977, 561019, 561047, 561053, 561059, 561061, 561079, 561083, 561091, 561097, 561101, 561103,
561109, 561161, 561173, 561181, 561191, 561199, 561229, 561251, 561277, 561307, 561313, 561343, 561347,
561359, 561367, 561373, 561377, 561389, 561409, 561419, 561439, 561461, 561521, 561529, 561551, 561553,
561559, 561599, 561607, 561667, 561703, 561713, 561733, 561761, 561767, 561787, 561797, 561809, 561829,
561839, 561907, 561917, 561923, 561931, 561943, 561947, 561961, 561973, 561983, 561997, 562007, 562019,
562021, 562043, 562091, 562103, 562129, 562147, 562169, 562181, 562193, 562201, 562231, 562259, 562271,
562273, 562283, 562291, 562297, 562301, 562307, 562313, 562333, 562337, 562349, 562351, 562357, 562361,
562399, 562403, 562409, 562417, 562421, 562427, 562439, 562459, 562477, 562493, 562501, 562517, 562519,
562537, 562577, 562579, 562589, 562591, 562607, 562613, 562621, 562631, 562633, 562651, 562663, 562669,
562673, 562691, 562693, 562699, 562703, 562711, 562721, 562739, 562753, 562759, 562763, 562781, 562789,
562813, 562831, 562841, 562871, 562897, 562901, 562909, 562931, 562943, 562949, 562963, 562967, 562973,
562979, 562987, 562997, 563009, 563011, 563021, 563039, 563041, 563047, 563051, 563077, 563081, 563099,
563113, 563117, 563119, 563131, 563149, 563153, 563183, 563197, 563219, 563249, 563263, 563287, 563327,
563351, 563357, 563359, 563377, 563401, 563411, 563413, 563417, 563419, 563447, 563449, 563467, 563489,
563501, 563503, 563543, 563551, 563561, 563587, 563593, 563599, 563623, 563657, 563663, 563723, 563743,
563747, 563777, 563809, 563813, 563821, 563831, 563837, 563851, 563869, 563881, 563887, 563897, 563929,
563933, 563947, 563971, 563987, 563999, 564013, 564017, 564041, 564049, 564059, 564061, 564089, 564097,
564103, 564127, 564133, 564149, 564163, 564173, 564191, 564197, 564227, 564229, 564233, 564251, 564257,
564269, 564271, 564299, 564391, 564307, 564313, 564323, 564353, 564359, 564367, 564371, 564373, 564391,
564401, 564407, 564409, 564419, 564437, 564449, 564457, 564463, 564467, 564491, 564497, 564523, 564533,
564593, 564607, 564617, 564643, 564653, 564667, 564671, 564679, 564701, 564703, 564709, 564713, 564761,
564779, 564793, 564797, 564827, 564871, 564881, 564899, 564917, 564919, 564923, 564937, 564959, 564973,
564979, 564983, 564989, 564997, 565013, 565039, 565049, 565057, 565069, 565109, 565111, 565127, 565163,
565171, 565177, 565183, 565189, 565207, 565237, 565241, 565247, 565259, 565261, 565273, 565283, 565289,
565303, 565319, 565333, 565337, 565343, 565361, 565379, 565381, 565387, 565391, 565393, 565427, 565429,
565441, 565451, 565463, 565469, 565483, 565489, 565507, 565511, 565517, 565519, 565549, 565553, 565559,
565567, 565571, 565583, 565589, 565597, 565603, 565613, 565637, 565651, 565661, 565667, 565723, 565727,
565769, 565771, 565787, 565793, 565813, 565849, 565867, 565889, 565891, 565907, 565909, 565919, 565921,
565937, 565973, 565979, 565997, 566011, 566023, 566047, 566057, 566077, 566089, 566101, 566107, 566131,
566149, 566161, 566173, 566179, 566183, 566201, 566213, 566227, 566231, 566233, 566273, 566311, 566323,
566347, 566387, 566393, 566413, 566417, 566429, 566431, 566437, 566441, 566443, 566453, 566521, 566537,
566539, 566543, 566549, 566551, 566557, 566563, 566567, 566617, 566633, 566639, 566653, 566659, 566677,
566681, 566693, 566701, 566707, 566717, 566719, 566723, 566737, 566759, 566767, 566791, 566821, 566833,
566851, 566857, 566879, 566911, 566939, 566947, 566963, 566971, 566977, 566987, 566999, 567011, 567013,
567031, 567053, 567059, 567067, 567097, 567101, 567107, 567121, 567143, 567179, 567181, 567187, 567209,
567257, 567263, 567277, 567319, 567323, 567367, 567377, 567383, 567389, 567401, 567407, 567439, 567449,
567451, 567467, 567487, 567493, 567499, 567527, 567529, 567533, 567569, 567601, 567607, 567631, 567649,
567653, 567659, 567661, 567667, 567673, 567689, 567719, 567737, 567751, 567761, 567767, 567779, 567793,
567811, 567829, 567841, 567857, 567863, 567871, 567877, 567881, 567883, 567899, 567937, 567943, 567947,
567949, 567961, 567979, 567991, 567997, 568019, 568027, 568033, 568049, 568069, 568091, 568097, 568109,
568133, 568151, 568153, 568163, 568171, 568177, 568187, 568189, 568193, 568201, 568207, 568231, 568237,
568241, 568273, 568279, 568289, 568303, 568349, 568363, 568367, 568387, 568391, 568433, 568439, 568441,

568453, 568471, 568481, 568493, 568523, 568541, 568549, 568577, 568609, 568619, 568627, 568643, 568657,
568669, 568679, 568691, 568699, 568709, 568723, 568751, 568783, 568787, 568807, 568823, 568831, 568853,
568877, 568891, 568903, 568907, 568913, 568921, 568963, 568979, 568987, 568991, 568999, 569003, 569011,
569021, 569047, 569053, 569057, 569071, 569077, 569081, 569083, 569111, 569117, 569137, 569141, 569159,
569161, 569189, 569197, 569201, 569209, 569213, 569237, 569243, 569249, 569251, 569263, 569267, 569269,
569321, 569323, 569369, 569417, 569419, 569423, 569431, 569447, 569461, 569479, 569497, 569507, 569533,
569573, 569579, 569581, 569599, 569603, 569609, 569617, 569623, 569659, 569663, 569671, 569683, 569711,
569713, 569717, 569729, 569731, 569747, 569759, 569771, 569773, 569797, 569809, 569813, 569819, 569831,
569839, 569843, 569851, 569861, 569869, 569887, 569893, 569897, 569903, 569927, 569939, 569957, 569983,
570001, 570013, 570029, 570041, 570043, 570047, 570049, 570071, 570077, 570079, 570083, 570091, 570107,
570109, 570113, 570131, 570139, 570161, 570173, 570181, 570191, 570217, 570221, 570233, 570253, 570329,
570359, 570373, 570379, 570389, 570391, 570403, 570407, 570413, 570419, 570421, 570461, 570463, 570467,
570487, 570491, 570497, 570499, 570509, 570511, 570527, 570529, 570539, 570547, 570553, 570569, 570587,
570601, 570613, 570637, 570643, 570649, 570659, 570667, 570671, 570677, 570683, 570697, 570719, 570733,
570737, 570743, 570781, 570821, 570827, 570839, 570841, 570851, 570853, 570859, 570881, 570887, 570901,
570919, 570937, 570949, 570959, 570961, 570967, 570991, 571001, 571019, 571031, 571037, 571049, 571069,
571093, 571099, 571111, 571133, 571147, 571157, 571163, 571199, 571201, 571211, 571223, 571229, 571231,
571261, 571267, 571279, 571303, 571321, 571331, 571339, 571369, 571381, 571397, 571399, 571409, 571433,
571453, 571471, 571477, 571531, 571541, 571579, 571583, 571589, 571601, 571603, 571633, 571657, 571673,
571679, 571699, 571709, 571717, 571721, 571741, 571751, 571759, 571777, 571783, 571789, 571799, 571801,
571811, 571841, 571847, 571853, 571861, 571867, 571871, 571873, 571877, 571903, 571933, 571939, 571969,
571973, 572023, 572027, 572041, 572051, 572053, 572059, 572063, 572069, 572087, 572093, 572107, 572137,
572161, 572177, 572179, 572183, 572207, 572233, 572239, 572251, 572269, 572281, 572303, 572311, 572321,
572323, 572329, 572333, 572357, 572387, 572399, 572417, 572419, 572423, 572437, 572449, 572461, 572471,
572479, 572491, 572497, 572519, 572521, 572549, 572567, 572573, 572581, 572587, 572597, 572599, 572609,
572629, 572633, 572639, 572651, 572653, 572657, 572659, 572683, 572687, 572699, 572707, 572711, 572749,
572777, 572791, 572801, 572807, 572813, 572821, 572827, 572833, 572843, 572867, 572879, 572881, 572903,
572909, 572927, 572933, 572939, 572941, 572963, 572969, 572993, 573007, 573031, 573047, 573101, 573107,
573109, 573119, 573143, 573161, 573163, 573179, 573197, 573247, 573253, 573263, 573277, 573289, 573299,
573317, 573329, 573341, 573343, 573371, 573379, 573383, 573409, 573437, 573451, 573457, 573473, 573479,
573481, 573487, 573493, 573497, 573509, 573511, 573523, 573527, 573557, 573569, 573571, 573637, 573647,
573673, 573679, 573691, 573719, 573737, 573739, 573757, 573761, 573763, 573787, 573791, 573809, 573817,
573829, 573847, 573851, 573863, 573871, 573883, 573887, 573899, 573901, 573929, 573941, 573953, 573967,
573973, 573977, 574003, 574031, 574033, 574051, 574061, 574081, 574099, 574109, 574127, 574157, 574159,
574163, 574169, 574181, 574183, 574201, 574219, 574261, 574279, 574283, 574289, 574297, 574307, 574309,
574363, 574367, 574373, 574393, 574423, 574429, 574433, 574439, 574477, 574489, 574493, 574501, 574507,
574529, 574543, 574547, 574597, 574619, 574621, 574627, 574631, 574643, 574657, 574667, 574687, 574699,
574703, 574711, 574723, 574727, 574733, 574741, 574789, 574799, 574801, 574813, 574817, 574859, 574907,
574913, 574933, 574939, 574949, 574963, 574967, 574969, 575009, 575027, 575033, 575053, 575063, 575077,
575087, 575119, 575123, 575129, 575131, 575137, 575153, 575173, 575177, 575203, 575213, 575219, 575231,
575243, 575249, 575251, 575257, 575261, 575303, 575317, 575359, 575369, 575371, 575401, 575417, 575429,
575431, 575441, 575473, 575479, 575489, 575503, 575513, 575551, 575557, 575573, 575579, 575581, 575591,
575593, 575611, 575623, 575647, 575651, 575669, 575677, 575689, 575693, 575699, 575711, 575717, 575723,
575747, 575753, 575777, 575791, 575821, 575837, 575849, 575857, 575863, 575867, 575893, 575903, 575921,
575923, 575941, 575957, 575959, 575963, 575987, 576001, 576013, 576019, 576029, 576031, 576041, 576049,
576089, 576101, 576119, 576131, 576151, 576161, 576167, 576179, 576193, 576203, 576211, 576217, 576221,
576223, 576227, 576287, 576293, 576299, 576313, 576319, 576341, 576377, 576379, 576391, 576421, 576427,
576431, 576439, 576461, 576469, 576473, 576493, 576509, 576523, 576529, 576533, 576539, 576551, 576553,
576577, 576581, 576613, 576617, 576637, 576647, 576649, 576659, 576671, 576677, 576683, 576689, 576701,

576703, 576721, 576727, 576731, 576739, 576743, 576749, 576757, 576769, 576787, 576791, 576881, 576883,
576889, 576899, 576943, 576949, 576967, 576977, 577007, 577009, 577033, 577043, 577063, 577067, 577069,
577081, 577097, 577111, 577123, 577147, 577151, 577153, 577169, 577177, 577193, 577219, 577249, 577259,
577271, 577279, 577307, 577327, 577331, 577333, 577349, 577351, 577363, 577387, 577397, 577399, 577427,
577453, 577457, 577463, 577471, 577483, 577513, 577517, 577523, 577529, 577531, 577537, 577547, 577559,
577573, 577589, 577601, 577613, 577627, 577637, 577639, 577667, 577721, 577739, 577751, 577757, 577781,
577799, 577807, 577817, 577831, 577849, 577867, 577873, 577879, 577897, 577901, 577909, 577919, 577931,
577937, 577939, 577957, 577979, 577981, 578021, 578029, 578041, 578047, 578063, 578077, 578093, 578117,
578131, 578167, 578183, 578191, 578203, 578209, 578213, 578251, 578267, 578297, 578299, 578309, 578311,
578317, 578327, 578353, 578363, 578371, 578399, 578401, 578407, 578419, 578441, 578453, 578467, 578477,
578483, 578489, 578497, 578503, 578509, 578533, 578537, 578563, 578573, 578581, 578587, 578597, 578603,
578609, 578621, 578647, 578659, 578687, 578689, 578693, 578701, 578719, 578729, 578741, 578777, 578779,
578789, 578803, 578819, 578821, 578827, 578839, 578843, 578857, 578861, 578881, 578917, 578923, 578957,
578959, 578971, 578999, 579011, 579017, 579023, 579053, 579079, 579083, 579107, 579113, 579119, 579133,
579179, 579197, 579199, 579239, 579251, 579259, 579263, 579277, 579281, 579283, 579287, 579311, 579331,
579353, 579379, 579407, 579409, 579427, 579433, 579451, 579473, 579497, 579499, 579503, 579517, 579521,
579529, 579533, 579539, 579541, 579563, 579569, 579571, 579583, 579587, 579611, 579613, 579629, 579637,
579641, 579643, 579653, 579673, 579701, 579707, 579713, 579721, 579737, 579757, 579763, 579773, 579779,
579809, 579829, 579851, 579869, 579877, 579881, 579883, 579893, 579907, 579947, 579949, 579961, 579967,
579973, 579983, 580001, 580031, 580033, 580079, 580081, 580093, 580133, 580163, 580169, 580183, 580187,
580201, 580213, 580219, 580231, 580259, 580291, 580301, 580303, 580331, 580339, 580343, 580357, 580361,
580373, 580379, 580381, 580409, 580417, 580471, 580477, 580487, 580513, 580529, 580549, 580553, 580561,
580577, 580607, 580627, 580631, 580633, 580639, 580663, 580673, 580687, 580691, 580693, 580711, 580717,
580733, 580747, 580757, 580759, 580763, 580787, 580793, 580807, 580813, 580837, 580843, 580859, 580871,
580889, 580891, 580901, 580913, 580919, 580927, 580939, 580969, 580981, 580997, 581029, 581041, 581047,
581069, 581071, 581089, 581099, 581101, 581137, 581143, 581149, 581171, 581173, 581177, 581183, 581197,
581201, 581227, 581237, 581239, 581261, 581263, 581293, 581303, 581311, 581323, 581333, 581341, 581351,
581353, 581369, 581377, 581393, 581407, 581411, 581429, 581443, 581447, 581459, 581473, 581491, 581521,
581527, 581549, 581551, 581557, 581573, 581597, 581599, 581617, 581639, 581657, 581663, 581683, 581687,
581699, 581701, 581729, 581731, 581743, 581753, 581767, 581773, 581797, 581809, 581821, 581843, 581857,
581863, 581869, 581873, 581891, 581909, 581921, 581941, 581947, 581953, 581981, 581983, 582011, 582013,
582017, 582031, 582037, 582067, 582083, 582119, 582137, 582139, 582157, 582161, 582167, 582173, 582181,
582203, 582209, 582221, 582223, 582227, 582247, 582251, 582299, 582317, 582319, 582371, 582391, 582409,
582419, 582427, 582433, 582451, 582457, 582469, 582499, 582509, 582511, 582541, 582551, 582563, 582587,
582601, 582623, 582643, 582649, 582677, 582689, 582691, 582719, 582721, 582727, 582731, 582737, 582761,
582763, 582767, 582773, 582781, 582793, 582809, 582821, 582851, 582853, 582859, 582887, 582899, 582931,
582937, 582949, 582961, 582971, 582973, 582983, 583007, 583013, 583019, 583021, 583031, 583069, 583087,
583127, 583139, 583147, 583153, 583169, 583171, 583181, 583189, 583207, 583213, 583229, 583237, 583249,
583267, 583273, 583279, 583291, 583301, 583337, 583339, 583351, 583367, 583391, 583397, 583403, 583409,
583417, 583421, 583447, 583459, 583469, 583481, 583493, 583501, 583511, 583519, 583523, 583537, 583543,
583577, 583603, 583613, 583619, 583621, 583631, 583651, 583657, 583669, 583673, 583697, 583727, 583733,
583753, 583769, 583777, 583783, 583789, 583801, 583841, 583853, 583859, 583861, 583873, 583879, 583903,
583909, 583937, 583969, 583981, 583991, 583997, 584011, 584027, 584033, 584053, 584057, 584063, 584081,
584099, 584141, 584153, 584167, 584183, 584203, 584249, 584261, 584279, 584281, 584303, 584347, 584357,
584359, 584377, 584387, 584393, 584399, 584411, 584417, 584429, 584447, 584473, 584509, 584531,
584557, 584561, 584587, 584593, 584599, 584603, 584609, 584621, 584627, 584659, 584663, 584677, 584693,
584699, 584707, 584713, 584719, 584723, 584737, 584767, 584777, 584789, 584791, 584809, 584849, 584863,
584869, 584873, 584879, 584897, 584911, 584917, 584923, 584951, 584963, 584971, 584981, 584993, 584999,
585019, 585023, 585031, 585037, 585041, 585043, 585049, 585061, 585071, 585073, 585077, 585107, 585113,

585119, 585131, 585149, 585163, 585199, 585217, 585251, 585269, 585271, 585283, 585289, 585313, 585317,
585337, 585341, 585367, 585383, 585391, 585413, 585437, 585443, 585461, 585467, 585493, 585503, 585517,
585547, 585551, 585569, 585577, 585581, 585587, 585593, 585601, 585619, 585643, 585653, 585671, 585677,
585691, 585721, 585727, 585733, 585737, 585743, 585749, 585757, 585779, 585791, 585799, 585839, 585841,
585847, 585853, 585857, 585863, 585877, 585881, 585883, 585889, 585899, 585911, 585913, 585917, 585919,
585953, 585989, 585997, 586009, 586037, 586051, 586057, 586067, 586073, 586087, 586111, 586121, 586123,
586129, 586139, 586147, 586153, 586189, 586213, 586237, 586273, 586277, 586291, 586301, 586309, 586319,
586349, 586361, 586363, 586367, 586387, 586403, 586429, 586433, 586457, 586459, 586463, 586471, 586493,
586499, 586501, 586541, 586543, 586567, 586571, 586577, 586589, 586601, 586603, 586609, 586627, 586631,
586633, 586667, 586679, 586693, 586711, 586723, 586741, 586769, 586787, 586793, 586801, 586811, 586813,
586819, 586837, 586841, 586849, 586871, 586897, 586903, 586909, 586919, 586921, 586933, 586939, 586951,
586961, 586973, 586979, 586981, 587017, 587021, 587033, 587051, 587053, 587057, 587063, 587087, 587101,
587107, 587117, 587123, 587131, 587137, 587143, 587149, 587173, 587179, 587189, 587201, 587219, 587263,
587267, 587269, 587281, 587287, 587297, 587303, 587341, 587371, 587381, 587387, 587413, 587417, 587429,
587437, 587441, 587459, 587467, 587473, 587497, 587513, 587519, 587527, 587533, 587539, 587549, 587551,
587563, 587579, 587599, 587603, 587617, 587621, 587623, 587633, 587659, 587669, 587677, 587687, 587693,
587711, 587731, 587737, 587747, 587749, 587753, 587771, 587773, 587789, 587813, 587827, 587833, 587849,
587863, 587887, 587891, 587897, 587927, 587933, 587947, 587959, 587969, 587971, 587987, 587989, 587999,
588011, 588019, 588037, 588043, 588061, 588073, 588079, 588083, 588097, 588113, 588121, 588131, 588151,
588167, 588169, 588173, 588191, 588199, 588229, 588239, 588241, 588257, 588277, 588293, 588311, 588347,
588359, 588361, 588383, 588389, 588397, 588403, 588433, 588437, 588463, 588481, 588493, 588503, 588509,
588517, 588521, 588529, 588569, 588571, 588619, 588631, 588641, 588647, 588649, 588667, 588673, 588683,
588703, 588733, 588737, 588743, 588767, 588773, 588779, 588811, 588827, 588839, 588871, 588877, 588881,
588893, 588911, 588937, 588941, 588947, 588949, 588953, 588977, 589021, 589027, 589049, 589063, 589109,
589111, 589123, 589139, 589159, 589163, 589181, 589187, 589189, 589207, 589213, 589219, 589231, 589241,
589243, 589273, 589289, 589291, 589297, 589327, 589331, 589349, 589357, 589387, 589409, 589439, 589451,
589453, 589471, 589481, 589493, 589507, 589529, 589531, 589579, 589583, 589591, 589601, 589607, 589609,
589639, 589643, 589681, 589711, 589717, 589751, 589753, 589759, 589763, 589783, 589793, 589807, 589811,
589829, 589847, 589859, 589861, 589873, 589877, 589903, 589921, 589933, 589993, 589997, 590021, 590027,
590033, 590041, 590071, 590077, 590099, 590119, 590123, 590129, 590131, 590137, 590141, 590153, 590171,
590201, 590207, 590243, 590251, 590263, 590267, 590269, 590279, 590309, 590321, 590323, 590327, 590357,
590363, 590377, 590383, 590389, 590399, 590407, 590431, 590437, 590489, 590537, 590543, 590567, 590573,
590593, 590599, 590609, 590627, 590641, 590647, 590657, 590659, 590669, 590713, 590717, 590719, 590741,
590753, 590771, 590797, 590809, 590813, 590819, 590833, 590839, 590867, 590899, 590921, 590923, 590929,
590959, 590963, 590983, 590987, 591023, 591053, 591061, 591067, 591079, 591089, 591091, 591113, 591127,
591131, 591137, 591161, 591163, 591181, 591193, 591233, 591259, 591271, 591287, 591289, 591301, 591317,
591319, 591341, 591377, 591391, 591403, 591407, 591421, 591431, 591443, 591457, 591469, 591499, 591509,
591523, 591553, 591559, 591581, 591599, 591601, 591611, 591623, 591649, 591653, 591659, 591673, 591691,
591709, 591739, 591743, 591749, 591751, 591757, 591779, 591791, 591827, 591841, 591847, 591863, 591881,
591887, 591893, 591901, 591937, 591959, 591973, 592019, 592027, 592049, 592057, 592061, 592073, 592087,
592099, 592121, 592129, 592133, 592139, 592157, 592199, 592217, 592219, 592223, 592237, 592261, 592289,
592303, 592307, 592309, 592321, 592337, 592343, 592351, 592357, 592367, 592369, 592387, 592391, 592393,
592429, 592451, 592453, 592463, 592469, 592483, 592489, 592507, 592517, 592531, 592547, 592561, 592577,
592589, 592597, 592601, 592609, 592621, 592639, 592643, 592649, 592661, 592663, 592681, 592693, 592723,
592727, 592741, 592747, 592759, 592763, 592793, 592843, 592849, 592853, 592861, 592873, 592877, 592897,
592903, 592919, 592931, 592939, 592967, 592973, 592987, 592993, 593003, 593029, 593041, 593051, 593059,
593071, 593081, 593083, 593111, 593119, 593141, 593143, 593149, 593171, 593179, 593183, 593207, 593209,
593213, 593227, 593231, 593233, 593251, 593261, 593273, 593291, 593293, 593297, 593321, 593323, 593353,
593381, 593387, 593399, 593401, 593407, 593429, 593447, 593449, 593473, 593479, 593491, 593497, 593501,

593507, 593513, 593519, 593531, 593539, 593573, 593587, 593597, 593603, 593627, 593629, 593633, 593641,
593647, 593651, 593689, 593707, 593711, 593767, 593777, 593783, 593839, 593851, 593863, 593869, 593899,
593903, 593933, 593951, 593969, 593977, 593987, 593993, 594023, 594037, 594047, 594091, 594103, 594107,
594119, 594137, 594151, 594157, 594161, 594163, 594179, 594193, 594203, 594211, 594227, 594241, 594271,
594281, 594283, 594287, 594299, 594311, 594313, 594329, 594359, 594367, 594379, 594397, 594401, 594403,
594421, 594427, 594449, 594457, 594467, 594469, 594499, 594511, 594521, 594523, 594533, 594551, 594563,
594569, 594571, 594577, 594617, 594637, 594641, 594653, 594667, 594679, 594697, 594709, 594721, 594739,
594749, 594751, 594773, 594793, 594821, 594823, 594827, 594829, 594857, 594889, 594899, 594911, 594917,
594929, 594931, 594953, 594959, 594961, 594977, 594989, 595003, 595037, 595039, 595043, 595057, 595069,
595073, 595081, 595087, 595093, 595097, 595117, 595123, 595129, 595139, 595141, 595157, 595159, 595181,
595183, 595201, 595207, 595229, 595247, 595253, 595261, 595267, 595271, 595277, 595291, 595303, 595313,
595319, 595333, 595339, 595351, 595363, 595373, 595379, 595381, 595411, 595451, 595453, 595481, 595513,
595519, 595523, 595547, 595549, 595571, 595577, 595579, 595613, 595627, 595687, 595703, 595709, 595711,
595717, 595733, 595741, 595801, 595807, 595817, 595843, 595873, 595877, 595927, 595939, 595943, 595949,
595951, 595957, 595961, 595963, 595967, 595981, 596009, 596021, 596027, 596047, 596053, 596059, 596069,
596081, 596083, 596093, 596117, 596119, 596143, 596147, 596159, 596179, 596209, 596227, 596231, 596243,
596251, 596257, 596261, 596273, 596279, 596291, 596293, 596317, 596341, 596363, 596369, 596399, 596419,
596423, 596461, 596489, 596503, 596507, 596537, 596569, 596573, 596579, 596587, 596593, 596599, 596611,
596623, 596633, 596653, 596663, 596669, 596671, 596693, 596707, 596737, 596741, 596749, 596767, 596779,
596789, 596803, 596821, 596831, 596839, 596851, 596857, 596861, 596863, 596879, 596899, 596917, 596927,
596929, 596933, 596941, 596963, 596977, 596983, 596987, 597031, 597049, 597053, 597059, 597073, 597127,
597131, 597133, 597137, 597169, 597209, 597221, 597239, 597253, 597263, 597269, 597271, 597301, 597307,
597349, 597353, 597361, 597367, 597383, 597391, 597403, 597407, 597409, 597419, 597433, 597437, 597451,
597473, 597497, 597521, 597523, 597539, 597551, 597559, 597577, 597581, 597589, 597593, 597599, 597613,
597637, 597643, 597659, 597671, 597673, 597677, 597679, 597689, 597697, 597757, 597761, 597767, 597769,
597781, 597803, 597823, 597827, 597833, 597853, 597859, 597869, 597889, 597899, 597901, 597923, 597929,
597967, 597997, 598007, 598049, 598051, 598057, 598079, 598093, 598099, 598123, 598127, 598141, 598151,
598159, 598163, 598187, 598189, 598193, 598219, 598229, 598261, 598303, 598307, 598333, 598363, 598369,
598379, 598387, 598399, 598421, 598427, 598439, 598447, 598457, 598463, 598487, 598489, 598501, 598537,
598541, 598571, 598613, 598643, 598649, 598651, 598657, 598669, 598681, 598687, 598691, 598711, 598721,
598727, 598729, 598777, 598783, 598789, 598799, 598817, 598841, 598853, 598867, 598877, 598883, 598891,
598903, 598931, 598933, 598963, 598967, 598973, 598981, 598987, 598999, 599003, 599009, 599021, 599023,
599069, 599087, 599117, 599143, 599147, 599149, 599153, 599191, 599213, 599231, 599243, 599251, 599273,
599281, 599303, 599309, 599321, 599341, 599353, 599359, 599371, 599383, 599387, 599399, 599407, 599413,
599419, 599429, 599477, 599479, 599491, 599513, 599519, 599537, 599551, 599561, 599591, 599597, 599603,
599611, 599623, 599629, 599657, 599663, 599681, 599693, 599699, 599701, 599713, 599719, 599741, 599759,
599779, 599783, 599803, 599831, 599843, 599857, 599869, 599891, 599899, 599927, 599933, 599939, 599941,
599959, 599983, 599993, 599999, 600011, 600043, 600053, 600071, 600073, 600091, 600101, 600109, 600167,
600169, 600203, 600217, 600221, 600233, 600239, 600241, 600247, 600269, 600283, 600289, 600293, 600307,
600311, 600317, 600319, 600337, 600359, 600361, 600367, 600371, 600401, 600403, 600407, 600421, 600433,
600449, 600451, 600463, 600469, 600487, 600517, 600529, 600557, 600569, 600577, 600601, 600623, 600631,
600641, 600659, 600673, 600689, 600697, 600701, 600703, 600727, 600751, 600791, 600823, 600827, 600833,
600841, 600857, 600877, 600881, 600883, 600889, 600893, 600931, 600947, 600949, 600959, 600961, 600973,
600979, 600983, 601021, 601031, 601037, 601039, 601043, 601061, 601067, 601079, 601093, 601127, 601147,
601187, 601189, 601193, 601201, 601207, 601219, 601231, 601241, 601247, 601259, 601267, 601283, 601291,
601297, 601309, 601313, 601319, 601333, 601339, 601357, 601379, 601397, 601411, 601423, 601439, 601451,
601457, 601487, 601507, 601541, 601543, 601589, 601591, 601607, 601631, 601651, 601669, 601687, 601697,
601717, 601747, 601751, 601759, 601763, 601771, 601801, 601807, 601813, 601819, 601823, 601831, 601849,
601873, 601883, 601889, 601897, 601903, 601943, 601949, 601961, 601969, 601981, 602029, 602033, 602039,

602047, 602057, 602081, 602083, 602087, 602093, 602099, 602111, 602137, 602141, 602143, 602153, 602179,
602197, 602201, 602221, 602227, 602233, 602257, 602267, 602269, 602279, 602297, 602309, 602311, 602317,
602321, 602333, 602341, 602351, 602377, 602383, 602401, 602411, 602431, 602453, 602461, 602477, 602479,
602489, 602501, 602513, 602521, 602543, 602551, 602593, 602597, 602603, 602621, 602627, 602639, 602647,
602677, 602687, 602689, 602711, 602713, 602717, 602729, 602743, 602753, 602759, 602773, 602779, 602801,
602821, 602831, 602839, 602867, 602873, 602887, 602891, 602909, 602929, 602947, 602951, 602971, 602977,
602983, 602999, 603011, 603013, 603023, 603047, 603077, 603091, 603101, 603103, 603131, 603133, 603149,
603173, 603191, 603203, 603209, 603217, 603227, 603257, 603283, 603311, 603319, 603349, 603389, 603391,
603401, 603431, 603443, 603457, 603467, 603487, 603503, 603521, 603523, 603529, 603541, 603553, 603557,
603563, 603569, 603607, 603613, 603623, 603641, 603667, 603679, 603689, 603719, 603731, 603739, 603749,
603761, 603769, 603781, 603791, 603793, 603817, 603821, 603833, 603847, 603851, 603853, 603859, 603881,
603893, 603899, 603901, 603907, 603913, 603917, 603919, 603923, 603931, 603937, 603947, 603949, 603989,
604001, 604007, 604013, 604031, 604057, 604063, 604069, 604073, 604171, 604189, 604223, 604237, 604243,
604249, 604259, 604277, 604291, 604309, 604313, 604319, 604339, 604343, 604349, 604361, 604369, 604379,
604397, 604411, 604427, 604433, 604441, 604477, 604481, 604517, 604529, 604547, 604559, 604579, 604589,
604603, 604609, 604613, 604619, 604649, 604651, 604661, 604697, 604699, 604711, 604727, 604729, 604733,
604753, 604759, 604781, 604787, 604801, 604811, 604819, 604823, 604829, 604837, 604859, 604861, 604867,
604883, 604907, 604931, 604939, 604949, 604957, 604973, 604997, 605009, 605021, 605023, 605039, 605051,
605069, 605071, 605113, 605117, 605123, 605147, 605167, 605173, 605177, 605191, 605221, 605233, 605237,
605239, 605249, 605257, 605261, 605309, 605323, 605329, 605333, 605347, 605369, 605393, 605401, 605411,
605413, 605443, 605471, 605477, 605497, 605503, 605509, 605531, 605533, 605543, 605551, 605573, 605593,
605597, 605599, 605603, 605609, 605617, 605629, 605639, 605641, 605687, 605707, 605719, 605779, 605789,
605809, 605837, 605849, 605861, 605867, 605873, 605879, 605887, 605893, 605909, 605921, 605933, 605947,
605953, 605977, 605987, 605993, 606017, 606029, 606031, 606037, 606041, 606049, 606059, 606077, 606079,
606083, 606091, 606113, 606121, 606131, 606173, 606181, 606223, 606241, 606247, 606251, 606299, 606301,
606311, 606313, 606323, 606341, 606379, 606383, 606413, 606433, 606443, 606449, 606493, 606497, 606503,
606521, 606527, 606539, 606559, 606569, 606581, 606587, 606589, 606607, 606643, 606649, 606653, 606659,
606673, 606721, 606731, 606733, 606737, 606743, 606757, 606791, 606811, 606829, 606833, 606839, 606847,
606857, 606863, 606899, 606913, 606919, 606943, 606959, 606961, 606967, 606971, 606997, 607001, 607003,
607007, 607037, 607043, 607049, 607063, 607067, 607081, 607091, 607093, 607097, 607109, 607127, 607129,
607147, 607151, 607153, 607157, 607163, 607181, 607199, 607213, 607219, 607249, 607253, 607261, 607301,
607303, 607307, 607309, 607319, 607331, 607337, 607339, 607349, 607357, 607363, 607417, 607421, 607423,
607471, 607493, 607517, 607531, 607549, 607573, 607583, 607619, 607627, 607667, 607669, 607681, 607697,
607703, 607721, 607723, 607727, 607741, 607769, 607813, 607819, 607823, 607837, 607843, 607861, 607883,
607889, 607909, 607921, 607931, 607933, 607939, 607951, 607961, 607967, 607991, 607993, 608011, 608029,
608033, 608087, 608089, 608099, 608117, 608123, 608129, 608131, 608147, 608161, 608177, 608191, 608207,
608213, 608269, 608273, 608297, 608299, 608303, 608339, 608347, 608357, 608359, 608369, 608371, 608383,
608389, 608393, 608401, 608411, 608423, 608429, 608431, 608459, 608471, 608483, 608497, 608519, 608521,
608527, 608581, 608591, 608593, 608609, 608611, 608633, 608653, 608659, 608669, 608677, 608693, 608701,
608737, 608743, 608749, 608759, 608767, 608789, 608819, 608831, 608843, 608851, 608857, 608863, 608873,
608887, 608897, 608899, 608903, 608941, 608947, 608953, 608977, 608987, 608989, 608999, 609043, 609047,
609067, 609071, 609079, 609101, 609107, 609113, 609143, 609149, 609163, 609173, 609179, 609199, 609209,
609221, 609227, 609233, 609241, 609253, 609269, 609277, 609283, 609289, 609307, 609313, 609337, 609359,
609361, 609373, 609379, 609391, 609397, 609403, 609407, 609421, 609437, 609443, 609461, 609487, 609503,
609509, 609517, 609527, 609533, 609541, 609571, 609589, 609593, 609599, 609601, 609607, 609613, 609617,
609619, 609641, 609673, 609683, 609701, 609709, 609743, 609751, 609757, 609779, 609781, 609803, 609809,
609821, 609859, 609877, 609887, 609907, 609911, 609913, 609923, 609929, 609979, 609989, 609991, 609997,
610031, 610063, 610081, 610123, 610157, 610163, 610187, 610193, 610199, 610217, 610219, 610229, 610243,
610271, 610279, 610289, 610301, 610327, 610331, 610339, 610391, 610409, 610417, 610429, 610439, 610447,

610457, 610469, 610501, 610523, 610541, 610543, 610553, 610559, 610567, 610579, 610583, 610619, 610633,
610639, 610651, 610661, 610667, 610681, 610699, 610703, 610721, 610733, 610739, 610741, 610763, 610781,
610783, 610787, 610801, 610817, 610823, 610829, 610837, 610843, 610847, 610849, 610867, 610877, 610879,
610891, 610913, 610919, 610921, 610933, 610957, 610969, 610993, 611011, 611027, 611033, 611057, 611069,
611071, 611081, 611101, 611111, 611113, 611131, 611137, 611147, 611189, 611207, 611213, 611257, 611263,
611279, 611293, 611297, 611323, 611333, 611389, 611393, 611411, 611419, 611441, 611449, 611453, 611459,
611467, 611483, 611497, 611531, 611543, 611549, 611551, 611557, 611561, 611587, 611603, 611621, 611641,
611657, 611671, 611693, 611707, 611729, 611753, 611791, 611801, 611803, 611827, 611833, 611837, 611839,
611873, 611879, 611887, 611903, 611921, 611927, 611939, 611951, 611953, 611957, 611969, 611977, 611993,
611999, 612011, 612023, 612037, 612041, 612043, 612049, 612061, 612067, 612071, 612083, 612107, 612109,
612113, 612133, 612137, 612149, 612169, 612173, 612181, 612193, 612217, 612223, 612229, 612259, 612263,
612301, 612307, 612317, 612319, 612331, 612341, 612349, 612371, 612373, 612377, 612383, 612401, 612407,
612439, 612481, 612497, 612511, 612553, 612583, 612589, 612611, 612613, 612637, 612643, 612649, 612671,
612679, 612713, 612719, 612727, 612737, 612751, 612763, 612791, 612797, 612809, 612811, 612817, 612823,
612841, 612847, 612853, 612869, 612877, 612889, 612923, 612929, 612947, 612967, 612971, 612977, 613007,
613009, 613013, 613049, 613061, 613097, 613099, 613141, 613153, 613163, 613169, 613177, 613181, 613189,
613199, 613213, 613219, 613229, 613231, 613243, 613247, 613253, 613267, 613279, 613289, 613297, 613337,
613357, 613363, 613367, 613381, 613421, 613427, 613439, 613441, 613447, 613451, 613463, 613469, 613471,
613493, 613499, 613507, 613523, 613549, 613559, 613573, 613577, 613597, 613607, 613609, 613633, 613637,
613651, 613661, 613667, 613673, 613699, 613733, 613741, 613747, 613759, 613763, 613807, 613813, 613817,
613829, 613841, 613849, 613861, 613883, 613889, 613903, 613957, 613967, 613969, 613981, 613993, 613999,
614041, 614051, 614063, 614071, 614093, 614101, 614113, 614129, 614143, 614147, 614153, 614167, 614177,
614179, 614183, 614219, 614267, 614279, 614291, 614293, 614297, 614321, 614333, 614377, 614387, 614413,
614417, 614437, 614477, 614483, 614503, 614527, 614531, 614543, 614561, 614563, 614569, 614609, 614611,
614617, 614623, 614633, 614639, 614657, 614659, 614671, 614683, 614687, 614693, 614701, 614717, 614729,
614741, 614743, 614749, 614753, 614759, 614773, 614827, 614843, 614849, 614851, 614863, 614881, 614893,
614909, 614917, 614927, 614963, 614981, 614983, 615019, 615031, 615047, 615053, 615067, 615101, 615103,
615107, 615137, 615151, 615161, 615187, 615229, 615233, 615253, 615259, 615269, 615289, 615299, 615313,
615337, 615341, 615343, 615367, 615379, 615389, 615401, 615403, 615413, 615427, 615431, 615437, 615449,
615473, 615479, 615491, 615493, 615497, 615509, 615521, 615539, 615557, 615577, 615599, 615607, 615617,
615623, 615661, 615677, 615679, 615709, 615721, 615731, 615739, 615743, 615749, 615751, 615761, 615767,
615773, 615793, 615799, 615821, 615827, 615829, 615833, 615869, 615883, 615887, 615907, 615919, 615941,
615949, 615971, 615997, 616003, 616027, 616051, 616069, 616073, 616079, 616103, 616111, 616117, 616129,
616139, 616141, 616153, 616157, 616169, 616171, 616181, 616207, 616211, 616219, 616223, 616229, 616243,
616261, 616277, 616289, 616307, 616313, 616321, 616327, 616361, 616367, 616387, 616391, 616393, 616409,
616411, 616433, 616439, 616459, 616463, 616481, 616489, 616501, 616507, 616513, 616519, 616523, 616529,
616537, 616547, 616579, 616589, 616597, 616639, 616643, 616669, 616673, 616703, 616717, 616723, 616729,
616741, 616757, 616769, 616783, 616787, 616789, 616793, 616799, 616829, 616841, 616843, 616849, 616871,
616877, 616897, 616909, 616933, 616943, 616951, 616961, 616991, 616997, 616999, 617011, 617027, 617039,
617051, 617053, 617059, 617077, 617087, 617107, 617119, 617129, 617131, 617147, 617153, 617161, 617189,
617191, 617231, 617233, 617237, 617249, 617257, 617269, 617273, 617293, 617311, 617327, 617333, 617339,
617341, 617359, 617363, 617369, 617387, 617401, 617411, 617429, 617447, 617453, 617467, 617471, 617473,
617479, 617509, 617521, 617531, 617537, 617579, 617587, 617647, 617651, 617657, 617677, 617681, 617689,
617693, 617699, 617707, 617717, 617719, 617723, 617731, 617759, 617761, 617767, 617777, 617791, 617801,
617809, 617819, 617843, 617857, 617873, 617879, 617887, 617917, 617951, 617959, 617963, 617971, 617983,
618029, 618031, 618041, 618049, 618053, 618083, 618119, 618131, 618161, 618173, 618199, 618227, 618229,
618253, 618257, 618269, 618271, 618287, 618301, 618311, 618323, 618329, 618337, 618347, 618349, 618361,
618377, 618407, 618413, 618421, 618437, 618439, 618463, 618509, 618521, 618547, 618559, 618571, 618577,
618581, 618587, 618589, 618593, 618619, 618637, 618643, 618671, 618679, 618703, 618707, 618719, 618799,

618823, 618833, 618841, 618847, 618857, 618859, 618869, 618883, 618913, 618929, 618941, 618971, 618979,
618991, 618997, 619007, 619009, 619019, 619027, 619033, 619057, 619061, 619067, 619079, 619111, 619117,
619139, 619159, 619169, 619181, 619187, 619189, 619207, 619247, 619253, 619261, 619273, 619277, 619279,
619303, 619309, 619313, 619331, 619363, 619373, 619391, 619397, 619471, 619477, 619511, 619537, 619543,
619561, 619573, 619583, 619589, 619603, 619607, 619613, 619621, 619657, 619669, 619681, 619687, 619693,
619711, 619739, 619741, 619753, 619763, 619771, 619793, 619807, 619811, 619813, 619819, 619831, 619841,
619849, 619867, 619897, 619909, 619921, 619967, 619979, 619981, 619987, 619999, 620003, 620029, 620033,
620051, 620099, 620111, 620117, 620159, 620161, 620171, 620183, 620197, 620201, 620227, 620233, 620237,
620239, 620251, 620261, 620297, 620303, 620311, 620317, 620329, 620351, 620359, 620363, 620377, 620383,
620393, 620401, 620413, 620429, 620437, 620441, 620461, 620467, 620491, 620507, 620519, 620531, 620549,
620561, 620567, 620569, 620579, 620603, 620623, 620639, 620647, 620657, 620663, 620671, 620689, 620693,
620717, 620731, 620743, 620759, 620771, 620773, 620777, 620813, 620821, 620827, 620831, 620849, 620869,
620887, 620909, 620911, 620929, 620933, 620947, 620957, 620981, 620999, 621007, 621013, 621017, 621029,
621031, 621043, 621059, 621083, 621097, 621113, 621133, 621139, 621143, 621217, 621223, 621227, 621239,
621241, 621259, 621289, 621301, 621317, 621337, 621343, 621347, 621353, 621359, 621371, 621389, 621419,
621427, 621431, 621443, 621451, 621461, 621473, 621521, 621527, 621541, 621583, 621611, 621617, 621619,
621629, 621631, 621641, 621671, 621679, 621697, 621701, 621703, 621721, 621739, 621749, 621757, 621769,
621779, 621799, 621821, 621833, 621869, 621871, 621883, 621893, 621913, 621923, 621937, 621941, 621983,
621997, 622009, 622019, 622043, 622049, 622051, 622067, 622073, 622091, 622103, 622109, 622123, 622129,
622133, 622151, 622157, 622159, 622177, 622187, 622189, 622241, 622243, 622247, 622249, 622277, 622301,
622313, 622331, 622333, 622337, 622351, 622367, 622397, 622399, 622423, 622477, 622481, 622483, 622493,
622513, 622519, 622529, 622547, 622549, 622561, 622571, 622577, 622603, 622607, 622613, 622619, 622621,
622637, 622639, 622663, 622669, 622709, 622723, 622729, 622751, 622777, 622781, 622793, 622813, 622849,
622861, 622879, 622889, 622901, 622927, 622943, 622957, 622967, 622987, 622997, 623003, 623009, 623017,
623023, 623041, 623057, 623059, 623071, 623107, 623171, 623209, 623221, 623261, 623263, 623269, 623279,
623281, 623291, 623299, 623303, 623321, 623327, 623341, 623351, 623353, 623383, 623387, 623393, 623401,
623417, 623423, 623431, 623437, 623477, 623521, 623531, 623537, 623563, 623591, 623617, 623621, 623633,
623641, 623653, 623669, 623671, 623677, 623681, 623683, 623699, 623717, 623719, 623723, 623729, 623743,
623759, 623767, 623771, 623803, 623839, 623851, 623867, 623869, 623879, 623881, 623893, 623923, 623929,
623933, 623947, 623957, 623963, 623977, 623983, 623989, 624007, 624031, 624037, 624047, 624049, 624067,
624089, 624097, 624119, 624133, 624139, 624149, 624163, 624191, 624199, 624203, 624209, 624229, 624233,
624241, 624251, 624259, 624271, 624277, 624311, 624313, 624319, 624329, 624331, 624347, 624391, 624401,
624419, 624443, 624451, 624467, 624469, 624479, 624487, 624497, 624509, 624517, 624521, 624539, 624541,
624577, 624593, 624599, 624601, 624607, 624643, 624649, 624667, 624683, 624707, 624709, 624721, 624727,
624731, 624737, 624763, 624769, 624787, 624791, 624797, 624803, 624809, 624829, 624839, 624847, 624851,
624859, 624917, 624961, 624973, 624977, 624983, 624997, 625007, 625033, 625057, 625063, 625087, 625103,
625109, 625111, 625129, 625133, 625169, 625171, 625181, 625187, 625199, 625213, 625231, 625237, 625253,
625267, 625279, 625283, 625307, 625319, 625343, 625351, 625367, 625369, 625397, 625409, 625451, 625477,
625483, 625489, 625507, 625517, 625529, 625543, 625589, 625591, 625609, 625621, 625627, 625631, 625637,
625643, 625657, 625661, 625663, 625697, 625699, 625763, 625777, 625789, 625811, 625819, 625831, 625837,
625861, 625871, 625883, 625909, 625913, 625927, 625939, 625943, 625969, 625979, 625997, 626009, 626011,
626033, 626051, 626063, 626113, 626117, 626147, 626159, 626173, 626177, 626189, 626191, 626201, 626207,
626239, 626251, 626261, 626317, 626323, 626333, 626341, 626347, 626363, 626377, 626389, 626393, 626443,
626477, 626489, 626519, 626533, 626539, 626581, 626597, 626599, 626609, 626611, 626617, 626621, 626623,
626627, 626629, 626663, 626683, 626687, 626693, 626701, 626711, 626713, 626723, 626741, 626749, 626761,
626771, 626783, 626797, 626809, 626833, 626837, 626861, 626887, 626917, 626921, 626929, 626947, 626953,
626959, 626963, 626987, 627017, 627041, 627059, 627071, 627073, 627083, 627089, 627091, 627101, 627119,
627131, 627139, 627163, 627169, 627191, 627197, 627217, 627227, 627251, 627257, 627269, 627271, 627293,
627301, 627329, 627349, 627353, 627377, 627379, 627383, 627391, 627433, 627449, 627479, 627481, 627491,

627511, 627541, 627547, 627559, 627593, 627611, 627617, 627619, 627637, 627643, 627659, 627661, 627667,
627673, 627709, 627721, 627733, 627749, 627773, 627787, 627791, 627797, 627799, 627811, 627841, 627859,
627901, 627911, 627919, 627943, 627947, 627953, 627961, 627973, 628013, 628021, 628037, 628049, 628051,
628057, 628063, 628093, 628097, 628127, 628139, 628171, 628183, 628189, 628193, 628207, 628213, 628217,
628219, 628231, 628261, 628267, 628289, 628301, 628319, 628357, 628363, 628373, 628379, 628391, 628399,
628423, 628427, 628447, 628477, 628487, 628493, 628499, 628547, 628561, 628583, 628591, 628651, 628673,
628679, 628681, 628687, 628699, 628709, 628721, 628753, 628757, 628759, 628781, 628783, 628787, 628799,
628801, 628811, 628819, 628841, 628861, 628877, 628909, 628913, 628921, 628937, 628939, 628973, 628993,
628997, 629003, 629009, 629011, 629023, 629029, 629059, 629081, 629113, 629137, 629143, 629171, 629177,
629203, 629243, 629249, 629263, 629281, 629311, 629339, 629341, 629351, 629371, 629381, 629383, 629401,
629411, 629417, 629429, 629449, 629467, 629483, 629491, 629509, 629513, 629537, 629567, 629569, 629591,
629593, 629609, 629611, 629617, 629623, 629653, 629683, 629687, 629689, 629701, 629711, 629723, 629737,
629743, 629747, 629767, 629773, 629779, 629803, 629807, 629819, 629843, 629857, 629861, 629873, 629891,
629897, 629899, 629903, 629921, 629927, 629929, 629939, 629963, 629977, 629987, 629989, 630017, 630023,
630029, 630043, 630067, 630101, 630107, 630127, 630151, 630163, 630167, 630169, 630181, 630193, 630197,
630229, 630247, 630263, 630281, 630299, 630307, 630319, 630349, 630353, 630391, 630433, 630451, 630467,
630473, 630481, 630493, 630521, 630523, 630529, 630559, 630577, 630583, 630587, 630589, 630593, 630607,
630613, 630659, 630677, 630689, 630701, 630709, 630713, 630719, 630733, 630737, 630797, 630803, 630823,
630827, 630841, 630863, 630871, 630893, 630899, 630901, 630907, 630911, 630919, 630941, 630967, 630997,
631003, 631013, 631039, 631061, 631121, 631133, 631139, 631151, 631153, 631157, 631171, 631181, 631187,
631223, 631229, 631247, 631249, 631259, 631271, 631273, 631291, 631307, 631339, 631357, 631361, 631387,
631391, 631399, 631409, 631429, 631453, 631457, 631459, 631469, 631471, 631483, 631487, 631507, 631513,
631529, 631531, 631537, 631549, 631559, 631573, 631577, 631583, 631597, 631613, 631619, 631643, 631667,
631679, 631681, 631711, 631717, 631723, 631733, 631739, 631751, 631753, 631789, 631817, 631819, 631843,
631847, 631853, 631859, 631861, 631867, 631889, 631901, 631903, 631913, 631927, 631931, 631937, 631979,
631987, 631991, 631993, 632029, 632041, 632053, 632083, 632087, 632089, 632101, 632117, 632123,
632141, 632147, 632153, 632189, 632209, 632221, 632227, 632231, 632251, 632257, 632267, 632273, 632297,
632299, 632321, 632323, 632327, 632329, 632347, 632351, 632353, 632363, 632371, 632381, 632389, 632393,
632447, 632459, 632473, 632483, 632497, 632501, 632503, 632521, 632557, 632561, 632591, 632609, 632623,
632627, 632629, 632647, 632669, 632677, 632683, 632699, 632713, 632717, 632743, 632747, 632773, 632777,
632813, 632839, 632843, 632851, 632857, 632881, 632897, 632911, 632923, 632939, 632941, 632971, 632977,
632987, 632993, 633001, 633013, 633037, 633053, 633067, 633079, 633091, 633133, 633151, 633161, 633187,
633197, 633209, 633221, 633253, 633257, 633263, 633271, 633287, 633307, 633317, 633337, 633359, 633377,
633379, 633383, 633401, 633407, 633427, 633449, 633461, 633463, 633467, 633469, 633473, 633487, 633497,
633559, 633569, 633571, 633583, 633599, 633613, 633623, 633629, 633649, 633653, 633667, 633739, 633751,
633757, 633767, 633781, 633791, 633793, 633797, 633799, 633803, 633823, 633833, 633877, 633883, 633923,
633931, 633937, 633943, 633953, 633961, 633967, 633991, 634003, 634013, 634031, 634061, 634079, 634091,
634097, 634103, 634141, 634157, 634159, 634169, 634177, 634181, 634187, 634199, 634211, 634223, 634237,
634241, 634247, 634261, 634267, 634273, 634279, 634301, 634307, 634313, 634327, 634331, 634343, 634367,
634373, 634397, 634421, 634441, 634471, 634483, 634493, 634499, 634511, 634519, 634523, 634531, 634541,
634567, 634573, 634577, 634597, 634603, 634609, 634643, 634649, 634651, 634679, 634681, 634687, 634703,
634709, 634717, 634727, 634741, 634747, 634757, 634759, 634793, 634807, 634817, 634841, 634853, 634859,
634861, 634871, 634891, 634901, 634903, 634927, 634937, 634939, 634943, 634969, 634979, 635003, 635021,
635039, 635051, 635057, 635087, 635119, 635147, 635149, 635197, 635203, 635207, 635249, 635251, 635263,
635267, 635279, 635287, 635291, 635293, 635309, 635317, 635333, 635339, 635347, 635351, 635353, 635359,
635363, 635387, 635389, 635413, 635423, 635431, 635441, 635449, 635461, 635471, 635483, 635507, 635519,
635527, 635533, 635563, 635567, 635599, 635603, 635617, 635639, 635653, 635659, 635689, 635707, 635711,
635729, 635731, 635737, 635777, 635801, 635809, 635813, 635821, 635837, 635849, 635867, 635879, 635891,
635893, 635909, 635917, 635923, 635939, 635959, 635969, 635977, 635981, 635983, 635989, 636017, 636023,

636043, 636059, 636061, 636071, 636073, 636107, 636109, 636133, 636137, 636149, 636193, 636211, 636217,
636241, 636247, 636257, 636263, 636277, 636283, 636287, 636301, 636313, 636319, 636331, 636343, 636353,
636359, 636403, 636407, 636409, 636421, 636469, 636473, 636499, 636533, 636539, 636541, 636547, 636553,
636563, 636569, 636613, 636619, 636631, 636653, 636673, 636697, 636719, 636721, 636731, 636739, 636749,
636761, 636763, 636773, 636781, 636809, 636817, 636821, 636829, 636851, 636863, 636877, 636917, 636919,
636931, 636947, 636953, 636967, 636983, 636997, 637001, 637003, 637067, 637073, 637079, 637097, 637129,
637139, 637157, 637163, 637171, 637199, 637201, 637229, 637243, 637271, 637277, 637283, 637291, 637297,
637309, 637319, 637321, 637327, 637337, 637339, 637349, 637369, 637379, 637409, 637421, 637423, 637447,
637459, 637463, 637471, 637489, 637499, 637513, 637519, 637529, 637531, 637543, 637573, 637597, 637601,
637603, 637607, 637627, 637657, 637669, 637691, 637699, 637709, 637711, 637717, 637723, 637727, 637729,
637751, 637771, 637781, 637783, 637787, 637817, 637829, 637831, 637841, 637873, 637883, 637909, 637933,
637937, 637939, 638023, 638047, 638051, 638059, 638063, 638081, 638117, 638123, 638147, 638159, 638161,
638171, 638177, 638179, 638201, 638233, 638263, 638269, 638303, 638317, 638327, 638347, 638359, 638371,
638423, 638431, 638437, 638453, 638459, 638467, 638489, 638501, 638527, 638567, 638581, 638587, 638621,
638629, 638633, 638663, 638669, 638689, 638699, 638717, 638719, 638767, 638801, 638819, 638839, 638857,
638861, 638893, 638923, 638933, 638959, 638971, 638977, 638993, 638999, 639007, 639011, 639043, 639049,
639053, 639083, 639091, 639137, 639143, 639151, 639157, 639167, 639169, 639181, 639211, 639253, 639257,
639259, 639263, 639269, 639299, 639307, 639311, 639329, 639337, 639361, 639371, 639391, 639433, 639439,
639451, 639487, 639491, 639493, 639511, 639517, 639533, 639547, 639563, 639571, 639577, 639589, 639599,
639601, 639631, 639637, 639647, 639671, 639677, 639679, 639689, 639697, 639701, 639703, 639713, 639719,
639731, 639739, 639757, 639833, 639839, 639851, 639853, 639857, 639907, 639911, 639937, 639941, 639949,
639959, 639983, 639997, 640007, 640009, 640019, 640027, 640039, 640043, 640049, 640061, 640069, 640099,
640109, 640121, 640127, 640139, 640151, 640153, 640163, 640193, 640219, 640223, 640229, 640231, 640247,
640249, 640259, 640261, 640267, 640279, 640303, 640307, 640333, 640363, 640369, 640411, 640421, 640457,
640463, 640477, 640483, 640499, 640529, 640531, 640579, 640583, 640589, 640613, 640621, 640631, 640649,
640663, 640667, 640669, 640687, 640691, 640727, 640733, 640741, 640771, 640777, 640793, 640837, 640847,
640853, 640859, 640873, 640891, 640901, 640907, 640919, 640933, 640943, 640949, 640957, 640963, 640967,
640973, 640993, 641051, 641057, 641077, 641083, 641089, 641093, 641101, 641129, 641131, 641143, 641167,
641197, 641203, 641213, 641227, 641239, 641261, 641279, 641287, 641299, 641317, 641327, 641371, 641387,
641411, 641413, 641419, 641437, 641441, 641453, 641467, 641471, 641479, 641491, 641513, 641519, 641521,
641549, 641551, 641579, 641581, 641623, 641633, 641639, 641681, 641701, 641713, 641747, 641749, 641761,
641789, 641791, 641803, 641813, 641819, 641821, 641827, 641833, 641843, 641863, 641867, 641873, 641881,
641891, 641897, 641909, 641923, 641929, 641959, 641969, 641981, 642011, 642013, 642049, 642071, 642077,
642079, 642113, 642121, 642133, 642149, 642151, 642157, 642163, 642197, 642199, 642211, 642217, 642223,
642233, 642241, 642247, 642253, 642281, 642359, 642361, 642373, 642403, 642407, 642419, 642427, 642457,
642487, 642517, 642527, 642529, 642533, 642547, 642557, 642563, 642581, 642613, 642623, 642673, 642683,
642701, 642737, 642739, 642769, 642779, 642791, 642797, 642799, 642809, 642833, 642853, 642869, 642871,
642877, 642881, 642899, 642907, 642931, 642937, 642947, 642953, 642973, 642977, 642997, 643009, 643021,
643031, 643039, 643043, 643051, 643061, 643073, 643081, 643087, 643099, 643121, 643129, 643183, 643187,
643199, 643213, 643217, 643231, 643243, 643273, 643301, 643303, 643369, 643373, 643403, 643421, 643429,
643439, 643453, 643457, 643463, 643469, 643493, 643507, 643523, 643547, 643553, 643567, 643583, 643589,
643619, 643633, 643639, 643649, 643651, 643661, 643681, 643691, 643693, 643697, 643703, 643723, 643729,
643751, 643781, 643847, 643849, 643859, 643873, 643879, 643883, 643889, 643919, 643927, 643949, 643957,
643961, 643969, 643991, 644009, 644029, 644047, 644051, 644053, 644057, 644089, 644101, 644107, 644117,
644123, 644129, 644131, 644141, 644143, 644153, 644159, 644173, 644191, 644197, 644201, 644227, 644239,
644257, 644261, 644291, 644297, 644327, 644341, 644353, 644359, 644363, 644377, 644381, 644383, 644401,
644411, 644431, 644443, 644447, 644489, 644491, 644507, 644513, 644519, 644531, 644549, 644557, 644563,
644569, 644593, 644597, 644599, 644617, 644629, 644647, 644653, 644669, 644671, 644687, 644701, 644717,
644729, 644731, 644747, 644753, 644767, 644783, 644789, 644797, 644801, 644837, 644843, 644857, 644863,

644867, 644869, 644881, 644899, 644909, 644911, 644923, 644933, 644951, 644977, 644999, 645011, 645013,
645019, 645023, 645037, 645041, 645049, 645067, 645077, 645083, 645091, 645097, 645131, 645137, 645149,
645179, 645187, 645233, 645257, 645313, 645329, 645347, 645353, 645367, 645383, 645397, 645409, 645419,
645431, 645433, 645443, 645467, 645481, 645493, 645497, 645499, 645503, 645521, 645527, 645529, 645571,
645577, 645581, 645583, 645599, 645611, 645629, 645641, 645647, 645649, 645661, 645683, 645691, 645703,
645713, 645727, 645737, 645739, 645751, 645763, 645787, 645803, 645833, 645839, 645851, 645857, 645877,
645889, 645893, 645901, 645907, 645937, 645941, 645973, 645979, 646003, 646013, 646027, 646039, 646067,
646073, 646099, 646103, 646147, 646157, 646159, 646169, 646181, 646183, 646189, 646193, 646199, 646237,
646253, 646259, 646267, 646271, 646273, 646291, 646301, 646307, 646309, 646339, 646379, 646397, 646403,
646411, 646421, 646423, 646433, 646453, 646519, 646523, 646537, 646543, 646549, 646571, 646573, 646577,
646609, 646619, 646631, 646637, 646643, 646669, 646687, 646721, 646757, 646771, 646781, 646823, 646831,
646837, 646843, 646859, 646873, 646879, 646883, 646889, 646897, 646909, 646913, 646927, 646937, 646957,
646979, 646981, 646991, 646993, 647011, 647033, 647039, 647047, 647057, 647069, 647081, 647099, 647111,
647113, 647117, 647131, 647147, 647161, 647189, 647201, 647209, 647219, 647261, 647263, 647293, 647303,
647321, 647327, 647333, 647341, 647357, 647359, 647363, 647371, 647399, 647401, 647417, 647429, 647441,
647453, 647477, 647489, 647503, 647509, 647527, 647531, 647551, 647557, 647579, 647587, 647593, 647609,
647617, 647627, 647641, 647651, 647659, 647663, 647687, 647693, 647719, 647723, 647741, 647743, 647747,
647753, 647771, 647783, 647789, 647809, 647821, 647837, 647839, 647851, 647861, 647891, 647893, 647909,
647917, 647951, 647953, 647963, 647987, 648007, 648019, 648029, 648041, 648047, 648059, 648061, 648073,
648079, 648097, 648101, 648107, 648119, 648133, 648173, 648181, 648191, 648199, 648211, 648217, 648229,
648239, 648257, 648259, 648269, 648283, 648289, 648293, 648317, 648331, 648341, 648343, 648371, 648377,
648379, 648383, 648391, 648433, 648437, 648449, 648481, 648509, 648563, 648607, 648617, 648619, 648629,
648631, 648649, 648653, 648671, 648677, 648689, 648709, 648719, 648731, 648763, 648779, 648803, 648841,
648859, 648863, 648871, 648887, 648889, 648911, 648917, 648931, 648937, 648953, 648961, 648971, 648997,
649001, 649007, 649039, 649063, 649069, 649073, 649079, 649081, 649087, 649093, 649123, 649141, 649147,
649151, 649157, 649183, 649217, 649261, 649273, 649277, 649279, 649283, 649291, 649307, 649321, 649361,
649379, 649381, 649403, 649421, 649423, 649427, 649457, 649469, 649471, 649483, 649487, 649499, 649501,
649507, 649511, 649529, 649541, 649559, 649567, 649573, 649577, 649613, 649619, 649631, 649633, 649639,
649643, 649651, 649657, 649661, 649697, 649709, 649717, 649739, 649751, 649769, 649771, 649777, 649783,
649787, 649793, 649799, 649801, 649813, 649829, 649843, 649849, 649867, 649871, 649877, 649879, 649897,
649907, 649921, 649937, 649969, 649981, 649991, 650011, 650017, 650059, 650071, 650081, 650099, 650107,
650179, 650183, 650189, 650213, 650227, 650261, 650269, 650281, 650291, 650317, 650327, 650329, 650347,
650359, 650387, 650401, 650413, 650449, 650477, 650479, 650483, 650519, 650537, 650543, 650549, 650563,
650567, 650581, 650591, 650599, 650609, 650623, 650627, 650669, 650701, 650759, 650761, 650779, 650813,
650821, 650827, 650833, 650851, 650861, 650863, 650869, 650873, 650911, 650917, 650927, 650933, 650953,
650971, 650987, 651017, 651019, 651029, 651043, 651067, 651071, 651097, 651103, 651109, 651127, 651139,
651143, 651169, 651179, 651181, 651191, 651193, 651221, 651223, 651239, 651247, 651251, 651257, 651271,
651281, 651289, 651293, 651323, 651331, 651347, 651361, 651397, 651401, 651437, 651439, 651461, 651473,
651481, 651487, 651503, 651509, 651517, 651587, 651617, 651641, 651647, 651649, 651667, 651683, 651689,
651697, 651727, 651731, 651733, 651767, 651769, 651793, 651803, 651809, 651811, 651821, 651839, 651841,
651853, 651857, 651863, 651869, 651877, 651881, 651901, 651913, 651943, 651971, 651997, 652019, 652033,
652039, 652063, 652079, 652081, 652087, 652117, 652121, 652153, 652189, 652207, 652217, 652229, 652237,
652241, 652243, 652261, 652279, 652283, 652291, 652319, 652321, 652331, 652339, 652343, 652357, 652361,
652369, 652373, 652381, 652411, 652417, 652429, 652447, 652451, 652453, 652493, 652499, 652507, 652541,
652543, 652549, 652559, 652567, 652573, 652577, 652591, 652601, 652607, 652609, 652621, 652627, 652651,
652657, 652667, 652699, 652723, 652727, 652733, 652739, 652741, 652747, 652753, 652759, 652787, 652811,
652831, 652837, 652849, 652853, 652871, 652903, 652909, 652913, 652921, 652931, 652933, 652937, 652943,
652957, 652969, 652991, 652997, 652999, 653033, 653057, 653083, 653111, 653113, 653117, 653143, 653153,
653197, 653203, 653207, 653209, 653243, 653267, 653273, 653281, 653311, 653321, 653339, 653357, 653363,

653431, 653461, 653473, 653491, 653501, 653503, 653507, 653519, 653537, 653539, 653561, 653563, 653579, 653593, 653617, 653621, 653623, 653641, 653647, 653651, 653659, 653687, 653693, 653707, 653711, 653713, 653743, 653749, 653761, 653777, 653789, 653797, 653801, 653819, 653831, 653879, 653881, 653893, 653899, 653903, 653927, 653929, 653941, 653951, 653963, 653969, 653977, 653993, 654001, 654011, 654019, 654023, 654029, 654047, 654053, 654067, 654089, 654107, 654127, 654149, 654161, 654163, 654167, 654169, 654187, 654191, 654209, 654221, 654223, 654229, 654233, 654257, 654293, 654301, 654307, 654323, 654343, 654349, 654371, 654397, 654413, 654421, 654427, 654439, 654491, 654499, 654509, 654527, 654529, 654539, 654541, 654553, 654571, 654587, 654593, 654601, 654611, 654613, 654623, 654629, 654671, 654679, 654697, 654701, 654727, 654739, 654743, 654749, 654767, 654779, 654781, 654799, 654803, 654817, 654821, 654827, 654839, 654853, 654877, 654889, 654917, 654923, 654931, 654943, 654967, 654991, 655001, 655003, 655013, 655021, 655033, 655037, 655043, 655069, 655087, 655103, 655111, 655121, 655157, 655181, 655211, 655219, 655223, 655229, 655241, 655243, 655261, 655267, 655273, 655283, 655289, 655301, 655331, 655337, 655351, 655357, 655373, 655379, 655387, 655399, 655439, 655453, 655471, 655489, 655507, 655511, 655517, 655531, 655541, 655547, 655559, 655561, 655579, 655583, 655597, 655601, 655637, 655643, 655649, 655651, 655657, 655687, 655693, 655717, 655723, 655727, 655757, 655807, 655847, 655849, 655859, 655883, 655901, 655909, 655913, 655927, 655943, 655961, 655987, 656023, 656039, 656063, 656077, 656113, 656119, 656129, 656141, 656147, 656153, 656171, 656221, 656237, 656263, 656267, 656273, 656291, 656297, 656303, 656311, 656321, 656323, 656329, 656333, 656347, 656371, 656377, 656389, 656407, 656423, 656429, 656459, 656471, 656479, 656483, 656519, 656527, 656561, 656587, 656597, 656599, 656603, 656609, 656651, 656657, 656671, 656681, 656683, 656687, 656701, 656707, 656737, 656741, 656749, 656753, 656771, 656783, 656791, 656809, 656819, 656833, 656839, 656891, 656917, 656923, 656939, 656951, 656959, 656977, 656989, 656993, 657017, 657029, 657047, 657049, 657061, 657071, 657079, 657089, 657091, 657113, 657121, 657127, 657131, 657187, 657193, 657197, 657233, 657257, 657269, 657281, 657289, 657299, 657311, 657313, 657323, 657347, 657361, 657383, 657403, 657413, 657431, 657439, 657451, 657469, 657473, 657491, 657493, 657497, 657499, 657523, 657529, 657539, 657557, 657581, 657583, 657589, 657607, 657617, 657649, 657653, 657659, 657661, 657703, 657707, 657719, 657743, 657779, 657793, 657809, 657827, 657841, 657863, 657893, 657911, 657929, 657931, 657947, 657959, 657973, 657983, 658001, 658043, 658051, 658057, 658069, 658079, 658111, 658117, 658123, 658127, 658139, 658153, 658159, 658169, 658187, 658199, 658211, 658219, 658247, 658253, 658261, 658277, 658279, 658303, 658309, 658319, 658321, 658327, 658349, 658351, 658367, 658379, 658391, 658403, 658417, 658433, 658447, 658453, 658477, 658487, 658507, 658547, 658549, 658573, 658579, 658589, 658591, 658601, 658607, 658613, 658633, 658639, 658643, 658649, 658663, 658681, 658703, 658751, 658753, 658783, 658807, 658817, 658831, 658837, 658841, 658871, 658873, 658883, 658897, 658907, 658913, 658919, 658943, 658961, 658963, 658969, 658979, 658991, 658997, 659011, 659023, 659047, 659059, 659063, 659069, 659077, 659101, 659137, 659159, 659171, 659173, 659177, 659189, 659221, 659231, 659237, 659251, 659279, 659299, 659317, 659327, 659333, 659353, 659371, 659419, 659423, 659437, 659453, 659467, 659473, 659497, 659501, 659513, 659521, 659531, 659539, 659563, 659569, 659591, 659597, 659609, 659611, 659621, 659629, 659639, 659653, 659657, 659669, 659671, 659689, 659693, 659713, 659723, 659741, 659759, 659761, 659783, 659819, 659831, 659843, 659849, 659863, 659873, 659881, 659899, 659917, 659941, 659947, 659951, 659963, 659983, 659999, 660001, 660013, 660029, 660047, 660053, 660061, 660067, 660071, 660073, 660097, 660103, 660119, 660131, 660137, 660157, 660167, 660181, 660197, 660199, 660217, 660227, 660241, 660251, 660271, 660277, 660281, 660299, 660329, 660337, 660347, 660349, 660367, 660377, 660379, 660391, 660403, 660409, 660449, 660493, 660503, 660509, 660521, 660529, 660547, 660557, 660559, 660563, 660589, 660593, 660599, 660601, 660607, 660617, 660619, 660643, 660659, 660661, 660683, 660719, 660727, 660731, 660733, 660757, 660769, 660787, 660791, 660799, 660809, 660811, 660817, 660833, 660851, 660853, 660887, 660893, 660899, 660901, 660917, 660923, 660941, 660949, 660973, 660983, 661009, 661019, 661027, 661049, 661061, 661091, 661093, 661097, 661099, 661103, 661109, 661117, 661121, 661139, 661183, 661187, 661189, 661201, 661217, 661231, 661237, 661253, 661259, 661267, 661321, 661327, 661343, 661361, 661373, 661393, 661417, 661421, 661439, 661459, 661477, 661481, 661483, 661513, 661517, 661541, 661547, 661553, 661603, 661607, 661613, 661621, 661663, 661673, 661679, 661697, 661721, 661741, 661769, 661777, 661823, 661849, 661873, 661877, 661879, 661883, 661889, 661897,

661909, 661931, 661939, 661949, 661951, 661961, 661987, 661993, 662003, 662021, 662029, 662047, 662059,
662063, 662083, 662107, 662111, 662141, 662143, 662149, 662177, 662203, 662227, 662231, 662251, 662261,
662267, 662281, 662287, 662309, 662323, 662327, 662339, 662351, 662353, 662357, 662369, 662401, 662407,
662443, 662449, 662477, 662483, 662491, 662513, 662527, 662531, 662537, 662539, 662551, 662567, 662591,
662617, 662639, 662647, 662657, 662671, 662681, 662689, 662693, 662713, 662719, 662743, 662771, 662773,
662789, 662797, 662819, 662833, 662839, 662843, 662867, 662897, 662899, 662917, 662939, 662941, 662947,
662951, 662953, 662957, 662999, 663001, 663007, 663031, 663037, 663049, 663053, 663071, 663097, 663127,
663149, 663161, 663163, 663167, 663191, 663203, 663209, 663239, 663241, 663263, 663269, 663281, 663283,
663301, 663319, 663331, 663349, 663359, 663371, 663407, 663409, 663437, 663463, 663517, 663529, 663539,
663541, 663547, 663557, 663563, 663569, 663571, 663581, 663583, 663587, 663589, 663599, 663601, 663631,
663653, 663659, 663661, 663683, 663709, 663713, 663737, 663763, 663787, 663797, 663821, 663823, 663827,
663853, 663857, 663869, 663881, 663893, 663907, 663937, 663959, 663961, 663967, 663973, 663977, 663979,
663983, 663991, 663997, 664009, 664019, 664043, 664061, 664067, 664091, 664099, 664109, 664117, 664121,
664123, 664133, 664141, 664151, 664177, 664193, 664199, 664211, 664243, 664253, 664271, 664273, 664289,
664319, 664331, 664357, 664369, 664379, 664381, 664403, 664421, 664427, 664441, 664459, 664471, 664507,
664511, 664529, 664537, 664549, 664561, 664571, 664579, 664583, 664589, 664597, 664603, 664613, 664619,
664621, 664633, 664661, 664663, 664667, 664669, 664679, 664687, 664691, 664693, 664711, 664739, 664757,
664771, 664777, 664789, 664793, 664799, 664843, 664847, 664849, 664879, 664891, 664933, 664949, 664967,
664973, 664997, 665011, 665017, 665029, 665039, 665047, 665051, 665053, 665069, 665089, 665111, 665113,
665117, 665123, 665131, 665141, 665153, 665177, 665179, 665201, 665207, 665213, 665221, 665233, 665239,
665251, 665267, 665279, 665293, 665299, 665303, 665311, 665351, 665359, 665369, 665381, 665387, 665419,
665429, 665447, 665479, 665501, 665503, 665507, 665527, 665549, 665557, 665563, 665569, 665573, 665591,
665603, 665617, 665629, 665633, 665659, 665677, 665713, 665719, 665723, 665747, 665761, 665773, 665783,
665789, 665801, 665803, 665813, 665843, 665857, 665897, 665921, 665923, 665947, 665953, 665981, 665983,
665993, 666016, 666019, 666023, 666031, 666041, 666067, 666073, 666079, 666089, 666091, 666109, 666119,
666139, 666143, 666167, 666173, 666187, 666191, 666203, 666229, 666269, 666277, 666301, 666329,
666353, 666403, 666427, 666431, 666433, 666437, 666439, 666461, 666467, 666493, 666511, 666527, 666529,
666541, 666557, 666559, 666599, 666607, 666637, 666643, 666647, 666649, 666667, 666671, 666683, 666697,
666707, 666727, 666733, 666737, 666749, 666751, 666769, 666773, 666811, 666821, 666823, 666829, 666857,
666871, 666889, 666901, 666929, 666937, 666959, 666979, 666983, 666989, 667013, 667019, 667021, 667081,
667091, 667103, 667123, 667127, 667129, 667141, 667171, 667181, 667211, 667229, 667241, 667243, 667273,
667283, 667309, 667321, 667333, 667351, 667361, 667363, 667367, 667379, 667417, 667421, 667423, 667427,
667441, 667463, 667477, 667487, 667501, 667507, 667519, 667531, 667547, 667549, 667553, 667559, 667561,
667577, 667631, 667643, 667649, 667657, 667673, 667687, 667691, 667697, 667699, 667727, 667741, 667753,
667769, 667781, 667801, 667817, 667819, 667829, 667837, 667859, 667861, 667867, 667883, 667903, 667921,
667949, 667963, 667987, 667991, 667999, 668009, 668029, 668033, 668047, 668051, 668069, 668089, 668093,
668111, 668141, 668153, 668159, 668179, 668201, 668203, 668209, 668221, 668243, 668273, 668303, 668347,
668407, 668417, 668471, 668509, 668513, 668527, 668531, 668533, 668539, 668543, 668567, 668579, 668581,
668599, 668609, 668611, 668617, 668623, 668671, 668677, 668687, 668699, 668713, 668719, 668737, 668741,
668747, 668761, 668791, 668803, 668813, 668821, 668851, 668867, 668869, 668873, 668879, 668903, 668929,
668939, 668947, 668959, 668963, 668989, 668999, 669023, 669029, 669049, 669077, 669089, 669091, 669107,
669113, 669121, 669127, 669133, 669167, 669173, 669181, 669241, 669247, 669271, 669283, 669287, 669289,
669301, 669311, 669329, 669359, 669371, 669377, 669379, 669391, 669401, 669419, 669433, 669437,
669451, 669463, 669479, 669481, 669527, 669551, 669577, 669607, 669611, 669637, 669649, 669659, 669661,
669667, 669673, 669677, 669679, 669689, 669701, 669707, 669733, 669763, 669787, 669791, 669839, 669847,
669853, 669857, 669859, 669863, 669869, 669887, 669901, 669913, 669923, 669931, 669937, 669943, 669947,
669971, 669989, 670001, 670031, 670037, 670039, 670049, 670051, 670097, 670099, 670129, 670139, 670147,
670177, 670193, 670199, 670211, 670217, 670223, 670231, 670237, 670249, 670261, 670279, 670297, 670303,
670321, 670333, 670343, 670349, 670363, 670379, 670399, 670409, 670447, 670457, 670471, 670487, 670489,

670493, 670507, 670511, 670517, 670541, 670543, 670559, 670577, 670583, 670597, 670613, 670619, 670627,
670639, 670669, 670673, 670693, 670711, 670727, 670729, 670739, 670763, 670777, 670781, 670811, 670823,
670849, 670853, 670867, 670877, 670897, 670903, 670919, 670931, 670951, 670963, 670987, 670991, 671003,
671017, 671029, 671039, 671059, 671063, 671081, 671087, 671093, 671123, 671131, 671141, 671159, 671161,
671189, 671201, 671219, 671233, 671249, 671257, 671261, 671269, 671287, 671299, 671303, 671323, 671339,
671353, 671357, 671369, 671383, 671401, 671417, 671431, 671443, 671467, 671471, 671477, 671501, 671519,
671533, 671537, 671557, 671581, 671591, 671603, 671609, 671633, 671647, 671651, 671681, 671701, 671717,
671729, 671743, 671753, 671777, 671779, 671791, 671831, 671837, 671851, 671887, 671893, 671903, 671911,
671917, 671921, 671933, 671939, 671941, 671947, 671969, 671971, 671981, 671999, 672019, 672029, 672041,
672043, 672059, 672073, 672079, 672097, 672103, 672107, 672127, 672131, 672137, 672143, 672151, 672167,
672169, 672181, 672193, 672209, 672223, 672227, 672229, 672251, 672271, 672283, 672289, 672293, 672311,
672317, 672323, 672341, 672349, 672377, 672379, 672439, 672443, 672473, 672493, 672499, 672521, 672557,
672577, 672587, 672593, 672629, 672641, 672643, 672653, 672667, 672703, 672743, 672757, 672767, 672779,
672781, 672787, 672799, 672803, 672811, 672817, 672823, 672827, 672863, 672869, 672871, 672883, 672901,
672913, 672937, 672943, 672949, 672953, 672967, 672977, 672983, 673019, 673039, 673063, 673069, 673073,
673091, 673093, 673109, 673111, 673117, 673121, 673129, 673157, 673193, 673199, 673201, 673207, 673223,
673241, 673247, 673271, 673273, 673291, 673297, 673313, 673327, 673339, 673349, 673381, 673391, 673397,
673399, 673403, 673411, 673427, 673429, 673441, 673447, 673451, 673457, 673459, 673469, 673487, 673499,
673513, 673529, 673549, 673553, 673567, 673573, 673579, 673609, 673613, 673619, 673637, 673639, 673643,
673649, 673667, 673669, 673747, 673769, 673781, 673787, 673793, 673801, 673811, 673817, 673837, 673879,
673891, 673921, 673943, 673951, 673961, 673979, 673991, 674017, 674057, 674059, 674071, 674083, 674099,
674117, 674123, 674131, 674159, 674161, 674173, 674183, 674189, 674227, 674231, 674239, 674249, 674263,
674269, 674273, 674299, 674321, 674347, 674357, 674363, 674371, 674393, 674419, 674431, 674449, 674461,
674483, 674501, 674533, 674537, 674551, 674563, 674603, 674647, 674669, 674677, 674683, 674693, 674699,
674701, 674711, 674717, 674719, 674731, 674741, 674749, 674759, 674761, 674767, 674771, 674789, 674813,
674827, 674831, 674833, 674837, 674851, 674857, 674867, 674879, 674903, 674929, 674941, 674953, 674957,
674977, 674987, 675029, 675067, 675071, 675079, 675083, 675097, 675109, 675113, 675131, 675133, 675151,
675161, 675163, 675173, 675179, 675187, 675197, 675221, 675239, 675247, 675251, 675253, 675263, 675271,
675299, 675313, 675319, 675341, 675347, 675391, 675407, 675413, 675419, 675449, 675457, 675463, 675481,
675511, 675539, 675541, 675551, 675553, 675559, 675569, 675581, 675593, 675601, 675607, 675611, 675617,
675629, 675643, 675713, 675739, 675743, 675751, 675781, 675797, 675817, 675823, 675827, 675839, 675841,
675859, 675863, 675877, 675881, 675889, 675923, 675929, 675931, 675959, 675973, 675977, 675979, 676007,
676009, 676031, 676037, 676043, 676051, 676057, 676061, 676069, 676099, 676103, 676111, 676129, 676147,
676171, 676211, 676217, 676219, 676241, 676253, 676259, 676279, 676289, 676297, 676337, 676339, 676349,
676363, 676373, 676387, 676391, 676409, 676411, 676421, 676427, 676463, 676469, 676493, 676523, 676573,
676589, 676597, 676601, 676649, 676661, 676679, 676703, 676717, 676721, 676727, 676733, 676747, 676751,
676763, 676771, 676807, 676829, 676859, 676861, 676883, 676891, 676903, 676909, 676919, 676927, 676931,
676937, 676943, 676961, 676967, 676979, 676981, 676987, 676993, 677011, 677021, 677029, 677041, 677057,
677077, 677081, 677107, 677111, 677113, 677119, 677147, 677167, 677177, 677213, 677227, 677231, 677233,
677239, 677309, 677311, 677321, 677323, 677333, 677357, 677371, 677387, 677423, 677441, 677447, 677459,
677461, 677471, 677473, 677531, 677533, 677539, 677543, 677561, 677563, 677587, 677627, 677639, 677647,
677657, 677681, 677683, 677687, 677717, 677737, 677767, 677779, 677783, 677791, 677813, 677827, 677857,
677891, 677927, 677947, 677953, 677959, 677983, 678023, 678037, 678047, 678061, 678077, 678101, 678103,
678133, 678157, 678169, 678179, 678191, 678199, 678203, 678211, 678217, 678221, 678229, 678253, 678289,
678299, 678329, 678341, 678343, 678367, 678371, 678383, 678401, 678407, 678409, 678413, 678421, 678437,
678451, 678463, 678479, 678481, 678493, 678499, 678533, 678541, 678553, 678563, 678577, 678581, 678593,
678599, 678607, 678611, 678631, 678637, 678641, 678647, 678649, 678653, 678659, 678719, 678721, 678731,
678739, 678749, 678757, 678761, 678763, 678767, 678773, 678779, 678809, 678823, 678829, 678833, 678859,
678871, 678883, 678901, 678907, 678941, 678943, 678949, 678959, 678971, 678989, 679033, 679037, 679039,

679051, 679067, 679087, 679111, 679123, 679127, 679153, 679157, 679169, 679171, 679183, 679207, 679219,
679223, 679229, 679249, 679277, 679279, 679297, 679309, 679319, 679333, 679361, 679363, 679369, 679373,
679381, 679403, 679409, 679417, 679423, 679433, 679451, 679463, 679487, 679501, 679517, 679519, 679531,
679537, 679561, 679597, 679603, 679607, 679633, 679639, 679669, 679681, 679691, 679699, 679709, 679733,
679741, 679747, 679751, 679753, 679781, 679793, 679807, 679823, 679829, 679837, 679843, 679859, 679867,
679879, 679883, 679891, 679897, 679907, 679909, 679919, 679933, 679951, 679957, 679961, 679969, 679981,
679993, 679999, 680003, 680027, 680039, 680059, 680077, 680081, 680083, 680107, 680123, 680129, 680159,
680161, 680177, 680189, 680203, 680209, 680213, 680237, 680249, 680263, 680291, 680293, 680297, 680299,
680321, 680327, 680341, 680347, 680353, 680377, 680387, 680399, 680401, 680411, 680417, 680431, 680441,
680443, 680453, 680489, 680503, 680507, 680509, 680531, 680539, 680567, 680569, 680587, 680597, 680611,
680623, 680633, 680651, 680657, 680681, 680707, 680749, 680759, 680767, 680783, 680803, 680809, 680831,
680857, 680861, 680873, 680879, 680881, 680917, 680929, 680959, 680971, 680987, 680989, 680993, 681001,
681011, 681019, 681041, 681047, 681049, 681061, 681067, 681089, 681091, 681113, 681127, 681137, 681151,
681167, 681179, 681221, 681229, 681251, 681253, 681257, 681259, 681271, 681293, 681311, 681337, 681341,
681361, 681367, 681371, 681403, 681407, 681409, 681419, 681427, 681449, 681451, 681481, 681487, 681493,
681497, 681521, 681523, 681539, 681557, 681563, 681589, 681607, 681613, 681623, 681631, 681647, 681673,
681677, 681689, 681719, 681727, 681731, 681763, 681773, 681781, 681787, 681809, 681823, 681833, 681839,
681841, 681883, 681899, 681913, 681931, 681943, 681949, 681971, 681977, 681979, 681983, 681997, 682001,
682009, 682037, 682049, 682063, 682069, 682079, 682141, 682147, 682151, 682153, 682183, 682207, 682219,
682229, 682237, 682247, 682259, 682277, 682289, 682291, 682303, 682307, 682321, 682327, 682333, 682337,
682361, 682373, 682411, 682417, 682421, 682427, 682439, 682447, 682463, 682471, 682483, 682489, 682511,
682519, 682531, 682547, 682597, 682607, 682637, 682657, 682673, 682679, 682697, 682699, 682723, 682729,
682733, 682739, 682751, 682763, 682777, 682789, 682811, 682819, 682901, 682933, 682943, 682951, 682967,
683003, 683021, 683041, 683047, 683071, 683083, 683087, 683119, 683129, 683143, 683149, 683159, 683201,
683231, 683251, 683257, 683273, 683299, 683303, 683337, 683323, 683341, 683351, 683357, 683377, 683381,
683401, 683407, 683437, 683447, 683453, 683461, 683471, 683477, 683479, 683483, 683489, 683503, 683513,
683567, 683591, 683597, 683603, 683651, 683653, 683681, 683687, 683693, 683699, 683701, 683713, 683719,
683731, 683737, 683747, 683759, 683777, 683783, 683789, 683807, 683819, 683821, 683831, 683833, 683843,
683857, 683861, 683863, 683873, 683887, 683899, 683909, 683911, 683923, 683933, 683939, 683957, 683983,
684007, 684017, 684037, 684053, 684091, 684109, 684113, 684119, 684121, 684127, 684157, 684163, 684191,
684217, 684221, 684239, 684269, 684287, 684289, 684293, 684311, 684329, 684337, 684347, 684349, 684373,
684379, 684407, 684419, 684427, 684433, 684443, 684451, 684469, 684473, 684493, 684527, 684547, 684557,
684559, 684569, 684581, 684587, 684599, 684617, 684637, 684643, 684647, 684683, 684713, 684727, 684731,
684751, 684757, 684767, 684769, 684773, 684791, 684793, 684799, 684809, 684829, 684841, 684857, 684869,
684889, 684923, 684949, 684961, 684973, 684977, 684989, 685001, 685019, 685031, 685039, 685051, 685057,
685063, 685073, 685081, 685093, 685099, 685103, 685109, 685123, 685141, 685169, 685177, 685199, 685231,
685247, 685249, 685271, 685297, 685301, 685319, 685337, 685339, 685361, 685367, 685369, 685381, 685393,
685417, 685427, 685429, 685453, 685459, 685471, 685493, 685511, 685513, 685519, 685537, 685541, 685547,
685591, 685609, 685613, 685621, 685631, 685637, 685649, 685669, 685679, 685697, 685717, 685723, 685733,
685739, 685747, 685753, 685759, 685781, 685793, 685819, 685849, 685859, 685907, 685939, 685963, 685969,
685973, 685987, 685991, 686003, 686009, 686011, 686027, 686029, 686039, 686041, 686051, 686057, 686087,
686089, 686099, 686117, 686131, 686141, 686143, 686149, 686173, 686177, 686197, 686201, 686209, 686267,
686269, 686293, 686317, 686321, 686333, 686339, 686353, 686359, 686363, 686417, 686423, 686437, 686449,
686453, 686473, 686479, 686503, 686513, 686519, 686551, 686563, 686593, 686611, 686639, 686669, 686671,
686687, 686723, 686729, 686731, 686737, 686761, 686773, 686789, 686797, 686801, 686837, 686843, 686863,
686879, 686891, 686893, 686897, 686911, 686947, 686963, 686969, 686971, 686977, 686989, 686993, 687007,
687013, 687017, 687019, 687023, 687031, 687041, 687061, 687073, 687083, 687101, 687107, 687109, 687121,
687131, 687139, 687151, 687161, 687163, 687179, 687223, 687233, 687277, 687289, 687299, 687307, 687311,
687317, 687331, 687341, 687343, 687359, 687383, 687389, 687397, 687403, 687413, 687431, 687433, 687437,

687443, 687457, 687461, 687473, 687481, 687499, 687517, 687521, 687523, 687541, 687551, 687559, 687581,
687593, 687623, 687637, 687641, 687647, 687679, 687683, 687691, 687707, 687721, 687737, 687749, 687767,
687773, 687779, 687787, 687809, 687823, 687829, 687839, 687847, 687893, 687901, 687917, 687923, 687931,
687949, 687961, 687977, 688003, 688013, 688027, 688031, 688063, 688067, 688073, 688087, 688097, 688111,
688133, 688139, 688147, 688159, 688187, 688201, 688217, 688223, 688249, 688253, 688277, 688297, 688309,
688333, 688339, 688357, 688379, 688393, 688397, 688403, 688411, 688423, 688433, 688447, 688451, 688453,
688477, 688511, 688531, 688543, 688561, 688573, 688591, 688621, 688627, 688631, 688637, 688657, 688661,
688669, 688679, 688697, 688717, 688729, 688733, 688741, 688747, 688757, 688763, 688777, 688783, 688799,
688813, 688861, 688867, 688871, 688889, 688907, 688939, 688951, 688957, 688969, 688979, 688999, 689021,
689033, 689041, 689063, 689071, 689077, 689081, 689089, 689093, 689107, 689113, 689131, 689141, 689167,
689201, 689219, 689233, 689237, 689257, 689261, 689267, 689279, 689291, 689309, 689317, 689321, 689341,
689357, 689369, 689383, 689389, 689393, 689411, 689431, 689441, 689459, 689461, 689467, 689509, 689551,
689561, 689581, 689587, 689597, 689599, 689603, 689621, 689629, 689641, 689693, 689699, 689713, 689723,
689761, 689771, 689779, 689789, 689797, 689803, 689807, 689827, 689831, 689851, 689867, 689869, 689873,
689879, 689891, 689893, 689903, 689917, 689921, 689929, 689951, 689957, 689959, 689963, 689981, 689987,
690037, 690059, 690073, 690089, 690103, 690119, 690127, 690139, 690143, 690163, 690187, 690233, 690259,
690269, 690271, 690281, 690293, 690323, 690341, 690367, 690377, 690397, 690407, 690419, 690427, 690433,
690439, 690449, 690467, 690491, 690493, 690509, 690511, 690533, 690541, 690553, 690583, 690589, 690607,
690611, 690629, 690661, 690673, 690689, 690719, 690721, 690757, 690787, 690793, 690817, 690839, 690841,
690869, 690871, 690887, 690889, 690919, 690929, 690953, 690997, 691001, 691037, 691051, 691063, 691079,
691109, 691111, 691121, 691129, 691147, 691151, 691153, 691181, 691183, 691189, 691193, 691199, 691231,
691241, 691267, 691289, 691297, 691309, 691333, 691337, 691343, 691349, 691363, 691381, 691399, 691409,
691433, 691451, 691463, 691489, 691499, 691531, 691553, 691573, 691583, 691589, 691591, 691631, 691637,
691651, 691661, 691681, 691687, 691693, 691697, 691709, 691721, 691723, 691727, 691729, 691739, 691759,
691763, 691787, 691799, 691813, 691829, 691837, 691841, 691843, 691871, 691877, 691891, 691897, 691903,
691907, 691919, 691921, 691931, 691949, 691973, 691979, 691991, 691997, 692009, 692017, 692053, 692059,
692063, 692071, 692089, 692099, 692117, 692141, 692147, 692149, 692161, 692191, 692221, 692239, 692249,
692269, 692273, 692281, 692287, 692297, 692299, 692309, 692327, 692333, 692347, 692353, 692371, 692387,
692389, 692399, 692401, 692407, 692413, 692423, 692431, 692441, 692453, 692459, 692467, 692513, 692521,
692537, 692539, 692543, 692567, 692581, 692591, 692621, 692641, 692647, 692651, 692663, 692689, 692707,
692711, 692717, 692729, 692743, 692753, 692761, 692771, 692779, 692789, 692821, 692851, 692863, 692893,
692917, 692927, 692929, 692933, 692957, 692963, 692969, 692983, 693019, 693037, 693041, 693061, 693079,
693089, 693097, 693103, 693127, 693137, 693149, 693157, 693167, 693169, 693179, 693223, 693257, 693283,
693317, 693323, 693337, 693353, 693359, 693373, 693397, 693401, 693403, 693409, 693421, 693431, 693437,
693487, 693493, 693503, 693523, 693527, 693529, 693533, 693569, 693571, 693601, 693607, 693619, 693629,
693659, 693661, 693677, 693683, 693689, 693691, 693697, 693701, 693727, 693731, 693733, 693739, 693743,
693757, 693779, 693793, 693799, 693809, 693827, 693829, 693851, 693859, 693871, 693877, 693881, 693943,
693961, 693967, 693989, 694019, 694033, 694039, 694061, 694069, 694079, 694081, 694087, 694091, 694123,
694189, 694193, 694201, 694207, 694223, 694259, 694261, 694271, 694273, 694277, 694313, 694319, 694327,
694333, 694339, 694349, 694357, 694361, 694367, 694373, 694381, 694387, 694391, 694409, 694427, 694457,
694471, 694481, 694483, 694487, 694511, 694513, 694523, 694541, 694549, 694559, 694567, 694571, 694591,
694597, 694609, 694619, 694633, 694649, 694651, 694717, 694721, 694747, 694763, 694781, 694783, 694789,
694829, 694831, 694867, 694871, 694873, 694879, 694901, 694919, 694951, 694957, 694979, 694987, 694997,
694999, 695003, 695017, 695021, 695047, 695059, 695069, 695081, 695087, 695089, 695099, 695111, 695117,
695131, 695141, 695171, 695207, 695239, 695243, 695257, 695263, 695269, 695281, 695293, 695297, 695309,
695323, 695327, 695329, 695347, 695369, 695371, 695377, 695389, 695407, 695411, 695441, 695447, 695467,
695477, 695491, 695503, 695509, 695561, 695567, 695573, 695581, 695593, 695599, 695603, 695621, 695627,
695641, 695659, 695663, 695677, 695687, 695689, 695701, 695719, 695743, 695749, 695771, 695777, 695791,
695801, 695809, 695839, 695843, 695867, 695873, 695879, 695881, 695899, 695917, 695927, 695939, 695999,

696019, 696053, 696061, 696067, 696077, 696079, 696083, 696107, 696109, 696119, 696149, 696181, 696239,
696253, 696257, 696263, 696271, 696281, 696313, 696317, 696323, 696343, 696349, 696359, 696361, 696373,
696379, 696403, 696413, 696427, 696433, 696457, 696481, 696491, 696497, 696503, 696517, 696523, 696533,
696547, 696569, 696607, 696611, 696617, 696623, 696629, 696653, 696659, 696679, 696691, 696719, 696721,
696737, 696743, 696757, 696763, 696793, 696809, 696811, 696823, 696827, 696833, 696851, 696853, 696887,
696889, 696893, 696907, 696929, 696937, 696961, 696989, 696991, 697009, 697013, 697019, 697033, 697049,
697063, 697069, 697079, 697087, 697093, 697111, 697121, 697127, 697133, 697141, 697157, 697181, 697201,
697211, 697217, 697259, 697261, 697267, 697271, 697303, 697327, 697351, 697373, 697379, 697381, 697387,
697397, 697399, 697409, 697423, 697441, 697447, 697453, 697457, 697481, 697507, 697511, 697513, 697519,
697523, 697553, 697579, 697583, 697591, 697601, 697603, 697637, 697643, 697673, 697681, 697687, 697691,
697693, 697703, 697727, 697729, 697733, 697757, 697759, 697787, 697819, 697831, 697877, 697891, 697897,
697909, 697913, 697937, 697951, 697967, 697973, 697979, 697993, 697999, 698017, 698021, 698039, 698051,
698053, 698077, 698083, 698111, 698171, 698183, 698239, 698249, 698251, 698261, 698263, 698273, 698287,
698293, 698297, 698311, 698329, 698339, 698359, 698371, 698387, 698393, 698413, 698417, 698419, 698437,
698447, 698471, 698483, 698491, 698507, 698521, 698527, 698531, 698539, 698543, 698557, 698567, 698591,
698641, 698653, 698669, 698701, 698713, 698723, 698729, 698773, 698779, 698821, 698827, 698849, 698891,
698899, 698903, 698923, 698939, 698977, 698983, 699001, 699007, 699037, 699053, 699059, 699073, 699077,
699089, 699113, 699119, 699133, 699151, 699157, 699169, 699187, 699191, 699197, 699211, 699217, 699221,
699241, 699253, 699271, 699287, 699289, 699299, 699319, 699323, 699343, 699367, 699373, 699379, 699383,
699401, 699427, 699437, 699443, 699449, 699463, 699469, 699493, 699511, 699521, 699527, 699529, 699539,
699541, 699557, 699571, 699581, 699617, 699631, 699641, 699649, 699697, 699709, 699719, 699733, 699757,
699761, 699767, 699791, 699793, 699817, 699823, 699863, 699931, 699943, 699947, 699953, 699961, 699967,
700001, 700027, 700057, 700067, 700079, 700081, 700087, 700099, 700103, 700109, 700127, 700129, 700171,
700199, 700201, 700211, 700223, 700229, 700237, 700241, 700277, 700279, 700303, 700307, 700319, 700331,
700339, 700361, 700363, 700367, 700387, 700391, 700393, 700423, 700429, 700433, 700459, 700471, 700499,
700523, 700537, 700561, 700571, 700573, 700577, 700591, 700597, 700627, 700633, 700639, 700643, 700673,
700681, 700703, 700717, 700751, 700759, 700781, 700789, 700801, 700811, 700831, 700837, 700849, 700871,
700877, 700883, 700897, 700907, 700919, 700933, 700937, 700949, 700963, 700993, 701009, 701011, 701023,
701033, 701047, 701089, 701117, 701147, 701159, 701177, 701179, 701209, 701219, 701221, 701227, 701257,
701279, 701291, 701299, 701329, 701341, 701357, 701359, 701377, 701383, 701399, 701401, 701413, 701417,
701419, 701443, 701447, 701453, 701473, 701479, 701489, 701497, 701507, 701509, 701527, 701531, 701549,
701579, 701581, 701593, 701609, 701611, 701621, 701627, 701629, 701653, 701669, 701671, 701681, 701699,
701711, 701719, 701731, 701741, 701761, 701783, 701791, 701819, 701837, 701863, 701881, 701903, 701951,
701957, 701963, 701969, 702007, 702011, 702017, 702067, 702077, 702101, 702113, 702127, 702131, 702137,
702139, 702173, 702179, 702193, 702199, 702203, 702211, 702239, 702257, 702269, 702281, 702283, 702311,
702313, 702323, 702329, 702337, 702341, 702347, 702349, 702353, 702379, 702391, 702407, 702413, 702431,
702433, 702439, 702451, 702469, 702497, 702503, 702511, 702517, 702523, 702529, 702539, 702551, 702557,
702587, 702589, 702599, 702607, 702613, 702623, 702671, 702679, 702683, 702701, 702707, 702721, 702731,
702733, 702743, 702773, 702787, 702803, 702809, 702817, 702827, 702847, 702851, 702853, 702869, 702881,
702887, 702893, 702913, 702937, 702983, 702991, 703013, 703033, 703039, 703081, 703117, 703121, 703123,
703127, 703139, 703141, 703169, 703193, 703211, 703217, 703223, 703229, 703231, 703243, 703249, 703267,
703277, 703301, 703309, 703321, 703327, 703331, 703349, 703357, 703379, 703393, 703411, 703441, 703447,
703459, 703463, 703471, 703489, 703499, 703531, 703537, 703559, 703561, 703631, 703643, 703657, 703663,
703673, 703679, 703691, 703699, 703709, 703711, 703721, 703733, 703753, 703763, 703789, 703819, 703837,
703849, 703861, 703873, 703883, 703897, 703903, 703907, 703943, 703949, 703957, 703981, 703991, 704003,
704009, 704017, 704023, 704027, 704029, 704059, 704069, 704087, 704101, 704111, 704117, 704131, 704141,
704153, 704161, 704177, 704183, 704189, 704213, 704219, 704233, 704243, 704251, 704269, 704279, 704281,
704287, 704299, 704303, 704309, 704321, 704357, 704393, 704399, 704419, 704441, 704447, 704449, 704453,
704461, 704477, 704507, 704521, 704527, 704549, 704551, 704567, 704569, 704579, 704581, 704593, 704603,

704617, 704647, 704657, 704663, 704681, 704687, 704713, 704719, 704731, 704747, 704761, 704771, 704777,
704779, 704783, 704797, 704801, 704807, 704819, 704833, 704839, 704849, 704857, 704861, 704863, 704867,
704897, 704929, 704933, 704947, 704983, 704989, 704993, 704999, 705011, 705013, 705017, 705031, 705043,
705053, 705073, 705079, 705097, 705113, 705119, 705127, 705137, 705161, 705163, 705167, 705169, 705181,
705191, 705197, 705209, 705247, 705259, 705269, 705277, 705293, 705307, 705317, 705389, 705403, 705409,
705421, 705427, 705437, 705461, 705491, 705493, 705499, 705521, 705533, 705559, 705613, 705631, 705643,
705689, 705713, 705737, 705751, 705763, 705769, 705779, 705781, 705787, 705821, 705827, 705829, 705833,
705841, 705863, 705871, 705883, 705899, 705919, 705937, 705949, 705967, 705973, 705989, 706001, 706003,
706009, 706019, 706033, 706039, 706049, 706051, 706067, 706099, 706109, 706117, 706133, 706141, 706151,
706157, 706159, 706183, 706193, 706201, 706207, 706213, 706229, 706253, 706267, 706283, 706291, 706297,
706301, 706309, 706313, 706337, 706357, 706369, 706373, 706403, 706417, 706427, 706463, 706481, 706487,
706499, 706507, 706523, 706547, 706561, 706597, 706603, 706613, 706621, 706631, 706633, 706661, 706669,
706679, 706703, 706709, 706729, 706733, 706747, 706751, 706753, 706757, 706763, 706787, 706793, 706801,
706829, 706837, 706841, 706847, 706883, 706897, 706907, 706913, 706919, 706921, 706943, 706961, 706973,
706987, 706999, 707011, 707027, 707029, 707053, 707071, 707099, 707111, 707117, 707131, 707143, 707153,
707159, 707177, 707191, 707197, 707219, 707249, 707261, 707279, 707293, 707299, 707321, 707341, 707359,
707383, 707407, 707429, 707431, 707437, 707459, 707467, 707501, 707527, 707543, 707561, 707563, 707573,
707627, 707633, 707647, 707653, 707669, 707671, 707677, 707683, 707689, 707711, 707717, 707723, 707747,
707753, 707767, 707789, 707797, 707801, 707813, 707827, 707831, 707849, 707857, 707869, 707873, 707887,
707911, 707923, 707929, 707933, 707939, 707951, 707953, 707957, 707969, 707981, 707983, 708007, 708011,
708017, 708023, 708031, 708041, 708047, 708049, 708053, 708061, 708091, 708109, 708119, 708131, 708137,
708139, 708161, 708163, 708179, 708199, 708221, 708223, 708229, 708251, 708269, 708283, 708287, 708293,
708311, 708329, 708343, 708347, 708353, 708359, 708361, 708371, 708403, 708437, 708457, 708473, 708479,
708481, 708493, 708497, 708517, 708527, 708559, 708563, 708569, 708583, 708593, 708599, 708601, 708641,
708647, 708667, 708689, 708703, 708733, 708751, 708803, 708823, 708839, 708857, 708859, 708893, 708899,
708907, 708913, 708923, 708937, 708943, 708959, 708979, 708989, 708991, 708997, 709043, 709057, 709097,
709117, 709123, 709139, 709141, 709151, 709153, 709157, 709201, 709211, 709217, 709231, 709237, 709271,
709273, 709279, 709283, 709307, 709321, 709337, 709349, 709351, 709381, 709409, 709417, 709421, 709433,
709447, 709451, 709453, 709469, 709507, 709519, 709531, 709537, 709547, 709561, 709589, 709603, 709607,
709609, 709649, 709651, 709663, 709673, 709679, 709691, 709693, 709703, 709729, 709739, 709741, 709769,
709777, 709789, 709799, 709817, 709823, 709831, 709843, 709847, 709853, 709861, 709871, 709879, 709901,
709909, 709913, 709921, 709927, 709957, 709963, 709967, 709981, 709991, 710009, 710023, 710027, 710051,
710053, 710081, 710089, 710119, 710189, 710207, 710219, 710221, 710257, 710261, 710273, 710293, 710299,
710321, 710323, 710327, 710341, 710351, 710371, 710377, 710383, 710389, 710399, 710441, 710443, 710449,
710459, 710473, 710483, 710491, 710503, 710513, 710519, 710527, 710531, 710557, 710561, 710569, 710573,
710599, 710603, 710609, 710621, 710623, 710627, 710641, 710663, 710683, 710693, 710713, 710777, 710779,
710791, 710813, 710837, 710839, 710849, 710851, 710863, 710867, 710873, 710887, 710903, 710909, 710911,
710917, 710929, 710933, 710951, 710959, 710971, 710977, 710987, 710989, 711001, 711017, 711019, 711023,
711041, 711049, 711089, 711097, 711121, 711131, 711133, 711143, 711163, 711173, 711181, 711187, 711209,
711223, 711259, 711287, 711299, 711307, 711317, 711329, 711353, 711371, 711397, 711409, 711427, 711437,
711463, 711479, 711497, 711499, 711509, 711517, 711523, 711539, 711563, 711577, 711583, 711589, 711617,
711629, 711649, 711653, 711679, 711691, 711701, 711707, 711709, 711713, 711727, 711731, 711749, 711751,
711757, 711793, 711811, 711817, 711829, 711839, 711847, 711859, 711877, 711889, 711899, 711913, 711923,
711929, 711937, 711947, 711959, 711967, 711973, 711983, 712007, 712021, 712051, 712067, 712093, 712109,
712121, 712133, 712157, 712169, 712171, 712183, 712199, 712219, 712237, 712279, 712289, 712301, 712303,
712319, 712321, 712331, 712339, 712357, 712409, 712417, 712427, 712429, 712433, 712447, 712477, 712483,
712489, 712493, 712499, 712507, 712511, 712531, 712561, 712571, 712573, 712601, 712603, 712631, 712651,
712669, 712681, 712687, 712693, 712697, 712711, 712717, 712739, 712781, 712807, 712819, 712837, 712841,
712843, 712847, 712883, 712889, 712891, 712909, 712913, 712927, 712939, 712951, 712961, 712967, 712973,

712981, 713021, 713039, 713059, 713077, 713107, 713117, 713129, 713147, 713149, 713159, 713171, 713177,
713183, 713189, 713191, 713227, 713233, 713239, 713243, 713261, 713267, 713281, 713287, 713309, 713311,
713329, 713347, 713351, 713353, 713357, 713381, 713389, 713399, 713407, 713411, 713417, 713467, 713477,
713491, 713497, 713501, 713509, 713533, 713563, 713569, 713597, 713599, 713611, 713627, 713653, 713663,
713681, 713737, 713743, 713747, 713753, 713771, 713807, 713827, 713831, 713833, 713861, 713863, 713873,
713891, 713903, 713917, 713927, 713939, 713941, 713957, 713981, 713987, 714029, 714037, 714061, 714073,
714107, 714113, 714139, 714143, 714151, 714163, 714169, 714199, 714223, 714227, 714247, 714257, 714283,
714341, 714349, 714361, 714377, 714443, 714463, 714479, 714481, 714487, 714503, 714509, 714517, 714521,
714529, 714551, 714557, 714563, 714569, 714577, 714601, 714619, 714673, 714677, 714691, 714719, 714739,
714751, 714773, 714781, 714787, 714797, 714809, 714827, 714839, 714841, 714851, 714853, 714869, 714881,
714887, 714893, 714907, 714911, 714919, 714943, 714947, 714949, 714971, 714991, 715019, 715031, 715049,
715063, 715069, 715073, 715087, 715109, 715123, 715151, 715153, 715157, 715159, 715171, 715189, 715193,
715223, 715229, 715237, 715243, 715249, 715259, 715289, 715301, 715303, 715313, 715339, 715357, 715361,
715373, 715397, 715417, 715423, 715439, 715441, 715453, 715457, 715489, 715499, 715523, 715537, 715549,
715567, 715571, 715577, 715579, 715613, 715621, 715639, 715643, 715651, 715657, 715679, 715681, 715699,
715727, 715739, 715753, 715777, 715789, 715801, 715811, 715817, 715823, 715843, 715849, 715859, 715867,
715873, 715877, 715879, 715889, 715903, 715909, 715919, 715927, 715943, 715961, 715963, 715969, 715973,
715991, 715999, 716003, 716033, 716063, 716087, 716117, 716123, 716137, 716143, 716161, 716171, 716173,
716249, 716257, 716279, 716291, 716299, 716321, 716351, 716383, 716389, 716399, 716411, 716413, 716447,
716449, 716453, 716459, 716477, 716479, 716483, 716491, 716501, 716531, 716543, 716549, 716563, 716581,
716591, 716621, 716629, 716633, 716659, 716663, 716671, 716687, 716693, 716707, 716713, 716731, 716741,
716743, 716747, 716783, 716789, 716809, 716819, 716827, 716857, 716861, 716869, 716897, 716899, 716917,
716929, 716951, 716953, 716959, 716981, 716987, 717001, 717011, 717047, 717089, 717091, 717103, 717109,
717113, 717127, 717133, 717139, 717149, 717151, 717161, 717191, 717229, 717259, 717271, 717289, 717293,
717317, 717323, 717331, 717341, 717397, 717413, 717419, 717427, 717443, 717449, 717463, 717491, 717511,
717527, 717529, 717533, 717539, 717551, 717559, 717581, 717589, 717593, 717631, 717653, 717659, 717667,
717679, 717683, 717697, 717719, 717751, 717797, 717803, 717811, 717817, 717841, 717851, 717883, 717887,
717917, 717919, 717923, 717967, 717979, 717989, 718007, 718043, 718049, 718051, 718087, 718093, 718121,
718139, 718163, 718169, 718171, 718183, 718187, 718241, 718259, 718271, 718303, 718321, 718331, 718337,
718343, 718349, 718357, 718379, 718381, 718387, 718391, 718411, 718423, 718427, 718433, 718453, 718457,
718463, 718493, 718511, 718513, 718541, 718547, 718559, 718579, 718603, 718621, 718633, 718657, 718661,
718691, 718703, 718717, 718723, 718741, 718747, 718759, 718801, 718807, 718813, 718841, 718847, 718871,
718897, 718901, 718919, 718931, 718937, 718943, 718973, 718999, 719009, 719011, 719027, 719041, 719057,
719063, 719071, 719101, 719119, 719143, 719149, 719153, 719167, 719177, 719179, 719183, 719189, 719197,
719203, 719227, 719237, 719239, 719267, 719281, 719297, 719333, 719351, 719353, 719377, 719393, 719413,
719419, 719441, 719447, 719483, 719503, 719533, 719557, 719567, 719569, 719573, 719597, 719599, 719633,
719639, 719659, 719671, 719681, 719683, 719689, 719699, 719713, 719717, 719723, 719731, 719749, 719753,
719773, 719779, 719791, 719801, 719813, 719821, 719833, 719839, 719893, 719903, 719911, 719941, 719947,
719951, 719959, 719981, 719989, 720007, 720019, 720023, 720053, 720059, 720089, 720091, 720101, 720127,
720133, 720151, 720173, 720179, 720193, 720197, 720211, 720221, 720229, 720241, 720253, 720257, 720281,
720283, 720289, 720299, 720301, 720311, 720319, 720359, 720361, 720367, 720373, 720397, 720403, 720407,
720413, 720439, 720481, 720491, 720497, 720527, 720547, 720569, 720571, 720607, 720611, 720617, 720619,
720653, 720661, 720677, 720683, 720697, 720703, 720743, 720763, 720767, 720773, 720779, 720791, 720793,
720829, 720847, 720857, 720869, 720877, 720887, 720899, 720901, 720913, 720931, 720943, 720947, 720961,
720971, 720983, 720991, 720997, 721003, 721013, 721037, 721043, 721051, 721057, 721079, 721087, 721109,
721111, 721117, 721129, 721139, 721141, 721159, 721163, 721169, 721177, 721181, 721199, 721207, 721213,
721219, 721223, 721229, 721243, 721261, 721267, 721283, 721291, 721307, 721319, 721321, 721333, 721337,
721351, 721363, 721379, 721381, 721387, 721397, 721439, 721451, 721481, 721499, 721529, 721547, 721561,
721571, 721577, 721597, 721613, 721619, 721621, 721631, 721661, 721663, 721687, 721697, 721703, 721709,

721733, 721739, 721783, 721793, 721843, 721849, 721859, 721883, 721891, 721909, 721921, 721951, 721961,
721979, 721991, 721997, 722011, 722023, 722027, 722047, 722063, 722069, 722077, 722093, 722119, 722123,
722147, 722149, 722153, 722159, 722167, 722173, 722213, 722237, 722243, 722257, 722273, 722287, 722291,
722299, 722311, 722317, 722321, 722333, 722341, 722353, 722363, 722369, 722377, 722389, 722411, 722417,
722431, 722459, 722467, 722479, 722489, 722509, 722521, 722537, 722539, 722563, 722581, 722599, 722611,
722633, 722639, 722663, 722669, 722713, 722723, 722737, 722749, 722783, 722791, 722797, 722807, 722819,
722833, 722849, 722881, 722899, 722903, 722921, 722933, 722963, 722971, 722977, 722983, 723029, 723031,
723043, 723049, 723053, 723067, 723071, 723089, 723101, 723103, 723109, 723113, 723119, 723127, 723133,
723157, 723161, 723167, 723169, 723181, 723193, 723209, 723221, 723227, 723257, 723259, 723263, 723269,
723271, 723287, 723293, 723319, 723337, 723353, 723361, 723379, 723391, 723407, 723409, 723413, 723421,
723439, 723451, 723467, 723473, 723479, 723491, 723493, 723529, 723551, 723553, 723559, 723563, 723587,
723589, 723601, 723607, 723617, 723623, 723661, 723721, 723727, 723739, 723761, 723791, 723797, 723799,
723803, 723823, 723829, 723839, 723851, 723857, 723859, 723893, 723901, 723907, 723913, 723917, 723923,
723949, 723959, 723967, 723973, 723977, 723997, 724001, 724007, 724021, 724079, 724093, 724099, 724111,
724117, 724121, 724123, 724153, 724187, 724211, 724219, 724259, 724267, 724277, 724291, 724303, 724309,
724313, 724331, 724393, 724403, 724433, 724441, 724447, 724453, 724459, 724469, 724481, 724487, 724499,
724513, 724517, 724519, 724531, 724547, 724553, 724567, 724573, 724583, 724597, 724601, 724609, 724621,
724627, 724631, 724639, 724643, 724651, 724721, 724723, 724729, 724733, 724747, 724751, 724769, 724777,
724781, 724783, 724807, 724813, 724837, 724847, 724853, 724879, 724901, 724903, 724939, 724949, 724961,
724967, 724991, 724993, 725009, 725041, 725057, 725071, 725077, 725099, 725111, 725113, 725119, 725147,
725149, 725159, 725161, 725189, 725201, 725209, 725273, 725293, 725303, 725317, 725321, 725323, 725327,
725341, 725357, 725359, 725371, 725381, 725393, 725399, 725423, 725437, 725447, 725449, 725479, 725507,
725519, 725531, 725537, 725579, 725587, 725597, 725603, 725639, 725653, 725663, 725671, 725687, 725723,
725731, 725737, 725749, 725789, 725801, 725807, 725827, 725861, 725863, 725867, 725891, 725897, 725909,
725929, 725939, 725953, 725981, 725983, 725993, 725999, 726007, 726013, 726023, 726043, 726071, 726091,
726097, 726101, 726107, 726109, 726137, 726139, 726149, 726157, 726163, 726169, 726181, 726191, 726221,
726287, 726289, 726301, 726307, 726331, 726337, 726367, 726371, 726377, 726379, 726391, 726413, 726419,
726431, 726457, 726463, 726469, 726487, 726497, 726521, 726527, 726533, 726559, 726589, 726599, 726601,
726611, 726619, 726623, 726629, 726641, 726647, 726659, 726679, 726689, 726697, 726701, 726707, 726751,
726779, 726787, 726797, 726809, 726811, 726839, 726841, 726853, 726893, 726899, 726911, 726917, 726923,
726941, 726953, 726983, 726989, 726991, 727003, 727009, 727019, 727021, 727049, 727061, 727063, 727079,
727121, 727123, 727157, 727159, 727169, 727183, 727189, 727201, 727211, 727241, 727247, 727249, 727261,
727267, 727271, 727273, 727289, 727297, 727313, 727327, 727343, 727351, 727369, 727399, 727409, 727427,
727451, 727459, 727471, 727483, 727487, 727499, 727501, 727541, 727561, 727577, 727589, 727613, 727621,
727633, 727667, 727673, 727691, 727703, 727711, 727717, 727729, 727733, 727747, 727759, 727763, 727777,
727781, 727799, 727807, 727817, 727823, 727843, 727847, 727877, 727879, 727891, 727933, 727939, 727949,
727981, 727997, 728003, 728017, 728027, 728047, 728069, 728087, 728113, 728129, 728131, 728173, 728191,
728207, 728209, 728261, 728267, 728269, 728281, 728293, 728303, 728317, 728333, 728369, 728381, 728383,
728417, 728423, 728437, 728471, 728477, 728489, 728521, 728527, 728537, 728551, 728557, 728561, 728573,
728579, 728627, 728639, 728647, 728659, 728681, 728687, 728699, 728701, 728713, 728723, 728729, 728731,
728743, 728747, 728771, 728809, 728813, 728831, 728837, 728839, 728843, 728851, 728867, 728869, 728873,
728881, 728891, 728899, 728911, 728921, 728927, 728929, 728941, 728947, 728953, 728969, 728971, 728993,
729019, 729023, 729037, 729041, 729059, 729073, 729139, 729143, 729173, 729187, 729191, 729199, 729203,
729217, 729257, 729269, 729271, 729293, 729301, 729329, 729331, 729359, 729367, 729371, 729373, 729389,
729403, 729413, 729451, 729457, 729473, 729493, 729497, 729503, 729511, 729527, 729551, 729557, 729559,
729569, 729571, 729577, 729587, 729601, 729607, 729613, 729637, 729643, 729649, 729661, 729671, 729679,
729689, 729713, 729719, 729737, 729749, 729761, 729779, 729787, 729791, 729821, 729851, 729871, 729877,
729907, 729913, 729919, 729931, 729941, 729943, 729947, 729977, 729979, 729991, 730003, 730021, 730033,
730049, 730069, 730091, 730111, 730139, 730157, 730187, 730199, 730217, 730237, 730253, 730277, 730283,

730297, 730321, 730339, 730363, 730397, 730399, 730421, 730447, 730451, 730459, 730469, 730487, 730537,
730553, 730559, 730567, 730571, 730573, 730589, 730591, 730603, 730619, 730633, 730637, 730663, 730669,
730679, 730727, 730747, 730753, 730757, 730777, 730781, 730783, 730789, 730799, 730811, 730819, 730823,
730837, 730843, 730853, 730867, 730879, 730889, 730901, 730909, 730913, 730943, 730969, 730973, 730993,
730999, 731033, 731041, 731047, 731053, 731057, 731113, 731117, 731141, 731173, 731183, 731189, 731191,
731201, 731209, 731219, 731233, 731243, 731249, 731251, 731257, 731261, 731267, 731287, 731299, 731327,
731333, 731359, 731363, 731369, 731389, 731413, 731447, 731483, 731501, 731503, 731509, 731531, 731539,
731567, 731587, 731593, 731597, 731603, 731611, 731623, 731639, 731651, 731681, 731683, 731711, 731713,
731719, 731729, 731737, 731741, 731761, 731767, 731779, 731803, 731807, 731821, 731827, 731831, 731839,
731851, 731869, 731881, 731893, 731909, 731911, 731921, 731923, 731933, 731957, 731981, 731999, 732023,
732029, 732041, 732073, 732077, 732079, 732097, 732101, 732133, 732157, 732169, 732181, 732187, 732191,
732197, 732209, 732211, 732217, 732229, 732233, 732239, 732257, 732271, 732283, 732287, 732293, 732299,
732311, 732323, 732331, 732373, 732439, 732449, 732461, 732467, 732491, 732493, 732497, 732509, 732521,
732533, 732541, 732601, 732617, 732631, 732653, 732673, 732689, 732703, 732709, 732713, 732731, 732749,
732761, 732769, 732799, 732817, 732827, 732829, 732833, 732841, 732863, 732877, 732889, 732911, 732923,
732943, 732959, 732967, 732971, 732997, 733003, 733009, 733067, 733097, 733099, 733111, 733123, 733127,
733133, 733141, 733147, 733157, 733169, 733177, 733189, 733227, 733241, 733273, 733277, 733283, 733289,
733301, 733307, 733321, 733331, 733333, 733339, 733351, 733373, 733387, 733391, 733393, 733399, 733409,
733427, 733433, 733459, 733477, 733489, 733511, 733517, 733519, 733559, 733561, 733591, 733619, 733639,
733651, 733687, 733697, 733741, 733751, 733753, 733757, 733793, 733807, 733813, 733823, 733829, 733841,
733847, 733849, 733867, 733871, 733879, 733883, 733919, 733921, 733937, 733939, 733949, 733963, 733973,
733981, 733991, 734003, 734017, 734021, 734047, 734057, 734087, 734113, 734131, 734143, 734159, 734171,
734177, 734189, 734197, 734203, 734207, 734221, 734233, 734263, 734267, 734273, 734291, 734303, 734329,
734347, 734381, 734389, 734401, 734411, 734423, 734429, 734431, 734443, 734471, 734473, 734477, 734479,
734497, 734537, 734543, 734549, 734557, 734567, 734627, 734647, 734653, 734659, 734663, 734687, 734693,
734707, 734717, 734729, 734737, 734743, 734759, 734771, 734803, 734807, 734813, 734819, 734837, 734849,
734869, 734879, 734887, 734897, 734911, 734933, 734941, 734953, 734957, 734959, 734971, 735001, 735019,
735043, 735061, 735067, 735071, 735073, 735083, 735107, 735109, 735113, 735139, 735143, 735157, 735169,
735173, 735181, 735187, 735193, 735209, 735211, 735239, 735247, 735263, 735271, 735283, 735307, 735311,
735331, 735337, 735341, 735359, 735367, 735373, 735389, 735391, 735419, 735421, 735431, 735439, 735443,
735451, 735461, 735467, 735473, 735479, 735491, 735529, 735533, 735557, 735571, 735617, 735649, 735653,
735659, 735673, 735689, 735697, 735719, 735731, 735733, 735739, 735751, 735781, 735809, 735821, 735829,
735853, 735871, 735877, 735883, 735901, 735919, 735937, 735949, 735953, 735979, 735983, 735997, 736007,
736013, 736027, 736037, 736039, 736051, 736061, 736063, 736091, 736093, 736097, 736111, 736121, 736147,
736159, 736181, 736187, 736243, 736247, 736249, 736259, 736273, 736277, 736279, 736357, 736361, 736363,
736367, 736369, 736381, 736387, 736399, 736403, 736409, 736429, 736433, 736441, 736447, 736469, 736471,
736511, 736577, 736607, 736639, 736657, 736679, 736691, 736699, 736717, 736721, 736741, 736787, 736793,
736817, 736823, 736843, 736847, 736867, 736871, 736889, 736903, 736921, 736927, 736937, 736951, 736961,
736973, 736987, 736993, 737017, 737039, 737041, 737047, 737053, 737059, 737083, 737089, 737111, 737119,
737129, 737131, 737147, 737159, 737179, 737183, 737203, 737207, 737251, 737263, 737279, 737281, 737287,
737291, 737293, 737309, 737327, 737339, 737351, 737353, 737411, 737413, 737423, 737431, 737479, 737483,
737497, 737501, 737507, 737509, 737531, 737533, 737537, 737563, 737567, 737573, 737591, 737593, 737617,
737629, 737641, 737657, 737663, 737683, 737687, 737717, 737719, 737729, 737747, 737753, 737767, 737773,
737797, 737801, 737809, 737819, 737843, 737857, 737861, 737873, 737887, 737897, 737921, 737927, 737929,
737969, 737981, 737999, 738011, 738029, 738043, 738053, 738071, 738083, 738107, 738109, 738121, 738151,
738163, 738173, 738197, 738211, 738217, 738223, 738247, 738263, 738301, 738313, 738317, 738319, 738341,
738349, 738373, 738379, 738383, 738391, 738401, 738403, 738421, 738443, 738457, 738469, 738487, 738499,
738509, 738523, 738539, 738547, 738581, 738583, 738589, 738623, 738643, 738677, 738707, 738713, 738721,
738743, 738757, 738781, 738791, 738797, 738811, 738827, 738839, 738847, 738851, 738863, 738877, 738889,

738917, 738919, 738923, 738937, 738953, 738961, 738977, 738989, 739003, 739021, 739027, 739031, 739051, 739061, 739069, 739087, 739099, 739103, 739111, 739117, 739121, 739153, 739163, 739171, 739183, 739187, 739199, 739201, 739217, 739241, 739253, 739273, 739283, 739301, 739303, 739307, 739327, 739331, 739337, 739351, 739363, 739369, 739373, 739379, 739391, 739393, 739397, 739399, 739433, 739439, 739463, 739469, 739493, 739507, 739511, 739513, 739523, 739549, 739553, 739579, 739601, 739603, 739621, 739631, 739633, 739637, 739649, 739693, 739699, 739723, 739751, 739759, 739771, 739777, 739787, 739799, 739813, 739829, 739847, 739853, 739859, 739861, 739909, 739931, 739943, 739951, 739957, 739967, 739969, 740011, 740021, 740023, 740041, 740053, 740059, 740087, 740099, 740123, 740141, 740143, 740153, 740161, 740171, 740189, 740191, 740227, 740237, 740279, 740287, 740303, 740321, 740323, 740329, 740351, 740359, 740371, 740387, 740423, 740429, 740461, 740473, 740477, 740483, 740513, 740521, 740527, 740533, 740549, 740561, 740581, 740591, 740599, 740603, 740651, 740653, 740659, 740671, 740681, 740687, 740693, 740711, 740713, 740717, 740737, 740749, 740801, 740849, 740891, 740893, 740897, 740903, 740923, 740939, 740951, 740969, 740989, 741001, 741007, 741011, 741031, 741043, 741053, 741061, 741071, 741077, 741079, 741101, 741119, 741121, 741127, 741131, 741137, 741163, 741187, 741193, 741227, 741229, 741233, 741253, 741283, 741337, 741341, 741343, 741347, 741373, 741401, 741409, 741413, 741431, 741457, 741467, 741469, 741473, 741479, 741491, 741493, 741509, 741541, 741547, 741563, 741569, 741593, 741599, 741641, 741661, 741667, 741677, 741679, 741683, 741691, 741709, 741721, 741781, 741787, 741803, 741809, 741827, 741833, 741847, 741857, 741859, 741869, 741877, 741883, 741913, 741929, 741941, 741967, 741973, 741991, 742009, 742031, 742037, 742057, 742069, 742073, 742111, 742117, 742127, 742151, 742153, 742193, 742199, 742201, 742211, 742213, 742219, 742229, 742241, 742243, 742253, 742277, 742283, 742289, 742307, 742327, 742333, 742351, 742369, 742381, 742393, 742409, 742439, 742457, 742499, 742507, 742513, 742519, 742531, 742537, 742541, 742549, 742559, 742579, 742591, 742607, 742619, 742657, 742663, 742673, 742681, 742697, 742699, 742711, 742717, 742723, 742757, 742759, 742783, 742789, 742801, 742817, 742891, 742897, 742909, 742913, 742943, 742949, 742967, 742981, 742991, 742993, 742999, 743027, 743047, 743059, 743069, 743089, 743111, 743123, 743129, 743131, 743137, 743143, 743159, 743161, 743167, 743173, 743177, 743179, 743203, 743209, 743221, 743251, 743263, 743269, 743273, 743279, 743297, 743321, 743333, 743339, 743363, 743377, 743401, 743423, 743447, 743507, 743549, 743551, 743573, 743579, 743591, 743609, 743657, 743669, 743671, 743689, 743693, 743711, 743731, 743747, 743777, 743779, 743791, 743803, 743819, 743833, 743837, 743849, 743851, 743881, 743891, 743917, 743921, 743923, 743933, 743947, 743987, 743989, 744019, 744043, 744071, 744077, 744083, 744113, 744127, 744137, 744179, 744187, 744199, 744203, 744221, 744239, 744251, 744253, 744283, 744301, 744313, 744353, 744371, 744377, 744389, 744391, 744397, 744407, 744409, 744431, 744451, 744493, 744503, 744511, 744539, 744547, 744559, 744599, 744607, 744637, 744641, 744649, 744659, 744661, 744677, 744701, 744707, 744721, 744727, 744739, 744761, 744767, 744791, 744811, 744817, 744823, 744829, 744833, 744859, 744893, 744911, 744917, 744941, 744949, 744959, 744977, 745001, 745013, 745027, 745033, 745037, 745051, 745067, 745103, 745117, 745133, 745141, 745181, 745187, 745189, 745201, 745231, 745243, 745247, 745249, 745273, 745301, 745307, 745337, 745343, 745357, 745369, 745379, 745391, 745397, 745471, 745477, 745517, 745529, 745531, 745543, 745567, 745573, 745601, 745609, 745621, 745631, 745649, 745673, 745697, 745699, 745709, 745711, 745727, 745733, 745741, 745747, 745751, 745753, 745757, 745817, 745837, 745859, 745873, 745903, 745931, 745933, 745939, 745951, 745973, 745981, 745993, 745999, 746017, 746023, 746033, 746041, 746047, 746069, 746099, 746101, 746107, 746117, 746129, 746153, 746167, 746171, 746177, 746183, 746191, 746197, 746203, 746209, 746227, 746231, 746233, 746243, 746267, 746287, 746303, 746309, 746329, 746353, 746363, 746371, 746411, 746413, 746429, 746477, 746479, 746483, 746497, 746503, 746507, 746509, 746531, 746533, 746561, 746563, 746597, 746653, 746659, 746671, 746677, 746723, 746737, 746743, 746747, 746749, 746773, 746777, 746791, 746797, 746807, 746813, 746839, 746843, 746869, 746873, 746891, 746899, 746903, 746939, 746951, 746957, 746959, 746969, 746981, 746989, 747037, 747049, 747053, 747073, 747107, 747113, 747139, 747157, 747161, 747199, 747203, 747223, 747239, 747259, 747277, 747283, 747287, 747319, 747323, 747343, 747361, 747377, 747391, 747401, 747407, 747421, 747427, 747449, 747451, 747457, 747463, 747493, 747497, 747499, 747521, 747529, 747547, 747557, 747563, 747583, 747587, 747599, 747611, 747619, 747647, 747673, 747679, 747713, 747731, 747737, 747743, 747763, 747781, 747811, 747827, 747829, 747833, 747839, 747841, 747853,

747863, 747869, 747871, 747889, 747917, 747919, 747941, 747953, 747977, 747979, 747991, 748003, 748019,
748021, 748039, 748057, 748091, 748093, 748133, 748169, 748183, 748199, 748207, 748211, 748217, 748219,
748249, 748271, 748273, 748283, 748301, 748331, 748337, 748339, 748343, 748361, 748379, 748387, 748441,
748453, 748463, 748471, 748481, 748487, 748499, 748513, 748523, 748541, 748567, 748589, 748597, 748603,
748609, 748613, 748633, 748637, 748639, 748669, 748687, 748691, 748703, 748711, 748717, 748723, 748729,
748763, 748777, 748789, 748801, 748807, 748817, 748819, 748823, 748829, 748831, 748849, 748861, 748871,
748877, 748883, 748889, 748921, 748933, 748963, 748973, 748981, 748987, 749011, 749027, 749051, 749069,
749081, 749083, 749093, 749129, 749137, 749143, 749149, 749153, 749167, 749171, 749183, 749197, 749209,
749219, 749237, 749249, 749257, 749267, 749279, 749297, 749299, 749323, 749339, 749347, 749351, 749383,
749393, 749401, 749423, 749429, 749431, 749443, 749449, 749453, 749461, 749467, 749471, 749543, 749557,
749587, 749641, 749653, 749659, 749677, 749701, 749711, 749729, 749741, 749747, 749761, 749773, 749779,
749803, 749807, 749809, 749843, 749851, 749863, 749891, 749893, 749899, 749909, 749923, 749927, 749939,
749941, 749971, 749993, 750019, 750037, 750059, 750077, 750083, 750097, 750119, 750121, 750131, 750133,
750137, 750151, 750157, 750161, 750163, 750173, 750179, 750203, 750209, 750223, 750229, 750287, 750311,
750313, 750353, 750383, 750401, 750413, 750419, 750437, 750457, 750473, 750487, 750509, 750517, 750521,
750553, 750571, 750599, 750613, 750641, 750653, 750661, 750667, 750679, 750691, 750707, 750713, 750719,
750721, 750749, 750769, 750787, 750791, 750797, 750803, 750809, 750817, 750829, 750853, 750857, 750863,
750917, 750929, 750943, 750961, 750977, 750983, 751001, 751007, 751021, 751027, 751057, 751061, 751087,
751103, 751123, 751133, 751139, 751141, 751147, 751151, 751181, 751183, 751189, 751193, 751199, 751207,
751217, 751237, 751259, 751273, 751277, 751291, 751297, 751301, 751307, 751319, 751321, 751327, 751343,
751351, 751357, 751363, 751367, 751379, 751411, 751423, 751447, 751453, 751463, 751481, 751523, 751529,
751549, 751567, 751579, 751609, 751613, 751627, 751631, 751633, 751637, 751643, 751661, 751669, 751691,
751711, 751717, 751727, 751739, 751747, 751753, 751759, 751763, 751787, 751799, 751813, 751823, 751841,
751853, 751867, 751871, 751879, 751901, 751909, 751913, 751921, 751943, 751957, 751969, 751987, 751997,
752009, 752023, 752033, 752053, 752083, 752093, 752107, 752111, 752117, 752137, 752149, 752177, 752183,
752189, 752197, 752201, 752203, 752207, 752251, 752263, 752273, 752281, 752287, 752291, 752293, 752299,
752303, 752351, 752359, 752383, 752413, 752431, 752447, 752449, 752459, 752483, 752489, 752503, 752513,
752519, 752527, 752569, 752581, 752593, 752603, 752627, 752639, 752651, 752681, 752683, 752699, 752701,
752707, 752747, 752771, 752789, 752797, 752803, 752809, 752819, 752821, 752831, 752833, 752861, 752867,
752881, 752891, 752903, 752911, 752929, 752933, 752977, 752993, 753001, 753007, 753019, 753023, 753031,
753079, 753091, 753127, 753133, 753139, 753143, 753161, 753187, 753191, 753197, 753229, 753257, 753307,
753329, 753341, 753353, 753367, 753373, 753383, 753409, 753421, 753427, 753437, 753439, 753461, 753463,
753497, 753499, 753527, 753547, 753569, 753583, 753587, 753589, 753611, 753617, 753619, 753631, 753647,
753659, 753677, 753679, 753689, 753691, 753707, 753719, 753721, 753737, 753743, 753751, 753773, 753793,
753799, 753803, 753811, 753821, 753839, 753847, 753859, 753931, 753937, 753941, 753947, 753959, 753979,
753983, 754003, 754027, 754037, 754043, 754057, 754067, 754073, 754081, 754093, 754099, 754109, 754111,
754121, 754123, 754133, 754153, 754157, 754181, 754183, 754207, 754211, 754217, 754223, 754241, 754249,
754267, 754279, 754283, 754289, 754297, 754301, 754333, 754337, 754343, 754367, 754373, 754379, 754381,
754399, 754417, 754421, 754427, 754451, 754463, 754483, 754489, 754513, 754531, 754549, 754573, 754577,
754583, 754597, 754627, 754639, 754651, 754703, 754709, 754711, 754717, 754723, 754739, 754751, 754771,
754781, 754811, 754829, 754861, 754877, 754891, 754903, 754907, 754921, 754931, 754937, 754939, 754967,
754969, 754973, 754979, 754981, 754991, 754993, 755009, 755033, 755057, 755071, 755077, 755081, 755087,
755107, 755137, 755143, 755147, 755171, 755173, 755203, 755213, 755233, 755239, 755257, 755267, 755273,
755309, 755311, 755317, 755329, 755333, 755351, 755357, 755371, 755387, 755393, 755399, 755401, 755413,
755437, 755441, 755449, 755473, 755483, 755509, 755539, 755551, 755561, 755567, 755569, 755593, 755597,
755617, 755627, 755663, 755681, 755707, 755717, 755719, 755737, 755759, 755767, 755771, 755789, 755791,
755809, 755813, 755861, 755863, 755869, 755879, 755899, 755903, 755959, 755969, 755977, 756011, 756023,
756043, 756053, 756097, 756101, 756127, 756131, 756139, 756149, 756167, 756179, 756191, 756199, 756227,
756247, 756251, 756253, 756271, 756281, 756289, 756293, 756319, 756323, 756331, 756373, 756403, 756419,

756421, 756433, 756443, 756463, 756467, 756527, 756533, 756541, 756563, 756571, 756593, 756601, 756607,
756629, 756641, 756649, 756667, 756673, 756683, 756689, 756703, 756709, 756719, 756727, 756739, 756773,
756799, 756829, 756839, 756853, 756869, 756881, 756887, 756919, 756923, 756961, 756967, 756971, 757019,
757039, 757063, 757067, 757109, 757111, 757151, 757157, 757171, 757181, 757201, 757241, 757243, 757247,
757259, 757271, 757291, 757297, 757307, 757319, 757327, 757331, 757343, 757363, 757381, 757387, 757403,
757409, 757417, 757429, 757433, 757457, 757481, 757487, 757507, 757513, 757517, 757543, 757553, 757577,
757579, 757583, 757607, 757633, 757651, 757661, 757693, 757699, 757709, 757711, 757727, 757751, 757753,
757763, 757793, 757807, 757811, 757819, 757829, 757879, 757903, 757909, 757927, 757937, 757943, 757951,
757993, 757997, 758003, 758029, 758041, 758053, 758071, 758083, 758099, 758101, 758111, 758137, 758141,
758159, 758179, 758189, 758201, 758203, 758227, 758231, 758237, 758243, 758267, 758269, 758273, 758279,
758299, 758323, 758339, 758341, 758357, 758363, 758383, 758393, 758411, 758431, 758441, 758449, 758453,
758491, 758501, 758503, 758519, 758521, 758551, 758561, 758573, 758579, 758599, 758617, 758629, 758633,
758671, 758687, 758699, 758707, 758711, 758713, 758729, 758731, 758741, 758743, 758753, 758767, 758783,
758789, 758819, 758827, 758837, 758851, 758867, 758887, 758893, 758899, 758929, 758941, 758957, 758963,
758969, 758971, 758987, 759001, 759019, 759029, 759037, 759047, 759053, 759089, 759103, 759113, 759131,
759149, 759167, 759173, 759179, 759181, 759193, 759223, 759229, 759263, 759287, 759293, 759301, 759313,
759329, 759359, 759371, 759377, 759397, 759401, 759431, 759433, 759457, 759463, 759467, 759491, 759503,
759523, 759547, 759553, 759557, 759559, 759569, 759571, 759581, 759589, 759599, 759617, 759623, 759631,
759637, 759641, 759653, 759659, 759673, 759691, 759697, 759701, 759709, 759719, 759727, 759739, 759757,
759763, 759797, 759799, 759821, 759833, 759881, 759893, 759911, 759923, 759929, 759947, 759953, 759959,
759961, 759973, 760007, 760043, 760063, 760079, 760093, 760103, 760117, 760129, 760141, 760147, 760153,
760163, 760169, 760183, 760187, 760211, 760229, 760231, 760237, 760241, 760261, 760267, 760273, 760289,
760297, 760301, 760321, 760343, 760367, 760373, 760411, 760423, 760433, 760447, 760453, 760457, 760477,
760489, 760499, 760511, 760519, 760531, 760537, 760549, 760553, 760561, 760567, 760579, 760607, 760619,
760621, 760637, 760649, 760657, 760693, 760723, 760729, 760759, 760769, 760783, 760807, 760813, 760841,
760843, 760847, 760871, 760891, 760897, 760901, 760913, 760927, 760933, 760939, 760951, 760961, 760993,
760997, 761003, 761009, 761023, 761051, 761069, 761087, 761113, 761119, 761129, 761153, 761161, 761177,
761179, 761183, 761203, 761207, 761213, 761227, 761249, 761251, 761261, 761263, 761291, 761297, 761347,
761351, 761357, 761363, 761377, 761381, 761389, 761393, 761399, 761407, 761417, 761429, 761437, 761441,
761443, 761459, 761471, 761477, 761483, 761489, 761521, 761531, 761533, 761543, 761561, 761567, 761591,
761597, 761603, 761611, 761623, 761633, 761669, 761671, 761681, 761689, 761711, 761713, 761731, 761773,
761777, 761779, 761807, 761809, 761833, 761861, 761863, 761869, 761879, 761897, 761927, 761939, 761963,
761977, 761983, 761993, 762001, 762007, 762017, 762031, 762037, 762049, 762053, 762061, 762101, 762121,
762187, 762211, 762227, 762233, 762239, 762241, 762253, 762257, 762277, 762319, 762329, 762367, 762371,
762373, 762379, 762389, 762397, 762401, 762407, 762409, 762479, 762491, 762499, 762529, 762539, 762547,
762557, 762563, 762571, 762577, 762583, 762599, 762647, 762653, 762659, 762667, 762721, 762737, 762743,
762761, 762779, 762791, 762809, 762821, 762823, 762847, 762871, 762877, 762893, 762899, 762901, 762913,
762917, 762919, 762959, 762967, 762973, 762989, 763001, 763013, 763027, 763031, 763039, 763043, 763067,
763073, 763093, 763111, 763123, 763141, 763157, 763159, 763183, 763201, 763223, 763237, 763261, 763267,
763271, 763303, 763307, 763339, 763349, 763369, 763381, 763391, 763403, 763409, 763417, 763423, 763429,
763447, 763457, 763471, 763481, 763493, 763513, 763523, 763549, 763559, 763573, 763579, 763583, 763597,
763601, 763613, 763619, 763621, 763627, 763649, 763663, 763673, 763699, 763739, 763751, 763753, 763757,
763771, 763787, 763801, 763811, 763823, 763843, 763859, 763879, 763883, 763897, 763901, 763907, 763913,
763921, 763927, 763937, 763943, 763957, 763967, 763999, 764003, 764011, 764017, 764021, 764041, 764051,
764053, 764059, 764081, 764089, 764111, 764131, 764143, 764149, 764171, 764189, 764209, 764233, 764249,
764251, 764261, 764273, 764293, 764317, 764321, 764327, 764339, 764341, 764369, 764381, 764399, 764431,
764447, 764459, 764471, 764501, 764521, 764539, 764551, 764563, 764587, 764591, 764593, 764611, 764623,
764627, 764629, 764657, 764683, 764689, 764717, 764719, 764723, 764783, 764789, 764809, 764837, 764839,
764849, 764857, 764887, 764891, 764893, 764899, 764903, 764947, 764969, 764971, 764977, 764989, 764993,

764999, 765007, 765031, 765041, 765043, 765047, 765059, 765091, 765097, 765103, 765109, 765131, 765137,
765139, 765143, 765151, 765169, 765181, 765199, 765203, 765209, 765211, 765227, 765229, 765241, 765251,
765257, 765283, 765287, 765293, 765307, 765313, 765319, 765329, 765353, 765379, 765383, 765389, 765409,
765437, 765439, 765461, 765467, 765487, 765497, 765503, 765521, 765533, 765539, 765577, 765581, 765587,
765613, 765619, 765623, 765649, 765659, 765673, 765707, 765727, 765749, 765763, 765767, 765773, 765781,
765823, 765827, 765847, 765851, 765857, 765859, 765881, 765889, 765893, 765899, 765907, 765913, 765931,
765949, 765953, 765971, 765983, 765991, 766021, 766039, 766049, 766067, 766079, 766091, 766097, 766109,
766111, 766127, 766163, 766169, 766177, 766187, 766211, 766223, 766229, 766231, 766237, 766247, 766261,
766273, 766277, 766301, 766313, 766321, 766333, 766357, 766361, 766369, 766373, 766387, 766393, 766399,
766421, 766439, 766453, 766457, 766471, 766477, 766487, 766501, 766511, 766531, 766541, 766543, 766553,
766559, 766583, 766609, 766637, 766639, 766651, 766679, 766687, 766721, 766739, 766757, 766763, 766769,
766793, 766807, 766811, 766813, 766817, 766861, 766867, 766873, 766877, 766891, 766901, 766907, 766937,
766939, 766943, 766957, 766967, 766999, 767017, 767029, 767051, 767071, 767089, 767093, 767101, 767111,
767131, 767147, 767153, 767161, 767167, 767203, 767243, 767279, 767287, 767293, 767309, 767317, 767321,
767323, 767339, 767357, 767359, 767381, 767399, 767423, 767443, 767471, 767489, 767509, 767513, 767521,
767527, 767537, 767539, 767549, 767551, 767587, 767597, 767603, 767617, 767623, 767633, 767647, 767677,
767681, 767707, 767729, 767747, 767749, 767759, 767761, 767773, 767783, 767813, 767827, 767831, 767843,
767857, 767863, 767867, 767869, 767881, 767909, 767951, 767957, 768013, 768029, 768041, 768049, 768059,
768073, 768101, 768107, 768127, 768133, 768139, 768161, 768167, 768169, 768191, 768193, 768197, 768199,
768203, 768221, 768241, 768259, 768263, 768301, 768319, 768323, 768329, 768343, 768347, 768353, 768359,
768371, 768373, 768377, 768389, 768401, 768409, 768419, 768431, 768437, 768457, 768461, 768479, 768491,
768503, 768541, 768563, 768571, 768589, 768613, 768623, 768629, 768631, 768641, 768643, 768653, 768671,
768727, 768751, 768767, 768773, 768787, 768793, 768799, 768811, 768841, 768851, 768853, 768857, 768869,
768881, 768923, 768931, 768941, 768953, 768979, 768983, 769003, 769007, 769019, 769033, 769039, 769057,
769073, 769081, 769091, 769117, 769123, 769147, 769151, 769159, 769169, 769207, 769231, 769243, 769247,
769259, 769261, 769273, 769289, 769297, 769309, 769319, 769339, 769357, 769387, 769411, 769421, 769423,
769429, 769453, 769459, 769463, 769469, 769487, 769541, 769543, 769547, 769553, 769577, 769579, 769589,
769591, 769597, 769619, 769627, 769661, 769663, 769673, 769687, 769723, 769729, 769733, 769739, 769751,
769781, 769789, 769799, 769807, 769837, 769871, 769903, 769919, 769927, 769943, 769961, 769963, 769973,
769987, 769997, 769999, 770027, 770039, 770041, 770047, 770053, 770057, 770059, 770069, 770101, 770111,
770113, 770123, 770129, 770167, 770177, 770179, 770183, 770191, 770207, 770227, 770233, 770239, 770261,
770281, 770291, 770309, 770311, 770353, 770359, 770387, 770401, 770417, 770437, 770447, 770449, 770459,
770503, 770519, 770527, 770533, 770537, 770551, 770557, 770573, 770579, 770587, 770591, 770597, 770611,
770639, 770641, 770647, 770657, 770663, 770669, 770741, 770761, 770767, 770771, 770789, 770801, 770813,
770837, 770839, 770843, 770863, 770867, 770873, 770881, 770897, 770909, 770927, 770929, 770951, 770971,
770981, 770993, 771011, 771013, 771019, 771031, 771037, 771047, 771049, 771073, 771079, 771091, 771109,
771143, 771163, 771179, 771181, 771209, 771217, 771227, 771233, 771269, 771283, 771289, 771293, 771299,
771301, 771349, 771359, 771389, 771401, 771403, 771427, 771431, 771437, 771439, 771461, 771473, 771481,
771499, 771503, 771509, 771517, 771527, 771553, 771569, 771583, 771587, 771607, 771619, 771623, 771629,
771637, 771643, 771653, 771679, 771691, 771697, 771703, 771739, 771763, 771769, 771781, 771809, 771853,
771863, 771877, 771887, 771889, 771899, 771917, 771937, 771941, 771961, 771971, 771973, 771997, 772001,
772003, 772019, 772061, 772073, 772081, 772091, 772097, 772127, 772139, 772147, 772159, 772169, 772181,
772207, 772229, 772231, 772273, 772279, 772297, 772313, 772333, 772339, 772349, 772367, 772379, 772381,
772391, 772393, 772403, 772439, 772441, 772451, 772459, 772477, 772493, 772517, 772537, 772567, 772571,
772573, 772591, 772619, 772631, 772649, 772657, 772661, 772663, 772669, 772691, 772697, 772703, 772721,
772757, 772771, 772789, 772843, 772847, 772853, 772859, 772867, 772903, 772907, 772909, 772913, 772921,
772949, 772963, 772987, 772991, 773021, 773023, 773027, 773029, 773039, 773057, 773063, 773081, 773083,
773093, 773117, 773147, 773153, 773159, 773207, 773209, 773231, 773239, 773249, 773251, 773273, 773287,
773299, 773317, 773341, 773363, 773371, 773387, 773393, 773407, 773417, 773447, 773453, 773473, 773491,

773497, 773501, 773533, 773537, 773561, 773567, 773569, 773579, 773599, 773603, 773609, 773611, 773657, 773659, 773681, 773683, 773693, 773713, 773719, 773723, 773767, 773777, 773779, 773803, 773821, 773831, 773837, 773849, 773863, 773867, 773869, 773879, 773897, 773909, 773933, 773939, 773951, 773953, 773987, 773989, 773999, 774001, 774017, 774023, 774047, 774071, 774073, 774083, 774107, 774119, 774127, 774131, 774133, 774143, 774149, 774161, 774173, 774181, 774199, 774217, 774223, 774229, 774233, 774239, 774283, 774289, 774313, 774317, 774337, 774343, 774377, 774427, 774439, 774463, 774467, 774491, 774511, 774523, 774541, 774551, 774577, 774583, 774589, 774593, 774601, 774629, 774643, 774661, 774667, 774671, 774679, 774691, 774703, 774733, 774757, 774773, 774779, 774791, 774797, 774799, 774803, 774811, 774821, 774833, 774853, 774857, 774863, 774901, 774919, 774929, 774931, 774959, 774997, 775007, 775037, 775043, 775057, 775063, 775079, 775087, 775091, 775097, 775121, 775147, 775153, 775157, 775163, 775189, 775193, 775237, 775241, 775259, 775267, 775273, 775309, 775343, 775349, 775361, 775363, 775367, 775393, 775417, 775441, 775451, 775477, 775507, 775513, 775517, 775531, 775553, 775573, 775601, 775603, 775613, 775627, 775633, 775639, 775661, 775669, 775681, 775711, 775729, 775739, 775741, 775757, 775777, 775787, 775807, 775811, 775823, 775861, 775871, 775889, 775919, 775933, 775937, 775939, 775949, 775963, 775987, 776003, 776029, 776047, 776057, 776059, 776077, 776099, 776117, 776119, 776137, 776143, 776159, 776173, 776177, 776179, 776183, 776201, 776219, 776221, 776233, 776249, 776257, 776267, 776287, 776317, 776327, 776357, 776389, 776401, 776429, 776449, 776453, 776467, 776471, 776483, 776497, 776507, 776513, 776521, 776551, 776557, 776561, 776563, 776569, 776599, 776627, 776651, 776683, 776693, 776719, 776729, 776753, 776759, 776801, 776813, 776819, 776837, 776851, 776861, 776869, 776879, 776887, 776899, 776921, 776947, 776969, 776977, 776983, 776987, 777001, 777011, 777013, 777031, 777041, 777071, 777097, 777103, 777109, 777137, 777143, 777151, 777167, 777169, 777173, 777181, 777187, 777191, 777199, 777209, 777221, 777241, 777247, 777251, 777269, 777277, 777313, 777317, 777349, 777353, 777373, 777383, 777389, 777391, 777419, 777421, 777431, 777433, 777437, 777451, 777463, 777473, 777479, 777541, 777551, 777571, 777583, 777589, 777617, 777619, 777641, 777643, 777661, 777671, 777677, 777683, 777731, 777737, 777743, 777761, 777769, 777781, 777787, 777817, 777839, 777857, 777859, 777863, 777871, 777877, 777901, 777911, 777919, 777977, 777979, 777989, 778013, 778027, 778049, 778051, 778061, 778079, 778081, 778091, 778097, 778109, 778111, 778121, 778123, 778153, 778163, 778187, 778201, 778213, 778223, 778237, 778241, 778247, 778301, 778307, 778313, 778319, 778333, 778357, 778361, 778363, 778391, 778397, 778403, 778409, 778417, 778439, 778469, 778507, 778511, 778513, 778523, 778529, 778537, 778541, 778553, 778559, 778567, 778579, 778597, 778633, 778643, 778663, 778667, 778681, 778693, 778697, 778699, 778709, 778717, 778727, 778733, 778759, 778763, 778769, 778777, 778793, 778819, 778831, 778847, 778871, 778873, 778879, 778903, 778907, 778913, 778927, 778933, 778951, 778963, 778979, 778993, 779003, 779011, 779021, 779039, 779063, 779069, 779081, 779101, 779111, 779131, 779137, 779159, 779173, 779189, 779221, 779231, 779239, 779249, 779267, 779327, 779329, 779341, 779347, 779351, 779353, 779357, 779377, 779413, 779477, 779489, 779507, 779521, 779531, 779543, 779561, 779563, 779573, 779579, 779591, 779593, 779599, 779609, 779617, 779621, 779657, 779659, 779663, 779693, 779699, 779707, 779731, 779747, 779749, 779761, 779767, 779771, 779791, 779797, 779827, 779837, 779869, 779873, 779879, 779887, 779899, 779927, 779939, 779971, 779981, 779983, 779993, 780029, 780037, 780041, 780047, 780049, 780061, 780119, 780127, 780163, 780173, 780179, 780191, 780193, 780211, 780223, 780233, 780253, 780257, 780287, 780323, 780343, 780347, 780371, 780379, 780383, 780389, 780397, 780401, 780421, 780433, 780457, 780469, 780499, 780523, 780553, 780583, 780587, 780601, 780613, 780631, 780649, 780667, 780671, 780679, 780683, 780697, 780707, 780719, 780721, 780733, 780799, 780803, 780809, 780817, 780823, 780833, 780841, 780851, 780853, 780869, 780877, 780887, 780889, 780917, 780931, 780953, 780961, 780971, 780973, 780991, 781003, 781007, 781021, 781043, 781051, 781063, 781069, 781087, 781111, 781117, 781127, 781129, 781139, 781163, 781171, 781199, 781211, 781217, 781229, 781243, 781247, 781271, 781283, 781301, 781307, 781309, 781321, 781327, 781351, 781357, 781367, 781369, 781387, 781397, 781399, 781409, 781423, 781427, 781433, 781453, 781481, 781483, 781493, 781511, 781513, 781519, 781523, 781531, 781559, 781567, 781589, 781601, 781607, 781619, 781631, 781633, 781661, 781673, 781681, 781721, 781733, 781741, 781771, 781799, 781801, 781817, 781819, 781853, 781861, 781867, 781883, 781889, 781897, 781919, 781951, 781961, 781967, 781969, 781973, 781987, 781997, 781999, 782003, 782009, 782011, 782053, 782057, 782071, 782083,

782087, 782107, 782113, 782123, 782129, 782137, 782141, 782147, 782149, 782183, 782189, 782191, 782209,
782219, 782231, 782251, 782263, 782267, 782297, 782311, 782329, 782339, 782371, 782381, 782387, 782389,
782393, 782429, 782443, 782461, 782473, 782489, 782497, 782501, 782519, 782539, 782581, 782611, 782641,
782659, 782669, 782671, 782687, 782689, 782707, 782711, 782723, 782777, 782783, 782791, 782839, 782849,
782861, 782891, 782911, 782921, 782941, 782963, 782981, 782983, 782993, 783007, 783011, 783019, 783023,
783043, 783077, 783089, 783119, 783121, 783131, 783137, 783143, 783149, 783151, 783191, 783193, 783197,
783227, 783247, 783257, 783259, 783269, 783283, 783317, 783323, 783329, 783337, 783359, 783361, 783373,
783379, 783407, 783413, 783421, 783473, 783487, 783527, 783529, 783533, 783553, 783557, 783569, 783571,
783599, 783613, 783619, 783641, 783647, 783661, 783677, 783689, 783691, 783701, 783703, 783707, 783719,
783721, 783733, 783737, 783743, 783749, 783763, 783767, 783779, 783781, 783787, 783791, 783793, 783799,
783803, 783829, 783869, 783877, 783931, 783953, 784009, 784039, 784061, 784081, 784087, 784097, 784103,
784109, 784117, 784129, 784153, 784171, 784181, 784183, 784211, 784213, 784219, 784229, 784243, 784249,
784283, 784307, 784309, 784313, 784321, 784327, 784349, 784351, 784367, 784373, 784379, 784387, 784409,
784411, 784423, 784447, 784451, 784457, 784463, 784471, 784481, 784489, 784501, 784513, 784541, 784543,
784547, 784561, 784573, 784577, 784583, 784603, 784627, 784649, 784661, 784687, 784697, 784717, 784723,
784727, 784753, 784789, 784799, 784831, 784837, 784841, 784859, 784867, 784897, 784913, 784919, 784939,
784957, 784961, 784981, 785003, 785017, 785033, 785053, 785093, 785101, 785107, 785119, 785123, 785129,
785143, 785153, 785159, 785167, 785203, 785207, 785219, 785221, 785227, 785249, 785269, 785287, 785293,
785299, 785303, 785311, 785321, 785329, 785333, 785341, 785347, 785353, 785357, 785363, 785377, 785413,
785423, 785431, 785459, 785461, 785483, 785501, 785503, 785527, 785537, 785549, 785569, 785573, 785579,
785591, 785597, 785623, 785627, 785641, 785651, 785671, 785693, 785717, 785731, 785737, 785753, 785773,
785777, 785779, 785801, 785803, 785809, 785839, 785857, 785861, 785879, 785903, 785921, 785923, 785947,
785951, 785963, 786001, 786013, 786017, 786031, 786047, 786053, 786059, 786061, 786077, 786109, 786127,
786151, 786167, 786173, 786179, 786197, 786211, 786223, 786241, 786251, 786271, 786307, 786311, 786319,
786329, 786337, 786349, 786371, 786407, 786419, 786431, 786433, 786449, 786469, 786491, 786547, 786551,
786553, 786587, 786589, 786613, 786629, 786659, 786661, 786673, 786691, 786697, 786701, 786703, 786707,
786719, 786739, 786763, 786803, 786823, 786829, 786833, 786859, 786881, 786887, 786889, 786901, 786931,
786937, 786941, 786949, 786959, 786971, 786979, 786983, 787021, 787043, 787051, 787057, 787067, 787069,
787079, 787091, 787099, 787123, 787139, 787153, 787181, 787187, 787207, 787217, 787243, 787261, 787277,
787289, 787309, 787331, 787333, 787337, 787357, 787361, 787427, 787429, 787433, 787439, 787447, 787469,
787477, 787483, 787489, 787513, 787517, 787519, 787529, 787537, 787541, 787547, 787573, 787601, 787609,
787621, 787639, 787649, 787667, 787697, 787711, 787747, 787751, 787757, 787769, 787771, 787777, 787783,
787793, 787807, 787811, 787817, 787823, 787837, 787879, 787883, 787903, 787907, 787939, 787973, 787981,
787993, 787999, 788009, 788023, 788027, 788033, 788041, 788071, 788077, 788087, 788089, 788093, 788107,
788129, 788153, 788159, 788167, 788173, 788189, 788209, 788213, 788231, 788261, 788267, 788287, 788309,
788317, 788321, 788351, 788353, 788357, 788363, 788369, 788377, 788383, 788393, 788399, 788413,
788419, 788429, 788449, 788467, 788479, 788497, 788521, 788527, 788531, 788537, 788549, 788561, 788563,
788569, 788603, 788621, 788651, 788659, 788677, 788687, 788701, 788719, 788761, 788779, 788789, 788813,
788819, 788849, 788863, 788867, 788869, 788873, 788891, 788897, 788903, 788927, 788933, 788941, 788947,
788959, 788971, 788993, 788999, 789001, 789017, 789029, 789031, 789067, 789077, 789091, 789097, 789101,
789109, 789121, 789133, 789137, 789149, 789169, 789181, 789221, 789227, 789251, 789311, 789323, 789331,
789343, 789367, 789377, 789389, 789391, 789407, 789419, 789443, 789473, 789491, 789493, 789511, 789527,
789533, 789557, 789571, 789577, 789587, 789589, 789611, 789623, 789631, 789653, 789671, 789673, 789683,
789689, 789709, 789713, 789721, 789731, 789739, 789749, 789793, 789823, 789829, 789847, 789851, 789857,
789883, 789941, 789959, 789961, 789967, 789977, 789979, 790000, 790021, 790033, 790043, 790051, 790057,
790063, 790087, 790093, 790099, 790121, 790169, 790171, 790189, 790199, 790201, 790219, 790241, 790261,
790271, 790277, 790289, 790291, 790327, 790331, 790333, 790351, 790369, 790379, 790397, 790403, 790417,
790421, 790429, 790451, 790459, 790481, 790501, 790513, 790519, 790523, 790529, 790547, 790567, 790583,
790589, 790607, 790613, 790633, 790637, 790649, 790651, 790693, 790697, 790703, 790709, 790733, 790739,

790747, 790753, 790781, 790793, 790817, 790819, 790831, 790843, 790861, 790871, 790879, 790883, 790897,
790927, 790957, 790961, 790967, 790969, 790991, 790997, 791003, 791009, 791017, 791029, 791047, 791053,
791081, 791093, 791099, 791111, 791117, 791137, 791159, 791191, 791201, 791209, 791227, 791233, 791251,
791257, 791261, 791291, 791309, 791311, 791317, 791321, 791347, 791363, 791377, 791387, 791411, 791419,
791431, 791443, 791447, 791473, 791489, 791519, 791543, 791561, 791563, 791569, 791573, 791599, 791627,
791629, 791657, 791663, 791677, 791699, 791773, 791783, 791789, 791797, 791801, 791803, 791827, 791849,
791851, 791887, 791891, 791897, 791899, 791909, 791927, 791929, 791933, 791951, 791969, 791971, 791993,
792023, 792031, 792037, 792041, 792049, 792061, 792067, 792073, 792101, 792107, 792109, 792119, 792131,
792151, 792163, 792179, 792223, 792227, 792229, 792241, 792247, 792257, 792263, 792277, 792283, 792293,
792299, 792301, 792307, 792317, 792359, 792371, 792377, 792383, 792397, 792413, 792443, 792461, 792479,
792481, 792487, 792521, 792529, 792551, 792553, 792559, 792563, 792581, 792593, 792601, 792613, 792629,
792637, 792641, 792643, 792647, 792667, 792679, 792689, 792691, 792697, 792703, 792709, 792713, 792731,
792751, 792769, 792793, 792797, 792821, 792871, 792881, 792893, 792907, 792919, 792929, 792941, 792959,
792973, 792983, 792989, 792991, 793043, 793069, 793099, 793103, 793123, 793129, 793139, 793159, 793181,
793187, 793189, 793207, 793229, 793253, 793279, 793297, 793301, 793327, 793333, 793337, 793343, 793379,
793399, 793439, 793447, 793453, 793487, 793489, 793493, 793511, 793517, 793519, 793537, 793547, 793553,
793561, 793591, 793601, 793607, 793621, 793627, 793633, 793669, 793673, 793691, 793699, 793711, 793717,
793721, 793733, 793739, 793757, 793769, 793777, 793787, 793789, 793813, 793841, 793843, 793853, 793867,
793889, 793901, 793927, 793931, 793939, 793957, 793967, 793979, 793981, 793999, 794009, 794011, 794023,
794033, 794039, 794041, 794063, 794071, 794077, 794089, 794111, 794113, 794119, 794137, 794141, 794149,
794153, 794161, 794173, 794179, 794191, 794201, 794203, 794207, 794221, 794231, 794239, 794249, 794327,
794341, 794363, 794383, 794389, 794399, 794407, 794413, 794449, 794471, 794473, 794477, 794483, 794491,
794509, 794531, 794537, 794543, 794551, 794557, 794569, 794579, 794587, 794593, 794641, 794653, 794657,
794659, 794669, 794693, 794711, 794741, 794743, 794749, 794779, 794831, 794879, 794881, 794887, 794921,
794923, 794953, 794957, 794993, 794999, 795001, 795007, 795023, 795071, 795077, 795079, 795083, 795097,
795101, 795103, 795121, 795127, 795139, 795149, 795161, 795187, 795203, 795211, 795217, 795233, 795239,
795251, 795253, 795299, 795307, 795323, 795329, 795337, 795343, 795349, 795427, 795449, 795461, 795467,
795479, 795493, 795503, 795517, 795527, 795533, 795539, 795551, 795581, 795589, 795601, 795643, 795647,
795649, 795653, 795659, 795661, 795667, 795679, 795703, 795709, 795713, 795727, 795737, 795761, 795763,
795791, 795793, 795797, 795799, 795803, 795827, 795829, 795871, 795877, 795913, 795917, 795931, 795937,
795941, 795943, 795947, 795979, 795983, 795997, 796001, 796009, 796063, 796067, 796091, 796121, 796139,
796141, 796151, 796171, 796177, 796181, 796189, 796193, 796217, 796247, 796259, 796267, 796291, 796303,
796307, 796337, 796339, 796361, 796363, 796373, 796379, 796387, 796391, 796409, 796447, 796451, 796459,
796487, 796493, 796517, 796531, 796541, 796553, 796561, 796567, 796571, 796583, 796591, 796619, 796633,
796657, 796673, 796687, 796693, 796699, 796709, 796711, 796751, 796759, 796769, 796777, 796781, 796799,
796801, 796813, 796819, 796847, 796849, 796853, 796867, 796871, 796877, 796889, 796921, 796931, 796933,
796937, 796951, 796967, 796969, 796981, 797003, 797009, 797021, 797029, 797033, 797039, 797051, 797053,
797057, 797063, 797077, 797119, 797131, 797143, 797161, 797171, 797201, 797207, 797273, 797281, 797287,
797309, 797311, 797333, 797353, 797359, 797383, 797389, 797399, 797417, 797429, 797473, 797497, 797507,
797509, 797539, 797549, 797551, 797557, 797561, 797567, 797569, 797579, 797581, 797591, 797593, 797611,
797627, 797633, 797647, 797681, 797689, 797701, 797711, 797729, 797743, 797747, 797767, 797773, 797813,
797833, 797851, 797869, 797887, 797897, 797911, 797917, 797933, 797947, 797957, 797977, 797987, 798023,
798043, 798059, 798067, 798071, 798079, 798089, 798097, 798101, 798121, 798131, 798139, 798143, 798151,
798173, 798179, 798191, 798197, 798199, 798221, 798223, 798227, 798251, 798257, 798263, 798271, 798293,
798319, 798331, 798373, 798383, 798397, 798403, 798409, 798443, 798451, 798461, 798481, 798487, 798503,
798517, 798521, 798527, 798533, 798569, 798599, 798613, 798641, 798647, 798649, 798667, 798691, 798697,
798701, 798713, 798727, 798737, 798751, 798757, 798773, 798781, 798799, 798823, 798871, 798887, 798911,
798923, 798929, 798937, 798943, 798961, 799003, 799021, 799031, 799061, 799063, 799091, 799093, 799103,
799147, 799151, 799171, 799217, 799219, 799223, 799259, 799291, 799301, 799303, 799307, 799313, 799333,

799343, 799361, 799363, 799369, 799417, 799427, 799441, 799453, 799471, 799481, 799483, 799489, 799507,
799523, 799529, 799543, 799553, 799573, 799609, 799613, 799619, 799621, 799633, 799637, 799651, 799657,
799661, 799679, 799723, 799727, 799739, 799741, 799753, 799759, 799789, 799801, 799807, 799817, 799837,
799853, 799859, 799873, 799891, 799921, 799949, 799961, 799979, 799991, 799993, 799999, 800011, 800029,
800053, 800057, 800077, 800083, 800089, 800113, 800117, 800119, 800123, 800131, 800143, 800159, 800161,
800171, 800209, 800213, 800221, 800231, 800237, 800243, 800281, 800287, 800291, 800311, 800329, 800333,
800351, 800357, 800399, 800407, 800417, 800419, 800441, 800447, 800473, 800477, 800483, 800497, 800509,
800519, 800521, 800533, 800537, 800539, 800549, 800557, 800573, 800587, 800593, 800599, 800621, 800623,
800647, 800651, 800659, 800663, 800669, 800677, 800687, 800693, 800707, 800711, 800729, 800731, 800741,
800743, 800759, 800773, 800783, 800801, 800861, 800873, 800879, 800897, 800903, 800909, 800923, 800953,
800959, 800971, 800977, 800993, 800999, 801001, 801007, 801011, 801019, 801037, 801061, 801077, 801079,
801103, 801107, 801127, 801137, 801179, 801187, 801197, 801217, 801247, 801277, 801289, 801293, 801301,
801331, 801337, 801341, 801349, 801371, 801379, 801403, 801407, 801419, 801421, 801461, 801469, 801487,
801503, 801517, 801539, 801551, 801557, 801569, 801571, 801607, 801611, 801617, 801631, 801641, 801677,
801683, 801701, 801707, 801709, 801733, 801761, 801791, 801809, 801811, 801817, 801833, 801841, 801859,
801883, 801947, 801949, 801959, 801973, 801989, 802007, 802019, 802027, 802031, 802037, 802073, 802103,
802121, 802127, 802129, 802133, 802141, 802147, 802159, 802163, 802177, 802181, 802183, 802189, 802231,
802253, 802279, 802283, 802297, 802331, 802339, 802357, 802387, 802421, 802441, 802453, 802463, 802471,
802499, 802511, 802523, 802531, 802573, 802583, 802589, 802597, 802603, 802609, 802643, 802649, 802651,
802661, 802667, 802709, 802721, 802729, 802733, 802751, 802759, 802777, 802783, 802787, 802793, 802799,
802811, 802829, 802831, 802873, 802909, 802913, 802933, 802951, 802969, 802979, 802987, 803027, 803041,
803053, 803057, 803059, 803087, 803093, 803119, 803141, 803171, 803189, 803207, 803227, 803237, 803251,
803269, 803273, 803287, 803311, 803323, 803333, 803347, 803359, 803389, 803393, 803399, 803417, 803441,
803443, 803447, 803449, 803461, 803479, 803483, 803497, 803501, 803513, 803519, 803549, 803587, 803591,
803609, 803611, 803623, 803629, 803651, 803659, 803669, 803687, 803717, 803729, 803731, 803741, 803749,
803813, 803819, 803849, 803857, 803867, 803893, 803897, 803911, 803921, 803927, 803939, 803963, 803977,
803987, 803989, 804007, 804017, 804031, 804043, 804059, 804073, 804077, 804091, 804107, 804113, 804119,
804127, 804157, 804161, 804179, 804191, 804197, 804203, 804211, 804239, 804259, 804281, 804283, 804313,
804317, 804329, 804337, 804341, 804367, 804371, 804383, 804409, 804443, 804449, 804473, 804493, 804497,
804511, 804521, 804523, 804541, 804553, 804571, 804577, 804581, 804589, 804607, 804611, 804613, 804619,
804653, 804689, 804697, 804703, 804709, 804743, 804751, 804757, 804761, 804767, 804803, 804823, 804829,
804833, 804847, 804857, 804877, 804889, 804893, 804901, 804913, 804919, 804929, 804941, 804943, 804983,
804989, 804997, 805019, 805027, 805031, 805033, 805037, 805061, 805067, 805073, 805081, 805097, 805099,
805109, 805111, 805121, 805153, 805159, 805177, 805187, 805213, 805219, 805223, 805241, 805249, 805267,
805271, 805279, 805289, 805297, 805309, 805313, 805327, 805331, 805333, 805339, 805369, 805381, 805397,
805403, 805421, 805451, 805463, 805471, 805487, 805499, 805501, 805507, 805517, 805523, 805531, 805537,
805559, 805573, 805583, 805589, 805633, 805639, 805687, 805703, 805711, 805723, 805729, 805741, 805757,
805789, 805799, 805807, 805811, 805843, 805853, 805859, 805867, 805873, 805877, 805891, 805901, 805913,
805933, 805967, 805991, 806009, 806011, 806017, 806023, 806027, 806033, 806041, 806051, 806059, 806087,
806107, 806111, 806129, 806137, 806153, 806159, 806177, 806203, 806213, 806233, 806257, 806261, 806263,
806269, 806291, 806297, 806317, 806329, 806363, 806369, 806371, 806381, 806383, 806389, 806447, 806453,
806467, 806483, 806503, 806513, 806521, 806543, 806549, 806579, 806581, 806609, 806639, 806657, 806671,
806719, 806737, 806761, 806783, 806789, 806791, 806801, 806807, 806821, 806857, 806893, 806903, 806917,
806929, 806941, 806947, 806951, 806977, 806999, 807011, 807017, 807071, 807077, 807083, 807089, 807097,
807113, 807119, 807127, 807151, 807181, 807187, 807193, 807197, 807203, 807217, 807221, 807241, 807251,
807259, 807281, 807299, 807337, 807371, 807379, 807383, 807403, 807407, 807409, 807419, 807427, 807463,
807473, 807479, 807487, 807491, 807493, 807509, 807511, 807523, 807539, 807559, 807571, 807607, 807613,
807629, 807637, 807647, 807689, 807707, 807731, 807733, 807749, 807757, 807787, 807797, 807809, 807817,
807869, 807871, 807901, 807907, 807923, 807931, 807941, 807943, 807949, 807973, 807997, 808019, 808021,

808039, 808081, 808097, 808111, 808147, 808153, 808169, 808177, 808187, 808211, 808217, 808229, 808237,
808261, 808267, 808307, 808309, 808343, 808349, 808351, 808361, 808363, 808369, 808373, 808391, 808399,
808417, 808421, 808439, 808441, 808459, 808481, 808517, 808523, 808553, 808559, 808579, 808589, 808597,
808601, 808603, 808627, 808637, 808651, 808679, 808681, 808693, 808699, 808721, 808733, 808739, 808747,
808751, 808771, 808777, 808789, 808793, 808837, 808853, 808867, 808919, 808937, 808957, 808961, 808981,
808991, 808993, 809023, 809041, 809051, 809063, 809087, 809093, 809101, 809141, 809143, 809147, 809173,
809177, 809189, 809201, 809203, 809213, 809231, 809239, 809243, 809261, 809269, 809273, 809297, 809309,
809323, 809339, 809357, 809359, 809377, 809383, 809399, 809401, 809407, 809423, 809437, 809443, 809447,
809453, 809461, 809491, 809507, 809521, 809527, 809563, 809569, 809579, 809581, 809587, 809603, 809629,
809701, 809707, 809719, 809729, 809737, 809741, 809747, 809749, 809759, 809771, 809779, 809797, 809801,
809803, 809821, 809827, 809833, 809839, 809843, 809869, 809891, 809903, 809909, 809917, 809929, 809981,
809983, 809993, 810013, 810023, 810049, 810053, 810059, 810071, 810079, 810091, 810109, 810137, 810149,
810151, 810191, 810193, 810209, 810223, 810239, 810253, 810259, 810269, 810281, 810307, 810319, 810343,
810349, 810353, 810361, 810367, 810377, 810379, 810389, 810391, 810401, 810409, 810419, 810427, 810437,
810443, 810457, 810473, 810487, 810493, 810503, 810517, 810533, 810539, 810541, 810547, 810553, 810571,
810581, 810583, 810587, 810643, 810653, 810659, 810671, 810697, 810737, 810757, 810763, 810769, 810791,
810809, 810839, 810853, 810871, 810881, 810893, 810907, 810913, 810923, 810941, 810949, 810961, 810967,
810973, 810989, 811037, 811039, 811067, 811081, 811099, 811123, 811127, 811147, 811157, 811163, 811171,
811183, 811193, 811199, 811207, 811231, 811241, 811253, 811259, 811273, 811277, 811289, 811297, 811337,
811351, 811379, 811387, 811411, 811429, 811441, 811457, 811469, 811493, 811501, 811511, 811519, 811523,
811553, 811561, 811583, 811607, 811619, 811627, 811637, 811649, 811651, 811667, 811691, 811697, 811703,
811709, 811729, 811747, 811753, 811757, 811763, 811771, 811777, 811799, 811819, 811861, 811871, 811879,
811897, 811919, 811931, 811933, 811957, 811961, 811981, 811991, 811997, 812011, 812033, 812047, 812051,
812057, 812081, 812101, 812129, 812137, 812167, 812173, 812179, 812183, 812191, 812213, 812221, 812233,
812249, 812257, 812267, 812281, 812297, 812299, 812309, 812341, 812347, 812351, 812353, 812359, 812363,
812381, 812387, 812393, 812401, 812431, 812443, 812467, 812473, 812477, 812491, 812501, 812503, 812519,
812527, 812587, 812597, 812599, 812627, 812633, 812639, 812641, 812671, 812681, 812689, 812699, 812701,
812711, 812717, 812731, 812759, 812761, 812807, 812849, 812857, 812869, 812921, 812939, 812963, 812969,
813013, 813017, 813023, 813041, 813049, 813061, 813083, 813089, 813091, 813097, 813107, 813121, 813133,
813157, 813167, 813199, 813203, 813209, 813217, 813221, 813227, 813251, 813269, 813277, 813283, 813287,
813299, 813301, 813311, 813343, 813361, 813367, 813377, 813383, 813401, 813419, 813427, 813443, 813493,
813499, 813503, 813511, 813529, 813541, 813559, 813577, 813583, 813601, 813613, 813623, 813647, 813677,
813697, 813707, 813721, 813749, 813767, 813797, 813811, 813817, 813829, 813833, 813847, 813863, 813871,
813893, 813907, 813931, 813961, 813971, 813991, 813997, 814003, 814007, 814013, 814019, 814031, 814043,
814049, 814061, 814063, 814067, 814069, 814081, 814097, 814127, 814129, 814139, 814171, 814183, 814193,
814199, 814211, 814213, 814237, 814241, 814243, 814279, 814309, 814327, 814337, 814367, 814379, 814381,
814393, 814399, 814403, 814423, 814447, 814469, 814477, 814493, 814501, 814531, 814537, 814543, 814559,
814577, 814579, 814601, 814603, 814609, 814631, 814633, 814643, 814687, 814699, 814717, 814741, 814747,
814763, 814771, 814783, 814789, 814799, 814823, 814829, 814841, 814859, 814873, 814883, 814889, 814901,
814903, 814927, 814937, 814939, 814943, 814949, 814991, 815029, 815033, 815047, 815053, 815063, 815123,
815141, 815149, 815159, 815173, 815197, 815209, 815231, 815251, 815257, 815261, 815273, 815279, 815291,
815317, 815333, 815341, 815351, 815389, 815401, 815411, 815413, 815417, 815431, 815453, 815459, 815471,
815491, 815501, 815519, 815527, 815533, 815539, 815543, 815569, 815587, 815599, 815621, 815623, 815627,
815653, 815663, 815669, 815671, 815681, 815687, 815693, 815713, 815729, 815809, 815819, 815821, 815831,
815851, 815869, 815891, 815897, 815923, 815933, 815939, 815953, 815963, 815977, 815989, 816019, 816037,
816043, 816047, 816077, 816091, 816103, 816113, 816121, 816131, 816133, 816157, 816161, 816163, 816169,
816191, 816203, 816209, 816217, 816223, 816227, 816239, 816251, 816271, 816317, 816329, 816341, 816353,
816367, 816377, 816401, 816427, 816443, 816451, 816469, 816499, 816521, 816539, 816547, 816559, 816581,
816587, 816589, 816593, 816649, 816653, 816667, 816689, 816691, 816703, 816709, 816743, 816763, 816769,

816779, 816811, 816817, 816821, 816839, 816841, 816847, 816857, 816859, 816869, 816883, 816887, 816899,
816911, 816917, 816919, 816929, 816941, 816947, 816961, 816971, 817013, 817027, 817039, 817049, 817051,
817073, 817081, 817087, 817093, 817111, 817123, 817127, 817147, 817151, 817153, 817163, 817169, 817183,
817211, 817237, 817273, 817277, 817279, 817291, 817303, 817319, 817321, 817331, 817337, 817357, 817379,
817403, 817409, 817433, 817457, 817463, 817483, 817519, 817529, 817549, 817561, 817567, 817603, 817637,
817651, 817669, 817679, 817697, 817709, 817711, 817721, 817723, 817727, 817757, 817769, 817777, 817783,
817787, 817793, 817823, 817837, 817841, 817867, 817871, 817877, 817889, 817891, 817897, 817907, 817913,
817919, 817933, 817951, 817979, 817987, 818011, 818017, 818021, 818093, 818099, 818101, 818113, 818123,
818143, 818171, 818173, 818189, 818219, 818231, 818239, 818249, 818281, 818287, 818291, 818303, 818309,
818327, 818339, 818341, 818347, 818353, 818359, 818371, 818383, 818393, 818399, 818413, 818429, 818453,
818473, 818509, 818561, 818569, 818579, 818581, 818603, 818621, 818659, 818683, 818687, 818689, 818707,
818717, 818723, 818813, 818819, 818821, 818827, 818837, 818887, 818897, 818947, 818959, 818963, 818969,
818977, 818999, 819001, 819017, 819029, 819031, 819037, 819061, 819073, 819083, 819101, 819131, 819149,
819157, 819167, 819173, 819187, 819229, 819239, 819241, 819251, 819253, 819263, 819271, 819289, 819307,
819311, 819317, 819319, 819367, 819373, 819389, 819391, 819407, 819409, 819419, 819431, 819437, 819443,
819449, 819457, 819463, 819473, 819487, 819491, 819493, 819499, 819503, 819509, 819523, 819563, 819583,
819593, 819607, 819617, 819619, 819629, 819647, 819653, 819659, 819673, 819691, 819701, 819719, 819737,
819739, 819761, 819769, 819773, 819781, 819787, 819799, 819811, 819823, 819827, 819829, 819853, 819899,
819911, 819913, 819937, 819943, 819977, 819989, 819991, 820037, 820051, 820067, 820073, 820093, 820109,
820117, 820129, 820133, 820163, 820177, 820187, 820201, 820213, 820223, 820231, 820241, 820243, 820247,
820271, 820273, 820279, 820319, 820321, 820331, 820333, 820343, 820349, 820361, 820367, 820399, 820409,
820411, 820427, 820429, 820441, 820459, 820481, 820489, 820537, 820541, 820559, 820577, 820597, 820609,
820619, 820627, 820637, 820643, 820649, 820657, 820679, 820681, 820691, 820711, 820723, 820733, 820747,
820753, 820759, 820763, 820789, 820793, 820837, 820873, 820891, 820901, 820907, 820909, 820921, 820927,
820957, 820969, 820991, 820997, 821003, 821027, 821039, 821053, 821057, 821063, 821069, 821081, 821089,
821099, 821101, 821113, 821131, 821143, 821191, 821153, 821167, 821173, 821207, 821209, 821263, 821281,
821291, 821297, 821311, 821329, 821333, 821377, 821383, 821411, 821441, 821449, 821459, 821461, 821467,
821477, 821479, 821489, 821497, 821507, 821519, 821551, 821573, 821603, 821641, 821647, 821651, 821663,
821677, 821741, 821747, 821753, 821759, 821771, 821801, 821803, 821809, 821819, 821827, 821833, 821851,
821857, 821861, 821869, 821879, 821897, 821911, 821939, 821941, 821971, 821993, 821999, 822007, 822011,
822013, 822037, 822049, 822067, 822079, 822113, 822131, 822139, 822161, 822163, 822167, 822169, 822191,
822197, 822221, 822223, 822229, 822233, 822253, 822259, 822277, 822293, 822299, 822313, 822317, 822323,
822329, 822343, 822347, 822361, 822379, 822383, 822389, 822391, 822407, 822431, 822433, 822517, 822539,
822541, 822551, 822553, 822557, 822571, 822581, 822587, 822589, 822599, 822607, 822611, 822631, 822667,
822671, 822673, 822683, 822691, 822697, 822713, 822721, 822727, 822739, 822743, 822761, 822763, 822781,
822791, 822793, 822803, 822821, 822823, 822839, 822853, 822881, 822883, 822889, 822893, 822901, 822907,
822949, 822971, 822973, 822989, 823001, 823003, 823013, 823033, 823051, 823117, 823127, 823129, 823153,
823169, 823177, 823183, 823201, 823219, 823231, 823237, 823241, 823243, 823261, 823271, 823283, 823309,
823337, 823349, 823351, 823357, 823373, 823399, 823421, 823447, 823451, 823457, 823481, 823483, 823489,
823499, 823519, 823541, 823547, 823553, 823573, 823591, 823601, 823619, 823621, 823637, 823643, 823651,
823663, 823679, 823703, 823709, 823717, 823721, 823723, 823727, 823729, 823741, 823747, 823759, 823777,
823787, 823789, 823799, 823819, 823829, 823831, 823841, 823843, 823877, 823903, 823913, 823961, 823967,
823969, 823981, 823993, 823997, 824017, 824029, 824039, 824063, 824069, 824077, 824081, 824099, 824123,
824137, 824147, 824179, 824183, 824189, 824191, 824227, 824231, 824233, 824269, 824281, 824287, 824339,
824393, 824399, 824401, 824413, 824419, 824437, 824443, 824459, 824477, 824489, 824497, 824501, 824513,
824531, 824539, 824563, 824591, 824609, 824641, 824647, 824651, 824669, 824671, 824683, 824699, 824701,
824723, 824741, 824749, 824753, 824773, 824777, 824779, 824801, 824821, 824833, 824843, 824861, 824893,
824899, 824911, 824921, 824933, 824939, 824947, 824951, 824977, 824981, 824983, 825001, 825007, 825017,
825029, 825047, 825049, 825059, 825067, 825073, 825101, 825107, 825109, 825131, 825161, 825191, 825193,

825199, 825203, 825229, 825241, 825247, 825259, 825277, 825281, 825283, 825287, 825301, 825329, 825337,
825343, 825347, 825353, 825361, 825389, 825397, 825403, 825413, 825421, 825439, 825443, 825467, 825479,
825491, 825509, 825527, 825533, 825547, 825551, 825553, 825577, 825593, 825611, 825613, 825637, 825647,
825661, 825679, 825689, 825697, 825701, 825709, 825733, 825739, 825749, 825763, 825779, 825791, 825821,
825827, 825829, 825857, 825883, 825889, 825919, 825947, 825959, 825961, 825971, 825983, 825991, 825997,
826019, 826037, 826039, 826051, 826061, 826069, 826087, 826093, 826097, 826129, 826151, 826153, 826169,
826171, 826193, 826201, 826211, 826271, 826283, 826289, 826303, 826313, 826333, 826339, 826349, 826351,
826363, 826379, 826381, 826391, 826393, 826403, 826411, 826453, 826477, 826493, 826499, 826541, 826549,
826559, 826561, 826571, 826583, 826603, 826607, 826613, 826621, 826663, 826667, 826669, 826673, 826681,
826697, 826699, 826711, 826717, 826723, 826729, 826753, 826759, 826783, 826799, 826807, 826811, 826831,
826849, 826867, 826879, 826883, 826907, 826921, 826927, 826939, 826957, 826963, 826967, 826979, 826997,
827009, 827023, 827039, 827041, 827063, 827087, 827129, 827131, 827143, 827147, 827161, 827213, 827227,
827231, 827251, 827269, 827293, 827303, 827311, 827327, 827347, 827369, 827389, 827417, 827423, 827429,
827443, 827447, 827461, 827473, 827501, 827521, 827537, 827539, 827549, 827581, 827591, 827599, 827633,
827639, 827677, 827681, 827693, 827699, 827719, 827737, 827741, 827767, 827779, 827791, 827803, 827809,
827821, 827833, 827837, 827843, 827851, 827857, 827867, 827873, 827899, 827903, 827923, 827927, 827929,
827941, 827969, 827987, 827989, 828007, 828011, 828013, 828029, 828043, 828059, 828067, 828071, 828101,
828109, 828119, 828127, 828131, 828133, 828169, 828199, 828209, 828221, 828239, 828277, 828349, 828361,
828371, 828379, 828383, 828397, 828407, 828409, 828431, 828449, 828517, 828523, 828547, 828557, 828577,
828587, 828601, 828637, 828643, 828649, 828673, 828677, 828691, 828697, 828701, 828703, 828721, 828731,
828743, 828757, 828787, 828797, 828809, 828811, 828823, 828829, 828833, 828859, 828871, 828881, 828889,
828899, 828901, 828917, 828923, 828941, 828953, 828967, 828977, 829001, 829013, 829057, 829063, 829069,
829093, 829097, 829111, 829121, 829123, 829151, 829159, 829177, 829187, 829193, 829211, 829223, 829229,
829237, 829249, 829267, 829273, 829289, 829319, 829349, 829399, 829453, 829457, 829463, 829469, 829501,
829511, 829519, 829537, 829547, 829561, 829601, 829613, 829627, 829637, 829639, 829643, 829657, 829687,
829693, 829709, 829721, 829723, 829727, 829729, 829733, 829757, 829789, 829811, 829813, 829819, 829831,
829841, 829847, 829849, 829867, 829877, 829883, 829949, 829967, 829979, 829987, 829993, 830003, 830017,
830041, 830051, 830099, 830111, 830117, 830131, 830143, 830153, 830173, 830177, 830191, 830233, 830237,
830257, 830267, 830279, 830293, 830309, 830311, 830327, 830329, 830339, 830341, 830353, 830359, 830363,
830383, 830387, 830411, 830413, 830419, 830441, 830447, 830449, 830477, 830483, 830497, 830503, 830513,
830549, 830551, 830561, 830567, 830579, 830587, 830591, 830597, 830617, 830639, 830657, 830677, 830693,
830719, 830729, 830741, 830743, 830777, 830789, 830801, 830827, 830833, 830839, 830849, 830861, 830873,
830887, 830891, 830899, 830911, 830923, 830939, 830957, 830981, 830989, 831023, 831031, 831037, 831043,
831067, 831071, 831073, 831091, 831109, 831139, 831161, 831163, 831167, 831191, 831217, 831221, 831239,
831253, 831287, 831301, 831323, 831329, 831361, 831367, 831371, 831373, 831407, 831409, 831431, 831433,
831437, 831443, 831461, 831503, 831529, 831539, 831541, 831547, 831553, 831559, 831583, 831587, 831599,
831617, 831619, 831631, 831643, 831647, 831653, 831659, 831661, 831679, 831683, 831697, 831707, 831709,
831713, 831731, 831739, 831751, 831757, 831769, 831781, 831799, 831811, 831821, 831829, 831847, 831851,
831863, 831881, 831889, 831893, 831899, 831911, 831913, 831917, 831967, 831983, 832003, 832063, 832079,
832081, 832103, 832109, 832121, 832123, 832129, 832141, 832151, 832157, 832159, 832189, 832211, 832217,
832253, 832291, 832297, 832309, 832327, 832331, 832339, 832361, 832367, 832369, 832373, 832379, 832399,
832411, 832421, 832427, 832451, 832457, 832477, 832483, 832487, 832493, 832499, 832519, 832583, 832591,
832597, 832607, 832613, 832621, 832627, 832631, 832633, 832639, 832673, 832679, 832681, 832687, 832693,
832703, 832709, 832717, 832721, 832729, 832747, 832757, 832763, 832771, 832787, 832801, 832837, 832841,
832861, 832879, 832883, 832889, 832913, 832919, 832927, 832933, 832943, 832957, 832963, 832969, 832973,
832987, 833009, 833023, 833033, 833047, 833057, 833099, 833101, 833117, 833171, 833177, 833179, 833191,
833197, 833201, 833219, 833251, 833269, 833281, 833293, 833299, 833309, 833347, 833353, 833363, 833377,
833389, 833429, 833449, 833453, 833461, 833467, 833477, 833479, 833491, 833509, 833537, 833557, 833563,
833593, 833597, 833617, 833633, 833659, 833669, 833689, 833711, 833713, 833717, 833719, 833737, 833747,

833759, 833783, 833801, 833821, 833839, 833843, 833857, 833873, 833887, 833893, 833897, 833923, 833927,
833933, 833947, 833977, 833999, 834007, 834013, 834023, 834059, 834107, 834131, 834133, 834137, 834143,
834149, 834151, 834181, 834199, 834221, 834257, 834259, 834269, 834277, 834283, 834287, 834299, 834311,
834341, 834367, 834433, 834439, 834469, 834487, 834497, 834503, 834511, 834523, 834527, 834569, 834571,
834593, 834599, 834607, 834611, 834623, 834629, 834641, 834643, 834653, 834671, 834703, 834709, 834721,
834761, 834773, 834781, 834787, 834797, 834809, 834811, 834829, 834857, 834859, 834893, 834913, 834941,
834947, 834949, 834959, 834961, 834983, 834991, 835001, 835013, 835019, 835033, 835039, 835097, 835099,
835117, 835123, 835139, 835141, 835207, 835213, 835217, 835249, 835253, 835271, 835313, 835319, 835321,
835327, 835369, 835379, 835391, 835399, 835421, 835427, 835441, 835451, 835453, 835459, 835469, 835489,
835511, 835531, 835553, 835559, 835591, 835603, 835607, 835609, 835633, 835643, 835661, 835663, 835673,
835687, 835717, 835721, 835733, 835739, 835759, 835789, 835811, 835817, 835819, 835823, 835831, 835841,
835847, 835859, 835897, 835909, 835927, 835931, 835937, 835951, 835957, 835973, 835979, 835987, 835993,
835997, 836047, 836063, 836071, 836107, 836117, 836131, 836137, 836149, 836153, 836159, 836161, 836183,
836189, 836191, 836203, 836219, 836233, 836239, 836243, 836267, 836291, 836299, 836317, 836327, 836347,
836351, 836369, 836377, 836387, 836413, 836449, 836471, 836477, 836491, 836497, 836501, 836509, 836567,
836569, 836573, 836609, 836611, 836623, 836657, 836663, 836677, 836683, 836699, 836701, 836707, 836713,
836729, 836747, 836749, 836753, 836761, 836789, 836807, 836821, 836833, 836839, 836861, 836863, 836873,
836879, 836881, 836917, 836921, 836939, 836951, 836971, 837017, 837043, 837047, 837059, 837071, 837073,
837077, 837079, 837107, 837113, 837139, 837149, 837157, 837191, 837203, 837257, 837271, 837283, 837293,
837307, 837313, 837359, 837367, 837373, 837377, 837379, 837409, 837413, 837439, 837451, 837461, 837467,
837497, 837503, 837509, 837521, 837533, 837583, 837601, 837611, 837619, 837631, 837659, 837667, 837673,
837677, 837679, 837721, 837731, 837737, 837773, 837779, 837797, 837817, 837833, 837847, 837853, 837887,
837923, 837929, 837931, 837937, 837943, 837979, 838003, 838021, 838037, 838039, 838043, 838063, 838069,
838091, 838093, 838099, 838133, 838139, 838141, 838153, 838157, 838169, 838171, 838193, 838207, 838247,
838249, 838349, 838351, 838363, 838367, 838379, 838391, 838393, 838399, 838403, 838421, 838429, 838441,
838447, 838459, 838463, 838471, 838483, 838517, 838547, 838553, 838561, 838571, 838583, 838589, 838597,
838601, 838609, 838613, 838631, 838633, 838657, 838667, 838687, 838693, 838711, 838751, 838757, 838769,
838771, 838777, 838781, 838807, 838813, 838837, 838853, 838889, 838897, 838909, 838913, 838919, 838927,
838931, 838939, 838949, 838951, 838963, 838969, 838991, 838993, 839009, 839029, 839051, 839071, 839087,
839117, 839131, 839161, 839203, 839207, 839221, 839227, 839261, 839269, 839303, 839323, 839327, 839351,
839353, 839369, 839381, 839413, 839429, 839437, 839441, 839453, 839459, 839471, 839473, 839483, 839491,
839497, 839519, 839539, 839551, 839563, 839599, 839603, 839609, 839611, 839617, 839621, 839633, 839651,
839653, 839669, 839693, 839723, 839731, 839767, 839771, 839791, 839801, 839809, 839831, 839837, 839873,
839879, 839887, 839897, 839899, 839903, 839911, 839921, 839957, 839959, 839963, 839981, 839999, 840023,
840053, 840061, 840067, 840083, 840109, 840137, 840139, 840149, 840163, 840179, 840181, 840187, 840197,
840223, 840239, 840241, 840253, 840269, 840277, 840289, 840299, 840319, 840331, 840341, 840347, 840353,
840439, 840451, 840457, 840467, 840473, 840479, 840491, 840523, 840547, 840557, 840571, 840589, 840601,
840611, 840643, 840661, 840683, 840703, 840709, 840713, 840727, 840733, 840743, 840757, 840761, 840767,
840817, 840821, 840823, 840839, 840841, 840859, 840863, 840907, 840911, 840923, 840929, 840941, 840943,
840967, 840979, 840989, 840991, 841003, 841013, 841019, 841021, 841063, 841069, 841079, 841081, 841091,
841097, 841103, 841147, 841157, 841189, 841193, 841207, 841213, 841219, 841223, 841231, 841237, 841241,
841259, 841273, 841277, 841283, 841289, 841297, 841307, 841327, 841333, 841349, 841369, 841391, 841397,
841411, 841427, 841447, 841457, 841459, 841541, 841549, 841559, 841573, 841597, 841601, 841637, 841651,
841661, 841663, 841691, 841697, 841727, 841741, 841751, 841793, 841801, 841849, 841859, 841873, 841879,
841889, 841913, 841921, 841927, 841931, 841933, 841979, 841987, 842003, 842021, 842041, 842047, 842063,
842071, 842077, 842081, 842087, 842089, 842111, 842113, 842141, 842147, 842159, 842161, 842167, 842173,
842183, 842203, 842209, 842249, 842267, 842279, 842291, 842293, 842311, 842321, 842323, 842339, 842341,
842351, 842353, 842371, 842383, 842393, 842399, 842407, 842417, 842419, 842423, 842447, 842449, 842473,
842477, 842483, 842489, 842497, 842507, 842519, 842521, 842531, 842551, 842581, 842587, 842599, 842617,

842623, 842627, 842657, 842701, 842729, 842747, 842759, 842767, 842771, 842791, 842801, 842813, 842819,
842857, 842869, 842879, 842887, 842923, 842939, 842951, 842957, 842969, 842977, 842981, 842987, 842993,
843043, 843067, 843079, 843091, 843103, 843113, 843127, 843131, 843137, 843173, 843179, 843181, 843209,
843211, 843229, 843253, 843257, 843289, 843299, 843301, 843307, 843331, 843347, 843361, 843371, 843377,
843379, 843383, 843397, 843443, 843449, 843457, 843461, 843473, 843487, 843497, 843503, 843527, 843539,
843553, 843559, 843587, 843589, 843607, 843613, 843629, 843643, 843649, 843677, 843679, 843701, 843737,
843757, 843763, 843779, 843781, 843793, 843797, 843811, 843823, 843833, 843841, 843881, 843883, 843889,
843901, 843907, 843911, 844001, 844013, 844043, 844061, 844069, 844087, 844093, 844111, 844117, 844121,
844127, 844139, 844141, 844153, 844157, 844163, 844183, 844187, 844199, 844201, 844243, 844247, 844253,
844279, 844289, 844297, 844309, 844321, 844351, 844369, 844421, 844427, 844429, 844433, 844439, 844447,
844453, 844457, 844463, 844469, 844483, 844489, 844499, 844507, 844511, 844513, 844517, 844523, 844549,
844553, 844601, 844603, 844609, 844619, 844621, 844631, 844639, 844643, 844651, 844709, 844717, 844733,
844757, 844763, 844769, 844771, 844777, 844841, 844847, 844861, 844867, 844891, 844897, 844903, 844913,
844927, 844957, 844999, 845003, 845017, 845021, 845027, 845041, 845069, 845083, 845099, 845111, 845129,
845137, 845167, 845179, 845183, 845197, 845203, 845209, 845219, 845231, 845237, 845261, 845279, 845287,
845303, 845309, 845333, 845347, 845357, 845363, 845371, 845381, 845387, 845431, 845441, 845447, 845459,
845489, 845491, 845531, 845567, 845599, 845623, 845653, 845657, 845659, 845683, 845717, 845723, 845729,
845749, 845753, 845771, 845777, 845809, 845833, 845849, 845863, 845879, 845881, 845893, 845909, 845921,
845927, 845941, 845951, 845969, 845981, 845983, 845987, 845989, 846037, 846059, 846061, 846067, 846113,
846137, 846149, 846161, 846179, 846187, 846217, 846229, 846233, 846247, 846259, 846271, 846323, 846341,
846343, 846353, 846359, 846361, 846383, 846389, 846397, 846401, 846403, 846407, 846421, 846427, 846437,
846457, 846487, 846493, 846499, 846529, 846563, 846577, 846589, 846647, 846661, 846667, 846673, 846689,
846721, 846733, 846739, 846749, 846751, 846757, 846779, 846823, 846841, 846851, 846869, 846871, 846877,
846913, 846917, 846919, 846931, 846943, 846949, 846953, 846961, 846973, 846977, 846983, 846997, 847009,
847031, 847037, 847043, 847051, 847069, 847073, 847079, 847097, 847103, 847109, 847129, 847139, 847151,
847157, 847163, 847169, 847193, 847201, 847213, 847219, 847237, 847247, 847271, 847277, 847279, 847283,
847309, 847321, 847339, 847361, 847367, 847373, 847393, 847423, 847453, 847477, 847493, 847499, 847507,
847519, 847531, 847537, 847543, 847549, 847577, 847589, 847601, 847607, 847621, 847657, 847663, 847673,
847681, 847687, 847697, 847703, 847727, 847729, 847741, 847787, 847789, 847813, 847817, 847853, 847871,
847883, 847901, 847919, 847933, 847937, 847949, 847967, 847969, 847991, 847993, 847997, 848017, 848051,
848087, 848101, 848119, 848123, 848131, 848143, 848149, 848173, 848201, 848203, 848213, 848227, 848251,
848269, 848273, 848297, 848321, 848359, 848363, 848383, 848387, 848399, 848417, 848423, 848429, 848443,
848461, 848467, 848473, 848489, 848531, 848537, 848557, 848567, 848579, 848591, 848593, 848599, 848611,
848629, 848633, 848647, 848651, 848671, 848681, 848699, 848707, 848713, 848737, 848747, 848761, 848779,
848789, 848791, 848797, 848803, 848807, 848839, 848843, 848849, 848851, 848857, 848879, 848893, 848909,
848921, 848923, 848927, 848933, 848941, 848959, 848983, 848993, 849019, 849047, 849049, 849061, 849083,
849097, 849103, 849119, 849127, 849131, 849143, 849161, 849179, 849197, 849203, 849217, 849221, 849223,
849241, 849253, 849271, 849301, 849311, 849347, 849349, 849353, 849383, 849391, 849419, 849427, 849461,
849467, 849481, 849523, 849533, 849539, 849571, 849581, 849587, 849593, 849599, 849601, 849649, 849691,
849701, 849703, 849721, 849727, 849731, 849733, 849743, 849763, 849767, 849773, 849829, 849833, 849839,
849857, 849869, 849883, 849917, 849923, 849931, 849943, 849967, 849973, 849991, 849997, 850009, 850021,
850027, 850033, 850043, 850049, 850061, 850063, 850081, 850093, 850121, 850133, 850139, 850147, 850177,
850181, 850189, 850207, 850211, 850229, 850243, 850247, 850253, 850261, 850271, 850273, 850301, 850303,
850331, 850337, 850349, 850351, 850373, 850387, 850393, 850397, 850403, 850417, 850427, 850433, 850439,
850453, 850457, 850481, 850529, 850537, 850567, 850571, 850613, 850631, 850637, 850673, 850679, 850691,
850711, 850727, 850753, 850781, 850807, 850823, 850849, 850853, 850879, 850891, 850897, 850933, 850943,
850951, 850973, 850979, 851009, 851017, 851033, 851041, 851051, 851057, 851087, 851093, 851113, 851117,
851131, 851153, 851159, 851171, 851177, 851197, 851203, 851209, 851231, 851239, 851251, 851261, 851267,
851273, 851293, 851297, 851303, 851321, 851327, 851351, 851359, 851363, 851381, 851387, 851393, 851401,

851413, 851419, 851423, 851449, 851471, 851491, 851507, 851519, 851537, 851549, 851569, 851573, 851597,
851603, 851623, 851633, 851639, 851647, 851659, 851671, 851677, 851689, 851723, 851731, 851749, 851761,
851797, 851801, 851803, 851813, 851821, 851831, 851839, 851843, 851863, 851881, 851891, 851899, 851953,
851957, 851971, 852011, 852013, 852031, 852037, 852079, 852101, 852121, 852139, 852143, 852149, 852151,
852167, 852179, 852191, 852197, 852199, 852211, 852233, 852239, 852253, 852259, 852263, 852287, 852289,
852301, 852323, 852347, 852367, 852391, 852409, 852427, 852437, 852457, 852463, 852521, 852557, 852559,
852563, 852569, 852581, 852583, 852589, 852613, 852617, 852623, 852641, 852661, 852671, 852673, 852689,
852749, 852751, 852757, 852763, 852769, 852793, 852799, 852809, 852827, 852829, 852833, 852847, 852851,
852857, 852871, 852881, 852889, 852893, 852913, 852937, 852953, 852959, 852989, 852997, 853007, 853031,
853033, 853049, 853057, 853079, 853091, 853103, 853123, 853133, 853159, 853187, 853189, 853211, 853217,
853241, 853283, 853289, 853291, 853319, 853339, 853357, 853387, 853403, 853427, 853429, 853439, 853477,
853481, 853493, 853529, 853543, 853547, 853571, 853577, 853597, 853637, 853663, 853667, 853669, 853687,
853693, 853703, 853717, 853733, 853739, 853759, 853763, 853793, 853799, 853807, 853813, 853819, 853823,
853837, 853843, 853873, 853889, 853901, 853903, 853913, 853933, 853949, 853969, 853981, 853999, 854017,
854033, 854039, 854041, 854047, 854053, 854083, 854089, 854093, 854099, 854111, 854123, 854129, 854141,
854149, 854159, 854171, 854213, 854257, 854263, 854299, 854303, 854323, 854327, 854333, 854351, 854353,
854363, 854383, 854387, 854407, 854417, 854419, 854423, 854431, 854443, 854459, 854461, 854467, 854479,
854527, 854533, 854569, 854587, 854593, 854599, 854617, 854621, 854629, 854647, 854683, 854713, 854729,
854747, 854771, 854801, 854807, 854849, 854869, 854881, 854897, 854899, 854921, 854923, 854927, 854929,
854951, 854957, 854963, 854993, 854999, 855031, 855059, 855061, 855067, 855079, 855089, 855119, 855131,
855143, 855187, 855191, 855199, 855203, 855221, 855229, 855241, 855269, 855271, 855277, 855293, 855307,
855311, 855317, 855331, 855359, 855373, 855377, 855391, 855397, 855401, 855419, 855427, 855431, 855461,
855467, 855499, 855511, 855521, 855527, 855581, 855601, 855607, 855619, 855641, 855667, 855671, 855683,
855697, 855709, 855713, 855719, 855721, 855727, 855731, 855733, 855737, 855739, 855781, 855787, 855821,
855851, 855857, 855863, 855887, 855889, 855901, 855919, 855923, 855937, 855947, 855983, 855989, 855997,
856021, 856043, 856057, 856061, 856073, 856081, 856099, 856111, 856117, 856133, 856139, 856147, 856153,
856169, 856181, 856187, 856213, 856237, 856241, 856249, 856277, 856279, 856301, 856309, 856333, 856343,
856351, 856369, 856381, 856391, 856393, 856411, 856417, 856421, 856441, 856459, 856469, 856483, 856487,
856507, 856519, 856529, 856547, 856549, 856553, 856567, 856571, 856627, 856637, 856649, 856693, 856697,
856699, 856703, 856711, 856717, 856721, 856733, 856759, 856787, 856789, 856799, 856811, 856813, 856831,
856841, 856847, 856853, 856897, 856901, 856903, 856909, 856927, 856939, 856943, 856949, 856969, 856993,
857009, 857011, 857027, 857029, 857039, 857047, 857053, 857069, 857081, 857083, 857099, 857107, 857137,
857161, 857167, 857201, 857203, 857221, 857249, 857267, 857273, 857281, 857287, 857309, 857321, 857333,
857341, 857347, 857357, 857369, 857407, 857411, 857419, 857431, 857453, 857459, 857471, 857513, 857539,
857551, 857567, 857569, 857573, 857579, 857581, 857629, 857653, 857669, 857671, 857687, 857707,
857711, 857713, 857723, 857737, 857741, 857743, 857749, 857809, 857821, 857827, 857839, 857851, 857867,
857873, 857897, 857903, 857929, 857951, 857953, 857957, 857959, 857963, 857977, 857981, 858001, 858029,
858043, 858073, 858083, 858101, 858103, 858113, 858127, 858149, 858161, 858167, 858217, 858223, 858233,
858239, 858241, 858251, 858259, 858269, 858281, 858293, 858301, 858307, 858311, 858317, 858373, 858397,
858427, 858433, 858457, 858463, 858467, 858479, 858497, 858503, 858527, 858563, 858577, 858589, 858623,
858631, 858673, 858691, 858701, 858707, 858709, 858713, 858749, 858757, 858763, 858769, 858787, 858817,
858821, 858833, 858841, 858859, 858877, 858883, 858899, 858911, 858919, 858931, 858943, 858953, 858961,
858989, 858997, 859003, 859031, 859037, 859049, 859051, 859057, 859081, 859091, 859093, 859109, 859121,
859181, 859189, 859213, 859223, 859249, 859259, 859267, 859273, 859277, 859279, 859297, 859321, 859361,
859363, 859373, 859381, 859393, 859423, 859433, 859447, 859459, 859477, 859493, 859513, 859553, 859559,
859561, 859567, 859577, 859601, 859603, 859609, 859619, 859633, 859657, 859667, 859669, 859679, 859681,
859697, 859709, 859751, 859783, 859787, 859799, 859801, 859823, 859841, 859849, 859853, 859861, 859891,
859913, 859919, 859927, 859933, 859939, 859973, 859981, 859987, 860009, 860011, 860029, 860051, 860059,
860063, 860071, 860077, 860087, 860089, 860107, 860113, 860117, 860143, 860239, 860257, 860267, 860291,

860297, 860309, 860311, 860317, 860323, 860333, 860341, 860351, 860357, 860369, 860381, 860383, 860393, 860399, 860413, 860417, 860423, 860441, 860479, 860501, 860507, 860513, 860533, 860543, 860569, 860579, 860581, 860593, 860599, 860609, 860623, 860641, 860647, 860663, 860689, 860701, 860747, 860753, 860759, 860779, 860789, 860791, 860809, 860813, 860819, 860843, 860861, 860887, 860891, 860911, 860917, 860921, 860927, 860929, 860939, 860941, 860957, 860969, 860971, 861001, 861013, 861019, 861031, 861037, 861043, 861053, 861059, 861079, 861083, 861089, 861109, 861121, 861131, 861139, 861163, 861167, 861191, 861199, 861221, 861239, 861293, 861299, 861317, 861347, 861353, 861361, 861391, 861433, 861437, 861439, 861491, 861493, 861499, 861541, 861547, 861551, 861559, 861563, 861571, 861589, 861599, 861613, 861617, 861647, 861659, 861691, 861701, 861703, 861719, 861733, 861739, 861743, 861761, 861797, 861799, 861803, 861823, 861829, 861853, 861857, 861871, 861877, 861881, 861899, 861901, 861907, 861929, 861937, 861941, 861947, 861977, 861979, 861997, 862009, 862013, 862031, 862033, 862061, 862067, 862097, 862117, 862123, 862129, 862139, 862157, 862159, 862171, 862177, 862181, 862187, 862207, 862219, 862229, 862231, 862241, 862249, 862259, 862261, 862273, 862283, 862289, 862297, 862307, 862319, 862331, 862343, 862369, 862387, 862397, 862399, 862409, 862417, 862423, 862441, 862447, 862471, 862481, 862483, 862487, 862493, 862501, 862541, 862553, 862559, 862567, 862571, 862573, 862583, 862607, 862627, 862633, 862649, 862651, 862669, 862703, 862727, 862739, 862769, 862777, 862783, 862789, 862811, 862819, 862861, 862879, 862907, 862909, 862913, 862919, 862921, 862943, 862957, 862973, 862987, 862991, 862997, 863003, 863017, 863047, 863081, 863087, 863119, 863123, 863131, 863143, 863153, 863179, 863197, 863231, 863251, 863279, 863287, 863299, 863309, 863323, 863363, 863377, 863393, 863479, 863491, 863497, 863509, 863521, 863537, 863539, 863561, 863593, 863609, 863633, 863641, 863671, 863689, 863693, 863711, 863729, 863743, 863749, 863767, 863771, 863783, 863801, 863803, 863833, 863843, 863851, 863867, 863869, 863879, 863887, 863897, 863899, 863909, 863917, 863921, 863959, 863983, 864007, 864011, 864013, 864029, 864037, 864047, 864049, 864053, 864077, 864079, 864091, 864103, 864107, 864119, 864121, 864131, 864137, 864151, 864167, 864169, 864191, 864203, 864211, 864221, 864223, 864251, 864277, 864289, 864299, 864301, 864307, 864319, 864323, 864341, 864359, 864361, 864379, 864407, 864419, 864427, 864439, 864449, 864491, 864503, 864509, 864511, 864533, 864541, 864551, 864581, 864583, 864587, 864613, 864623, 864629, 864631, 864641, 864673, 864679, 864691, 864707, 864733, 864737, 864757, 864781, 864793, 864803, 864811, 864817, 864883, 864887, 864901, 864911, 864917, 864947, 864953, 864959, 864967, 864979, 864989, 865001, 865003, 865043, 865049, 865057, 865061, 865069, 865087, 865091, 865103, 865121, 865153, 865159, 865177, 865201, 865211, 865213, 865217, 865231, 865247, 865253, 865259, 865261, 865301, 865307, 865313, 865321, 865327, 865339, 865343, 865349, 865357, 865363, 865379, 865409, 865457, 865477, 865481, 865483, 865493, 865499, 865511, 865537, 865577, 865591, 865597, 865609, 865619, 865637, 865639, 865643, 865661, 865681, 865687, 865717, 865721, 865729, 865741, 865747, 865751, 865757, 865769, 865771, 865783, 865801, 865807, 865817, 865819, 865829, 865847, 865859, 865867, 865871, 865877, 865889, 865933, 865937, 865957, 865979, 865993, 866003, 866009, 866011, 866029, 866051, 866053, 866057, 866081, 866083, 866087, 866093, 866101, 866119, 866123, 866161, 866183, 866197, 866213, 866221, 866231, 866279, 866293, 866309, 866311, 866329, 866353, 866389, 866399, 866417, 866431, 866443, 866461, 866471, 866477, 866513, 866519, 866573, 866581, 866623, 866629, 866639, 866641, 866653, 866683, 866689, 866693, 866707, 866713, 866717, 866737, 866743, 866759, 866777, 866783, 866819, 866843, 866849, 866851, 866857, 866869, 866909, 866917, 866927, 866933, 866941, 866953, 866963, 866969, 867001, 867007, 867011, 867023, 867037, 867059, 867067, 867079, 867091, 867121, 867131, 867143, 867151, 867161, 867173, 867203, 867211, 867227, 867233, 867253, 867257, 867259, 867263, 867271, 867281, 867301, 867319, 867337, 867343, 867371, 867389, 867397, 867401, 867409, 867413, 867431, 867443, 867457, 867463, 867467, 867487, 867509, 867511, 867541, 867547, 867553, 867563, 867571, 867577, 867589, 867617, 867619, 867623, 867631, 867641, 867653, 867677, 867679, 867689, 867701, 867719, 867733, 867743, 867773, 867781, 867793, 867803, 867817, 867827, 867829, 867857, 867871, 867887, 867913, 867943, 867947, 867959, 867991, 868019, 868033, 868039, 868051, 868069, 868073, 868081, 868103, 868111, 868121, 868123, 868151, 868157, 868171, 868177, 868199, 868211, 868229, 868249, 868267, 868271, 868277, 868291, 868313, 868327, 868331, 868337, 868349, 868369, 868379, 868381, 868397, 868409, 868423, 868451, 868453, 868459, 868487, 868489, 868493, 868529, 868531, 868537, 868559, 868561, 868577, 868583, 868603, 868613, 868639, 868663, 868669, 868691, 868697, 868727,

868739, 868741, 868771, 868783, 868787, 868793, 868799, 868801, 868817, 868841, 868849, 868867, 868873,
868877, 868883, 868891, 868909, 868937, 868939, 868943, 868951, 868957, 868993, 868997, 868999, 869017,
869021, 869039, 869053, 869059, 869069, 869081, 869119, 869131, 869137, 869153, 869173, 869179, 869203,
869233, 869249, 869251, 869257, 869273, 869291, 869293, 869299, 869303, 869317, 869321, 869339, 869369,
869371, 869381, 869399, 869413, 869419, 869437, 869443, 869461, 869467, 869471, 869489, 869501, 869521,
869543, 869551, 869563, 869579, 869587, 869597, 869599, 869657, 869663, 869683, 869689, 869707, 869717,
869747, 869753, 869773, 869777, 869779, 869807, 869809, 869819, 869849, 869863, 869879, 869887, 869893,
869899, 869909, 869927, 869951, 869959, 869983, 869989, 870007, 870013, 870031, 870047, 870049, 870059,
870083, 870097, 870109, 870127, 870131, 870137, 870151, 870161, 870169, 870173, 870197, 870211, 870223,
870229, 870239, 870241, 870253, 870271, 870283, 870301, 870323, 870329, 870341, 870367, 870391, 870403,
870407, 870413, 870431, 870433, 870437, 870461, 870479, 870491, 870497, 870517, 870533, 870547, 870577,
870589, 870593, 870601, 870613, 870629, 870641, 870643, 870679, 870691, 870703, 870731, 870739, 870743,
870773, 870787, 870809, 870811, 870823, 870833, 870847, 870853, 870871, 870889, 870901, 870907, 870911,
870917, 870929, 870931, 870953, 870967, 870977, 870983, 870997, 871001, 871021, 871027, 871037, 871061,
871103, 871147, 871159, 871163, 871177, 871181, 871229, 871231, 871249, 871259, 871271, 871289, 871303,
871337, 871349, 871393, 871439, 871459, 871463, 871477, 871513, 871517, 871531, 871553, 871571, 871589,
871597, 871613, 871621, 871639, 871643, 871649, 871657, 871679, 871681, 871687, 871727, 871763, 871771,
871789, 871817, 871823, 871837, 871867, 871883, 871901, 871919, 871931, 871957, 871963, 871973, 871987,
871993, 872017, 872023, 872033, 872041, 872057, 872071, 872077, 872089, 872099, 872107, 872129, 872141,
872143, 872149, 872159, 872161, 872173, 872177, 872189, 872203, 872227, 872231, 872237, 872243, 872251,
872257, 872269, 872281, 872317, 872323, 872351, 872353, 872369, 872381, 872383, 872387, 872393, 872411,
872419, 872429, 872437, 872441, 872453, 872471, 872477, 872479, 872533, 872549, 872561, 872563, 872567,
872587, 872609, 872611, 872621, 872623, 872647, 872657, 872659, 872671, 872687, 872731, 872737, 872747,
872749, 872761, 872789, 872791, 872843, 872863, 872923, 872947, 872951, 872953, 872959, 872999, 873017,
873043, 873049, 873073, 873079, 873083, 873091, 873109, 873113, 873121, 873133, 873139, 873157, 873209,
873247, 873251, 873263, 873293, 873317, 873319, 873331, 873343, 873349, 873359, 873403, 873407, 873419,
873421, 873427, 873437, 873461, 873463, 873469, 873497, 873527, 873529, 873539, 873541, 873553, 873569,
873571, 873617, 873619, 873641, 873643, 873659, 873667, 873671, 873689, 873707, 873709, 873721, 873727,
873739, 873767, 873773, 873781, 873787, 873863, 873877, 873913, 873959, 873979, 873989, 873991, 874001,
874009, 874037, 874063, 874087, 874091, 874099, 874103, 874109, 874117, 874121, 874127, 874151, 874193,
874213, 874217, 874229, 874249, 874267, 874271, 874277, 874301, 874303, 874331, 874337, 874343, 874351,
874373, 874387, 874397, 874403, 874409, 874427, 874457, 874459, 874477, 874487, 874537, 874543, 874547,
874567, 874583, 874597, 874619, 874637, 874639, 874651, 874661, 874673, 874681, 874693, 874697, 874711,
874721, 874723, 874729, 874739, 874763, 874771, 874777, 874799, 874807, 874813, 874823, 874831, 874847,
874859, 874873, 874879, 874889, 874891, 874919, 874957, 874967, 874987, 875011, 875027, 875033, 875089,
875107, 875113, 875117, 875129, 875141, 875183, 875201, 875209, 875213, 875233, 875239, 875243, 875261,
875263, 875267, 875269, 875297, 875299, 875317, 875323, 875327, 875333, 875339, 875341, 875363, 875377,
875389, 875393, 875417, 875419, 875429, 875443, 875447, 875477, 875491, 875503, 875509, 875513, 875519,
875521, 875543, 875579, 875591, 875593, 875617, 875621, 875627, 875629, 875647, 875659, 875663, 875681,
875683, 875689, 875701, 875711, 875717, 875731, 875741, 875759, 875761, 875773, 875779, 875783, 875803,
875821, 875837, 875851, 875893, 875923, 875929, 875933, 875947, 875969, 875981, 875983, 876011, 876013,
876017, 876019, 876023, 876041, 876067, 876077, 876079, 876097, 876103, 876107, 876121, 876131, 876137,
876149, 876181, 876191, 876193, 876199, 876203, 876229, 876233, 876257, 876263, 876287, 876301, 876307,
876311, 876329, 876331, 876341, 876349, 876371, 876373, 876431, 876433, 876443, 876479, 876481, 876497,
876523, 876529, 876569, 876581, 876593, 876607, 876611, 876619, 876643, 876647, 876653, 876661, 876677,
876719, 876721, 876731, 876749, 876751, 876761, 876769, 876787, 876791, 876797, 876817, 876823, 876833,
876851, 876853, 876871, 876893, 876913, 876929, 876947, 876971, 877003, 877027, 877043, 877057, 877073,
877091, 877109, 877111, 877117, 877133, 877169, 877181, 877187, 877199, 877213, 877223, 877237, 877267,
877291, 877297, 877301, 877313, 877321, 877333, 877343, 877351, 877361, 877367, 877379, 877397, 877399,

877403, 877411, 877423, 877463, 877469, 877531, 877543, 877567, 877573, 877577, 877601, 877609, 877619,
877621, 877651, 877661, 877699, 877739, 877771, 877783, 877817, 877823, 877837, 877843, 877853, 877867,
877871, 877873, 877879, 877883, 877907, 877909, 877937, 877939, 877949, 877997, 878011, 878021, 878023,
878039, 878041, 878077, 878083, 878089, 878099, 878107, 878113, 878131, 878147, 878153, 878159, 878167,
878173, 878183, 878191, 878197, 878201, 878221, 878239, 878279, 878287, 878291, 878299, 878309, 878359,
878377, 878387, 878411, 878413, 878419, 878443, 878453, 878467, 878489, 878513, 878539, 878551, 878567,
878573, 878593, 878597, 878609, 878621, 878629, 878641, 878651, 878659, 878663, 878677, 878681, 878699,
878719, 878737, 878743, 878749, 878777, 878783, 878789, 878797, 878821, 878831, 878833, 878837, 878851,
878863, 878869, 878873, 878893, 878929, 878939, 878953, 878957, 878987, 878989, 879001, 879007, 879023,
879031, 879061, 879089, 879097, 879103, 879113, 879119, 879133, 879143, 879167, 879169, 879181, 879199,
879227, 879239, 879247, 879259, 879269, 879271, 879283, 879287, 879299, 879331, 879341, 879343, 879353,
879371, 879391, 879401, 879413, 879449, 879457, 879493, 879523, 879533, 879539, 879553, 879581, 879583,
879607, 879617, 879623, 879629, 879649, 879653, 879661, 879667, 879673, 879679, 879689, 879691, 879701,
879707, 879709, 879713, 879721, 879743, 879797, 879799, 879817, 879821, 879839, 879859, 879863, 879881,
879917, 879919, 879941, 879953, 879961, 879973, 879979, 880001, 880007, 880021, 880027, 880031, 880043,
880057, 880067, 880069, 880091, 880097, 880109, 880127, 880133, 880151, 880153, 880199, 880211, 880219,
880223, 880247, 880249, 880259, 880283, 880301, 880303, 880331, 880337, 880343, 880349, 880361, 880367,
880409, 880421, 880423, 880427, 880483, 880487, 880513, 880519, 880531, 880541, 880543, 880553, 880559,
880571, 880573, 880589, 880603, 880661, 880667, 880673, 880681, 880687, 880699, 880703, 880709, 880723,
880727, 880729, 880751, 880793, 880799, 880801, 880813, 880819, 880823, 880853, 880861, 880871, 880883,
880903, 880907, 880909, 880939, 880949, 880951, 880961, 880981, 880993, 881003, 881009, 881017, 881029,
881057, 881071, 881077, 881099, 881119, 881141, 881143, 881147, 881159, 881171, 881173, 881191, 881197,
881207, 881219, 881233, 881249, 881269, 881273, 881311, 881317, 881327, 881333, 881351, 881357, 881369,
881393, 881407, 881411, 881417, 881437, 881449, 881471, 881473, 881477, 881479, 881509, 881527, 881533,
881537, 881539, 881591, 881597, 881611, 881641, 881663, 881669, 881681, 881707, 881711, 881729, 881743,
881779, 881813, 881833, 881849, 881897, 881899, 881911, 881917, 881939, 881953, 881963, 881983, 881987,
882017, 882019, 882029, 882031, 882047, 882061, 882067, 882071, 882083, 882103, 882139, 882157, 882169,
882173, 882179, 882187, 882199, 882239, 882241, 882247, 882251, 882253, 882263, 882289, 882313, 882359,
882367, 882377, 882389, 882391, 882433, 882439, 882449, 882451, 882461, 882481, 882491, 882517, 882529,
882551, 882571, 882577, 882587, 882593, 882599, 882617, 882631, 882653, 882659, 882697, 882701, 882703,
882719, 882727, 882733, 882751, 882773, 882779, 882823, 882851, 882863, 882877, 882881, 882883, 882907,
882913, 882923, 882943, 882953, 882961, 882967, 882979, 883013, 883049, 883061, 883073, 883087, 883093,
883109, 883111, 883117, 883121, 883163, 883187, 883193, 883213, 883217, 883229, 883231, 883237, 883241,
883247, 883249, 883273, 883279, 883307, 883327, 883331, 883339, 883343, 883357, 883391, 883397, 883409,
883411, 883423, 883429, 883433, 883451, 883471, 883483, 883489, 883517, 883537, 883549, 883577, 883579,
883613, 883621, 883627, 883639, 883661, 883667, 883691, 883697, 883699, 883703, 883721, 883733, 883739,
883763, 883777, 883781, 883783, 883807, 883871, 883877, 883889, 883921, 883933, 883963, 883969, 883973,
883979, 883991, 884003, 884011, 884029, 884057, 884069, 884077, 884087, 884111, 884129, 884131, 884159,
884167, 884171, 884183, 884201, 884227, 884231, 884243, 884251, 884267, 884269, 884287, 884293, 884309,
884311, 884321, 884341, 884353, 884363, 884369, 884371, 884417, 884423, 884437, 884441, 884453, 884483,
884489, 884491, 884497, 884501, 884537, 884573, 884579, 884591, 884593, 884617, 884651, 884669, 884693,
884699, 884717, 884743, 884789, 884791, 884803, 884813, 884827, 884831, 884857, 884881, 884899, 884921,
884951, 884959, 884977, 884981, 884987, 884999, 885023, 885041, 885061, 885083, 885091, 885097, 885103,
885107, 885127, 885133, 885161, 885163, 885169, 885187, 885217, 885223, 885233, 885239, 885251, 885257,
885263, 885289, 885301, 885307, 885331, 885359, 885371, 885383, 885389, 885397, 885403, 885421, 885427,
885449, 885473, 885487, 885497, 885503, 885509, 885517, 885529, 885551, 885553, 885589, 885607, 885611,
885623, 885679, 885713, 885721, 885727, 885733, 885737, 885769, 885791, 885793, 885803, 885811, 885821,
885823, 885839, 885869, 885881, 885883, 885889, 885893, 885919, 885923, 885931, 885943, 885947, 885959,
885961, 885967, 885971, 885977, 885991, 886007, 886013, 886019, 886021, 886031, 886043, 886069, 886097,

886117, 886129, 886163, 886177, 886181, 886183, 886189, 886199, 886241, 886243, 886247, 886271, 886283, 886307, 886313, 886337, 886339, 886349, 886367, 886381, 886387, 886421, 886427, 886429, 886433, 886453, 886463, 886469, 886471, 886493, 886511, 886517, 886519, 886537, 886541, 886547, 886549, 886583, 886591, 886607, 886609, 886619, 886643, 886651, 886663, 886667, 886741, 886747, 886751, 886759, 886777, 886793, 886799, 886807, 886819, 886859, 886867, 886891, 886909, 886913, 886967, 886969, 886973, 886979, 886981, 886987, 886993, 886999, 887017, 887057, 887059, 887069, 887093, 887101, 887113, 887141, 887143, 887153, 887171, 887177, 887191, 887203, 887233, 887261, 887267, 887269, 887291, 887311, 887323, 887333, 887377, 887387, 887399, 887401, 887423, 887441, 887449, 887459, 887479, 887483, 887503, 887533, 887543, 887567, 887569, 887573, 887581, 887599, 887617, 887629, 887633, 887641, 887651, 887657, 887659, 887669, 887671, 887681, 887693, 887701, 887707, 887717, 887743, 887749, 887759, 887819, 887827, 887837, 887839, 887849, 887867, 887903, 887911, 887921, 887923, 887941, 887947, 887987, 887989, 888001, 888011, 888047, 888059, 888061, 888077, 888091, 888103, 888109, 888133, 888143, 888157, 888161, 888163, 888179, 888203, 888211, 888247, 888257, 888263, 888271, 888287, 888313, 888319, 888323, 888359, 888361, 888373, 888389, 888397, 888409, 888413, 888427, 888431, 888443, 888451, 888457, 888469, 888479, 888493, 888499, 888533, 888541, 888557, 888623, 888631, 888637, 888653, 888659, 888661, 888683, 888689, 888691, 888721, 888737, 888751, 888761, 888773, 888779, 888781, 888793, 888799, 888809, 888827, 888857, 888869, 888871, 888887, 888917, 888919, 888931, 888959, 888961, 888967, 888983, 888989, 888997, 889001, 889027, 889037, 889039, 889043, 889051, 889069, 889081, 889087, 889123, 889139, 889171, 889177, 889211, 889237, 889247, 889261, 889271, 889279, 889289, 889309, 889313, 889327, 889337, 889349, 889351, 889363, 889367, 889373, 889391, 889411, 889429, 889439, 889453, 889481, 889489, 889501, 889519, 889579, 889589, 889597, 889631, 889639, 889657, 889673, 889687, 889697, 889699, 889703, 889727, 889747, 889769, 889783, 889829, 889871, 889873, 889877, 889879, 889891, 889901, 889907, 889909, 889921, 889937, 889951, 889957, 889963, 889997, 890003, 890011, 890027, 890053, 890063, 890083, 890107, 890111, 890117, 890119, 890129, 890147, 890159, 890161, 890177, 890221, 890231, 890237, 890287, 890291, 890303, 890317, 890333, 890371, 890377, 890419, 890429, 890437, 890441, 890459, 890467, 890501, 890531, 890543, 890551, 890563, 890597, 890609, 890653, 890657, 890671, 890683, 890707, 890711, 890717, 890737, 890761, 890789, 890797, 890803, 890809, 890821, 890833, 890843, 890861, 890863, 890867, 890881, 890887, 890893, 890927, 890933, 890941, 890957, 890963, 890969, 890993, 890999, 891001, 891017, 891047, 891049, 891061, 891067, 891091, 891101, 891103, 891133, 891151, 891161, 891173, 891179, 891223, 891239, 891251, 891277, 891287, 891311, 891323, 891329, 891349, 891377, 891379, 891389, 891391, 891409, 891421, 891427, 891439, 891481, 891487, 891491, 891493, 891509, 891521, 891523, 891551, 891557, 891559, 891563, 891571, 891577, 891587, 891593, 891601, 891617, 891629, 891643, 891647, 891659, 891661, 891677, 891679, 891707, 891743, 891749, 891763, 891767, 891797, 891799, 891809, 891817, 891823, 891827, 891829, 891851, 891859, 891887, 891889, 891893, 891899, 891907, 891923, 891929, 891967, 891983, 891991, 891997, 892019, 892027, 892049, 892057, 892079, 892091, 892093, 892097, 892103, 892123, 892141, 892153, 892159, 892169, 892189, 892219, 892237, 892249, 892253, 892261, 892267, 892271, 892291, 892321, 892351, 892357, 892387, 892391, 892421, 892433, 892439, 892457, 892471, 892481, 892513, 892523, 892531, 892547, 892553, 892559, 892579, 892597, 892603, 892609, 892627, 892643, 892657, 892663, 892667, 892709, 892733, 892747, 892757, 892763, 892777, 892781, 892783, 892817, 892841, 892849, 892861, 892877, 892901, 892919, 892933, 892951, 892973, 892987, 892999, 893003, 893023, 893029, 893033, 893041, 893051, 893059, 893093, 893099, 893107, 893111, 893117, 893119, 893131, 893147, 893149, 893161, 893183, 893213, 893219, 893227, 893237, 893257, 893261, 893281, 893317, 893339, 893341, 893351, 893359, 893363, 893381, 893383, 893407, 893413, 893419, 893429, 893441, 893449, 893479, 893489, 893509, 893521, 893549, 893567, 893591, 893603, 893609, 893653, 893657, 893671, 893681, 893701, 893719, 893723, 893743, 893777, 893797, 893821, 893839, 893857, 893863, 893873, 893881, 893897, 893903, 893917, 893929, 893933, 893939, 893989, 893999, 894011, 894037, 894059, 894067, 894073, 894097, 894109, 894119, 894137, 894139, 894151, 894161, 894167, 894181, 894191, 894193, 894203, 894209, 894211, 894221, 894227, 894233, 894239, 894247, 894259, 894277, 894281, 894287, 894301, 894329, 894343, 894371, 894391, 894403, 894407, 894409, 894419, 894427, 894431, 894449, 894451, 894503, 894511, 894521, 894527, 894541, 894547, 894559, 894581, 894589, 894611, 894613, 894637, 894643, 894667, 894689, 894709, 894713, 894721, 894731, 894749, 894763, 894779, 894791,

894793, 894811, 894869, 894871, 894893, 894917, 894923, 894947, 894973, 894997, 895003, 895007, 895009,
895039, 895049, 895051, 895079, 895087, 895127, 895133, 895151, 895157, 895159, 895171, 895189, 895211,
895231, 895241, 895243, 895247, 895253, 895277, 895283, 895291, 895309, 895313, 895319, 895333, 895343,
895351, 895357, 895361, 895387, 895393, 895421, 895423, 895457, 895463, 895469, 895471, 895507, 895529,
895553, 895571, 895579, 895591, 895613, 895627, 895633, 895649, 895651, 895667, 895669, 895673, 895681,
895691, 895703, 895709, 895721, 895729, 895757, 895771, 895777, 895787, 895789, 895799, 895801, 895813,
895823, 895841, 895861, 895879, 895889, 895901, 895903, 895913, 895927, 895933, 895957, 895987, 896003,
896009, 896047, 896069, 896101, 896107, 896111, 896113, 896123, 896143, 896167, 896191, 896201, 896263,
896281, 896293, 896297, 896299, 896323, 896327, 896341, 896347, 896353, 896369, 896381, 896417, 896443,
896447, 896449, 896453, 896479, 896491, 896509, 896521, 896531, 896537, 896543, 896549, 896557, 896561,
896573, 896587, 896617, 896633, 896647, 896669, 896677, 896681, 896717, 896719, 896723, 896771, 896783,
896803, 896837, 896867, 896879, 896897, 896921, 896927, 896947, 896953, 896963, 896983, 897007, 897011,
897019, 897049, 897053, 897059, 897067, 897077, 897101, 897103, 897119, 897133, 897137, 897157, 897163,
897191, 897223, 897229, 897241, 897251, 897263, 897269, 897271, 897301, 897307, 897317, 897319, 897329,
897349, 897359, 897373, 897401, 897433, 897443, 897461, 897467, 897469, 897473, 897497, 897499, 897517,
897527, 897553, 897557, 897563, 897571, 897577, 897581, 897593, 897601, 897607, 897629, 897647, 897649,
897671, 897691, 897703, 897707, 897709, 897727, 897751, 897779, 897781, 897817, 897829, 897847, 897877,
897881, 897887, 897899, 897907, 897931, 897947, 897971, 897983, 898013, 898019, 898033, 898063, 898067,
898069, 898091, 898097, 898109, 898129, 898133, 898147, 898153, 898171, 898181, 898189, 898199, 898211,
898213, 898223, 898231, 898241, 898243, 898253, 898259, 898279, 898283, 898291, 898307, 898319, 898327,
898361, 898369, 898409, 898421, 898423, 898427, 898439, 898459, 898477, 898481, 898483, 898493, 898519,
898523, 898543, 898549, 898553, 898561, 898607, 898613, 898621, 898661, 898663, 898669, 898673, 898691,
898717, 898727, 898753, 898763, 898769, 898787, 898813, 898819, 898823, 898853, 898867, 898873, 898889,
898897, 898921, 898927, 898951, 898981, 898987, 899009, 899051, 899057, 899069, 899123, 899149, 899153,
899159, 899161, 899177, 899179, 899183, 899189, 899209, 899221, 899233, 899237, 899263, 899273, 899291,
899309, 899321, 899387, 899401, 899413, 899429, 899447, 899467, 899473, 899477, 899491, 899519, 899531,
899537, 899611, 899617, 899659, 899671, 899681, 899687, 899693, 899711, 899719, 899749, 899753, 899761,
899779, 899791, 899807, 899831, 899849, 899851, 899863, 899881, 899891, 899893, 899903, 899917, 899939,
899971, 899981, 900001, 900007, 900019, 900037, 900061, 900089, 900091, 900103, 900121, 900139, 900143,
900149, 900157, 900161, 900169, 900187, 900217, 900233, 900241, 900253, 900259, 900283, 900287, 900293,
900307, 900329, 900331, 900349, 900397, 900409, 900443, 900461, 900481, 900491, 900511, 900539, 900551,
900553, 900563, 900569, 900577, 900583, 900587, 900589, 900593, 900607, 900623, 900649, 900659, 900671,
900673, 900689, 900701, 900719, 900737, 900743, 900751, 900761, 900763, 900773, 900797, 900803, 900817,
900821, 900863, 900869, 900917, 900929, 900931, 900937, 900959, 900971, 900973, 900997, 901007, 901009,
901013, 901063, 901067, 901079, 901093, 901097, 901111, 901133, 901141, 901169, 901171, 901177, 901183,
901193, 901207, 901211, 901213, 901247, 901249, 901253, 901273, 901279, 901309, 901333, 901339, 901367,
901399, 901403, 901423, 901427, 901429, 901441, 901447, 901451, 901457, 901471, 901489, 901499, 901501,
901513, 901517, 901529, 901547, 901567, 901591, 901613, 901643, 901657, 901679, 901687, 901709, 901717,
901739, 901741, 901751, 901781, 901787, 901811, 901819, 901841, 901861, 901891, 901907, 901909, 901919,
901931, 901937, 901963, 901973, 901993, 901997, 902009, 902017, 902029, 902039, 902047, 902053, 902087,
902089, 902119, 902137, 902141, 902179, 902191, 902201, 902227, 902261, 902263, 902281, 902299, 902303,
902311, 902333, 902347, 902351, 902357, 902389, 902401, 902413, 902437, 902449, 902471, 902477, 902483,
902501, 902507, 902521, 902563, 902569, 902579, 902591, 902597, 902599, 902611, 902639, 902653, 902659,
902669, 902677, 902687, 902719, 902723, 902753, 902761, 902767, 902771, 902777, 902789, 902807, 902821,
902827, 902849, 902873, 902903, 902933, 902953, 902963, 902971, 902977, 902981, 902987, 903017, 903029,
903037, 903073, 903079, 903103, 903109, 903143, 903151, 903163, 903179, 903197, 903211, 903223, 903251,
903257, 903269, 903311, 903323, 903337, 903347, 903359, 903367, 903389, 903391, 903403, 903407, 903421,
903443, 903449, 903451, 903457, 903479, 903493, 903527, 903541, 903547, 903563, 903569, 903607, 903613,
903641, 903649, 903673, 903677, 903691, 903701, 903709, 903751, 903757, 903761, 903781, 903803, 903827,

903841, 903871, 903883, 903899, 903913, 903919, 903949, 903967, 903979, 904019, 904027, 904049, 904067,
904069, 904073, 904087, 904093, 904097, 904103, 904117, 904121, 904147, 904157, 904181, 904193, 904201,
904207, 904217, 904219, 904261, 904283, 904289, 904297, 904303, 904357, 904361, 904369, 904399, 904441,
904459, 904483, 904489, 904499, 904511, 904513, 904517, 904523, 904531, 904559, 904573, 904577, 904601,
904619, 904627, 904633, 904637, 904643, 904661, 904663, 904667, 904679, 904681, 904693, 904697, 904721,
904727, 904733, 904759, 904769, 904777, 904781, 904789, 904793, 904801, 904811, 904823, 904847, 904861,
904867, 904873, 904879, 904901, 904903, 904907, 904919, 904931, 904933, 904987, 904997, 904999, 905011,
905053, 905059, 905071, 905083, 905087, 905111, 905123, 905137, 905143, 905147, 905161, 905167, 905171,
905189, 905197, 905207, 905209, 905213, 905227, 905249, 905269, 905291, 905297, 905299, 905329, 905339,
905347, 905381, 905413, 905449, 905453, 905461, 905477, 905491, 905497, 905507, 905551, 905581, 905587,
905599, 905617, 905621, 905629, 905647, 905651, 905659, 905677, 905683, 905687, 905693, 905701, 905713,
905719, 905759, 905761, 905767, 905783, 905803, 905819, 905833, 905843, 905897, 905909, 905917, 905923,
905951, 905959, 905963, 905999, 906007, 906011, 906013, 906023, 906029, 906043, 906089, 906107, 906119,
906121, 906133, 906179, 906187, 906197, 906203, 906211, 906229, 906233, 906259, 906263, 906289, 906293,
906313, 906317, 906329, 906331, 906343, 906349, 906371, 906377, 906383, 906391, 906403, 906421, 906427,
906431, 906461, 906473, 906481, 906487, 906497, 906517, 906523, 906539, 906541, 906557, 906589, 906601,
906613, 906617, 906641, 906649, 906673, 906679, 906691, 906701, 906707, 906713, 906727, 906749, 906751,
906757, 906767, 906779, 906793, 906809, 906817, 906823, 906839, 906847, 906869, 906881, 906901, 906911,
906923, 906929, 906931, 906943, 906949, 906973, 907019, 907021, 907031, 907063, 907073, 907099, 907111,
907133, 907139, 907141, 907163, 907169, 907183, 907199, 907211, 907213, 907217, 907223, 907229, 907237,
907259, 907267, 907279, 907297, 907301, 907321, 907331, 907363, 907367, 907369, 907391, 907393, 907397,
907399, 907427, 907433, 907447, 907457, 907469, 907471, 907481, 907493, 907507, 907513, 907549, 907561,
907567, 907583, 907589, 907637, 907651, 907657, 907663, 907667, 907691, 907693, 907703, 907717, 907723,
907727, 907733, 907757, 907759, 907793, 907807, 907811, 907813, 907831, 907843, 907849, 907871, 907891,
907909, 907913, 907927, 907957, 907967, 907969, 907997, 907999, 908003, 908041, 908053, 908057, 908071,
908081, 908101, 908113, 908129, 908137, 908153, 908179, 908183, 908197, 908213, 908221, 908233, 908249,
908287, 908317, 908321, 908353, 908359, 908363, 908377, 908381, 908417, 908419, 908441, 908449, 908459,
908471, 908489, 908491, 908503, 908513, 908521, 908527, 908533, 908539, 908543, 908549, 908573, 908581,
908591, 908597, 908603, 908617, 908623, 908627, 908653, 908669, 908671, 908711, 908723, 908731, 908741,
908749, 908759, 908771, 908797, 908807, 908813, 908819, 908821, 908849, 908851, 908857, 908861, 908863,
908879, 908881, 908893, 908909, 908911, 908927, 908953, 908959, 908993, 909019, 909023, 909031, 909037,
909043, 909047, 909061, 909071, 909089, 909091, 909107, 909113, 909119, 909133, 909151, 909173, 909203,
909217, 909239, 909241, 909247, 909253, 909281, 909287, 909289, 909299, 909301, 909317, 909319, 909329,
909331, 909341, 909343, 909371, 909379, 909383, 909401, 909409, 909437, 909451, 909457, 909463, 909481,
909521, 909529, 909539, 909541, 909547, 909577, 909599, 909611, 909613, 909631, 909637, 909679, 909683,
909691, 909697, 909731, 909737, 909743, 909761, 909767, 909773, 909787, 909791, 909803, 909809, 909829,
909833, 909859, 909863, 909877, 909889, 909899, 909901, 909907, 909911, 909917, 909971, 909973, 909977,
910003, 910031, 910051, 910069, 910093, 910097, 910099, 910103, 910109, 910121, 910127, 910139, 910141,
910171, 910177, 910199, 910201, 910207, 910213, 910219, 910229, 910277, 910279, 910307, 910361, 910369,
910421, 910447, 910451, 910453, 910457, 910471, 910519, 910523, 910561, 910577, 910583, 910603, 910619,
910621, 910627, 910631, 910643, 910661, 910691, 910709, 910711, 910747, 910751, 910771, 910781, 910787,
910799, 910807, 910817, 910849, 910853, 910883, 910909, 910939, 910957, 910981, 911003, 911011, 911023,
911033, 911039, 911063, 911077, 911087, 911089, 911101, 911111, 911129, 911147, 911159, 911161, 911167,
911171, 911173, 911179, 911201, 911219, 911227, 911231, 911233, 911249, 911269, 911291, 911293, 911303,
911311, 911321, 911327, 911341, 911357, 911359, 911363, 911371, 911413, 911419, 911437, 911453, 911459,
911503, 911507, 911527, 911549, 911593, 911597, 911621, 911633, 911657, 911663, 911671, 911681, 911683,
911689, 911707, 911719, 911723, 911737, 911749, 911773, 911777, 911783, 911819, 911831, 911837, 911839,
911851, 911861, 911873, 911879, 911893, 911899, 911903, 911917, 911947, 911951, 911957, 911959, 911969,
912007, 912031, 912047, 912049, 912053, 912061, 912083, 912089, 912103, 912167, 912173, 912187, 912193,

912211, 912217, 912227, 912239, 912251, 912269, 912287, 912337, 912343, 912349, 912367, 912391, 912397,
912403, 912409, 912413, 912449, 912451, 912463, 912467, 912469, 912481, 912487, 912491, 912497, 912511,
912521, 912523, 912533, 912539, 912559, 912581, 912631, 912647, 912649, 912727, 912763, 912773, 912797,
912799, 912809, 912823, 912829, 912839, 912851, 912853, 912859, 912869, 912871, 912911, 912929, 912941,
912953, 912959, 912971, 912973, 912979, 912991, 913013, 913027, 913037, 913039, 913063, 913067, 913103,
913139, 913151, 913177, 913183, 913217, 913247, 913259, 913279, 913309, 913321, 913327, 913331, 913337,
913373, 913397, 913417, 913421, 913433, 913441, 913447, 913457, 913483, 913487, 913513, 913571, 913573,
913579, 913589, 913637, 913639, 913687, 913709, 913723, 913739, 913753, 913771, 913799, 913811, 913853,
913873, 913889, 913907, 913921, 913933, 913943, 913981, 913999, 914021, 914027, 914041, 914047, 914117,
914131, 914161, 914189, 914191, 914213, 914219, 914237, 914239, 914257, 914269, 914279, 914293, 914321,
914327, 914339, 914351, 914357, 914359, 914363, 914369, 914371, 914429, 914443, 914449, 914461, 914467,
914477, 914491, 914513, 914519, 914521, 914533, 914561, 914569, 914579, 914581, 914591, 914597, 914609,
914611, 914629, 914647, 914657, 914701, 914713, 914723, 914731, 914737, 914777, 914783, 914789, 914791,
914801, 914813, 914819, 914827, 914843, 914857, 914861, 914867, 914873, 914887, 914891, 914897, 914941,
914951, 914971, 914981, 915007, 915017, 915029, 915041, 915049, 915053, 915067, 915071, 915113, 915139,
915143, 915157, 915181, 915191, 915197, 915199, 915203, 915221, 915223, 915247, 915251, 915253, 915259,
915283, 915301, 915311, 915353, 915367, 915379, 915391, 915437, 915451, 915479, 915487, 915527, 915533,
915539, 915547, 915557, 915587, 915589, 915601, 915611, 915613, 915623, 915631, 915641, 915659, 915683,
915697, 915703, 915727, 915731, 915737, 915757, 915763, 915769, 915799, 915839, 915851, 915869, 915881,
915911, 915917, 915919, 915947, 915949, 915961, 915973, 915991, 916031, 916033, 916049, 916057, 916061,
916073, 916099, 916103, 916109, 916121, 916127, 916129, 916141, 916169, 916177, 916183, 916187, 916189,
916213, 916217, 916219, 916259, 916261, 916273, 916291, 916319, 916337, 916339, 916361, 916367, 916387,
916411, 916417, 916441, 916451, 916457, 916463, 916469, 916471, 916477, 916501, 916507, 916511, 916537,
916561, 916571, 916583, 916613, 916621, 916633, 916649, 916651, 916679, 916703, 916733, 916771, 916781,
916787, 916831, 916837, 916841, 916859, 916871, 916879, 916907, 916913, 916931, 916933, 916939, 916961,
916973, 916999, 917003, 917039, 917041, 917051, 917053, 917083, 917089, 917093, 917101, 917117, 917117,
917123, 917141, 917153, 917159, 917173, 917179, 917209, 917219, 917227, 917237, 917239, 917243, 917251,
917281, 917291, 917317, 917327, 917333, 917353, 917363, 917381, 917407, 917443, 917459, 917461, 917471,
917503, 917513, 917519, 917549, 917557, 917573, 917591, 917593, 917611, 917617, 917629, 917633, 917641,
917659, 917669, 917687, 917689, 917713, 917729, 917737, 917753, 917759, 917767, 917771, 917773, 917783,
917789, 917803, 917809, 917827, 917831, 917837, 917843, 917849, 917869, 917887, 917893, 917923, 917927,
917951, 917971, 917993, 918011, 918019, 918041, 918067, 918079, 918089, 918103, 918109, 918131, 918139,
918143, 918149, 918157, 918161, 918173, 918193, 918199, 918209, 918223, 918257, 918259, 918263, 918283,
918301, 918319, 918329, 918341, 918347, 918353, 918361, 918371, 918389, 918397, 918431, 918433, 918439,
918443, 918469, 918481, 918497, 918529, 918539, 918563, 918581, 918583, 918587, 918613, 918641, 918647,
918653, 918677, 918679, 918683, 918733, 918737, 918751, 918763, 918767, 918779, 918787, 918793, 918823,
918829, 918839, 918857, 918877, 918889, 918899, 918913, 918943, 918947, 918949, 918959, 918971, 918989,
919013, 919019, 919021, 919031, 919033, 919063, 919067, 919081, 919109, 919111, 919129, 919147, 919153,
919169, 919183, 919189, 919223, 919229, 919231, 919249, 919253, 919267, 919301, 919313, 919319, 919337,
919349, 919351, 919381, 919393, 919409, 919417, 919421, 919423, 919427, 919447, 919511, 919519, 919531,
919559, 919571, 919591, 919613, 919621, 919631, 919679, 919691, 919693, 919703, 919729, 919757, 919759,
919769, 919781, 919799, 919811, 919817, 919823, 919859, 919871, 919883, 919901, 919903, 919913, 919927,
919937, 919939, 919949, 919951, 919969, 919979, 920011, 920021, 920039, 920053, 920107, 920123, 920137,
920147, 920149, 920167, 920197, 920201, 920203, 920209, 920219, 920233, 920263, 920267, 920273, 920279,
920281, 920291, 920323, 920333, 920357, 920371, 920377, 920393, 920399, 920407, 920411, 920419, 920441,
920443, 920467, 920473, 920477, 920497, 920509, 920519, 920539, 920561, 920609, 920641, 920651, 920653,
920677, 920687, 920701, 920707, 920729, 920741, 920743, 920753, 920761, 920783, 920789, 920791, 920807,
920827, 920833, 920849, 920863, 920869, 920891, 920921, 920947, 920951, 920957, 920963, 920971, 920999,
921001, 921007, 921013, 921029, 921031, 921073, 921079, 921091, 921121, 921133, 921143, 921149, 921157,

921169, 921191, 921197, 921199, 921203, 921223, 921233, 921241, 921257, 921259, 921287, 921293, 921331,
921353, 921373, 921379, 921407, 921409, 921457, 921463, 921467, 921491, 921497, 921499, 921517, 921523,
921563, 921581, 921589, 921601, 921611, 921629, 921637, 921643, 921647, 921667, 921677, 921703, 921733,
921737, 921743, 921749, 921751, 921761, 921779, 921787, 921821, 921839, 921841, 921871, 921887, 921889,
921901, 921911, 921913, 921919, 921931, 921959, 921989, 922021, 922027, 922037, 922039, 922043, 922057,
922067, 922069, 922073, 922079, 922081, 922087, 922099, 922123, 922169, 922211, 922217, 922223, 922237,
922247, 922261, 922283, 922289, 922291, 922303, 922309, 922321, 922331, 922333, 922351, 922357, 922367,
922391, 922423, 922451, 922457, 922463, 922487, 922489, 922499, 922511, 922513, 922517, 922531, 922549,
922561, 922601, 922613, 922619, 922627, 922631, 922637, 922639, 922643, 922667, 922679, 922681, 922699,
922717, 922729, 922739, 922741, 922781, 922807, 922813, 922853, 922861, 922897, 922907, 922931, 922973,
922993, 923017, 923023, 923029, 923047, 923051, 923053, 923107, 923123, 923129, 923137, 923141, 923147,
923171, 923177, 923179, 923183, 923201, 923203, 923227, 923233, 923239, 923249, 923309, 923311, 923333,
923341, 923347, 923369, 923371, 923387, 923399, 923407, 923411, 923437, 923441, 923449, 923453, 923467,
923471, 923501, 923509, 923513, 923539, 923543, 923551, 923561, 923567, 923579, 923581, 923591, 923599,
923603, 923617, 923641, 923653, 923687, 923693, 923701, 923711, 923719, 923743, 923773, 923789, 923809,
923833, 923849, 923851, 923861, 923869, 923903, 923917, 923929, 923939, 923947, 923953, 923959, 923963,
923971, 923977, 923983, 923987, 924019, 924023, 924031, 924037, 924041, 924043, 924059, 924073, 924083,
924097, 924101, 924109, 924139, 924151, 924173, 924191, 924197, 924241, 924269, 924281, 924283, 924299,
924323, 924337, 924359, 924361, 924383, 924397, 924401, 924403, 924419, 924421, 924431, 924437, 924463,
924493, 924499, 924503, 924523, 924527, 924529, 924551, 924557, 924601, 924617, 924641, 924643, 924659,
924661, 924683, 924697, 924709, 924713, 924719, 924727, 924731, 924743, 924751, 924757, 924769, 924773,
924779, 924793, 924809, 924811, 924827, 924829, 924841, 924871, 924877, 924881, 924907, 924929, 924961,
924967, 924997, 925019, 925027, 925033, 925039, 925051, 925063, 925073, 925079, 925081, 925087, 925097,
925103, 925109, 925117, 925121, 925147, 925153, 925159, 925163, 925181, 925189, 925193, 925217, 925237,
925241, 925271, 925273, 925279, 925291, 925307, 925339, 925349, 925369, 925373, 925387, 925391, 925399,
925409, 925423, 925447, 925469, 925487, 925499, 925501, 925513, 925517, 925523, 925559, 925577, 925579,
925597, 925607, 925619, 925621, 925637, 925649, 925663, 925669, 925679, 925697, 925721, 925733, 925741,
925783, 925789, 925823, 925831, 925843, 925849, 925891, 925901, 925913, 925921, 925937, 925943, 925949,
925961, 925979, 925987, 925997, 926017, 926027, 926033, 926077, 926087, 926089, 926099, 926111, 926113,
926129, 926131, 926153, 926161, 926171, 926179, 926183, 926203, 926227, 926239, 926251, 926273, 926293,
926309, 926327, 926351, 926353, 926357, 926377, 926389, 926399, 926411, 926423, 926437, 926461, 926467,
926489, 926503, 926507, 926533, 926537, 926557, 926561, 926567, 926581, 926587, 926617, 926623, 926633,
926657, 926659, 926669, 926671, 926689, 926701, 926707, 926741, 926747, 926767, 926777, 926797, 926803,
926819, 926843, 926851, 926867, 926879, 926899, 926903, 926921, 926957, 926963, 926971, 926977, 926983,
927001, 927007, 927013, 927049, 927077, 927083, 927089, 927097, 927137, 927149, 927161, 927167, 927187,
927191, 927229, 927233, 927259, 927287, 927301, 927313, 927317, 927323, 927361, 927373, 927397, 927403,
927431, 927439, 927491, 927497, 927517, 927529, 927533, 927541, 927557, 927569, 927587, 927629, 927631,
927643, 927649, 927653, 927671, 927677, 927683, 927709, 927727, 927743, 927763, 927769, 927779, 927791,
927803, 927821, 927833, 927841, 927847, 927853, 927863, 927869, 927961, 927967, 927973, 928001, 928043,
928051, 928063, 928079, 928097, 928099, 928111, 928139, 928141, 928153, 928157, 928159, 928163, 928177,
928223, 928231, 928253, 928267, 928271, 928273, 928289, 928307, 928313, 928331, 928337, 928351, 928399,
928409, 928423, 928427, 928429, 928453, 928457, 928463, 928469, 928471, 928513, 928547, 928559, 928561,
928597, 928607, 928619, 928621, 928637, 928643, 928649, 928651, 928661, 928679, 928689, 928703, 928769,
928771, 928787, 928793, 928799, 928813, 928817, 928819, 928849, 928859, 928871, 928883, 928903, 928913,
928927, 928933, 928979, 929003, 929009, 929011, 929023, 929029, 929051, 929057, 929059, 929063, 929069,
929077, 929083, 929087, 929113, 929129, 929141, 929153, 929161, 929171, 929197, 929207, 929209, 929239,
929251, 929261, 929281, 929293, 929303, 929311, 929323, 929333, 929381, 929389, 929393, 929399, 929417,
929419, 929431, 929459, 929483, 929497, 929501, 929507, 929527, 929549, 929557, 929561, 929573, 929581,
929587, 929609, 929623, 929627, 929629, 929639, 929641, 929647, 929671, 929693, 929717, 929737, 929741,

929743, 929749, 929777, 929791, 929807, 929809, 929813, 929843, 929861, 929869, 929881, 929891, 929897, 929941, 929953, 929963, 929977, 929983, 930011, 930043, 930071, 930073, 930077, 930079, 930089, 930101, 930113, 930119, 930157, 930173, 930179, 930187, 930191, 930197, 930199, 930211, 930229, 930269, 930277, 930283, 930287, 930289, 930301, 930323, 930337, 930379, 930389, 930409, 930437, 930467, 930469, 930481, 930491, 930499, 930509, 930547, 930551, 930569, 930571, 930583, 930593, 930617, 930619, 930637, 930653, 930667, 930689, 930707, 930719, 930737, 930749, 930763, 930773, 930779, 930817, 930827, 930841, 930847, 930859, 930863, 930889, 930911, 930931, 930973, 930977, 930989, 930991, 931003, 931013, 931067, 931087, 931097, 931123, 931127, 931129, 931153, 931163, 931169, 931181, 931193, 931199, 931213, 931237, 931241, 931267, 931289, 931303, 931309, 931313, 931319, 931351, 931363, 931387, 931417, 931421, 931487, 931499, 931517, 931529, 931537, 931543, 931571, 931573, 931577, 931597, 931621, 931639, 931657, 931691, 931709, 931727, 931729, 931739, 931747, 931751, 931757, 931781, 931783, 931789, 931811, 931837, 931849, 931859, 931873, 931877, 931883, 931901, 931907, 931913, 931921, 931933, 931943, 931949, 931967, 931981, 931999, 932003, 932021, 932039, 932051, 932081, 932101, 932117, 932119, 932131, 932149, 932153, 932177, 932189, 932203, 932207, 932209, 932219, 932221, 932227, 932231, 932257, 932303, 932317, 932333, 932341, 932353, 932357, 932413, 932417, 932419, 932431, 932441, 932447, 932471, 932473, 932483, 932497, 932513, 932521, 932537, 932549, 932557, 932563, 932567, 932579, 932587, 932593, 932597, 932609, 932647, 932651, 932663, 932677, 932681, 932683, 932749, 932761, 932779, 932783, 932801, 932803, 932819, 932839, 932863, 932879, 932887, 932917, 932923, 932927, 932941, 932947, 932951, 932963, 932969, 932983, 932999, 933001, 933019, 933047, 933059, 933061, 933067, 933073, 933151, 933157, 933173, 933199, 933209, 933217, 933221, 933241, 933259, 933263, 933269, 933293, 933301, 933313, 933319, 933329, 933349, 933389, 933397, 933403, 933407, 933421, 933433, 933463, 933479, 933497, 933523, 933551, 933553, 933563, 933601, 933607, 933613, 933643, 933649, 933671, 933677, 933703, 933707, 933739, 933761, 933781, 933787, 933797, 933809, 933811, 933817, 933839, 933847, 933851, 933853, 933883, 933893, 933923, 933931, 933943, 933949, 933953, 933967, 933973, 933979, 934001, 934009, 934033, 934039, 934049, 934051, 934057, 934067, 934069, 934079, 934111, 934117, 934121, 934127, 934151, 934159, 934187, 934223, 934229, 934243, 934253, 934259, 934277, 934291, 934301, 934319, 934343, 934387, 934393, 934399, 934403, 934429, 934441, 934463, 934469, 934481, 934487, 934489, 934499, 934517, 934523, 934537, 934543, 934547, 934561, 934567, 934579, 934597, 934603, 934607, 934613, 934639, 934669, 934673, 934693, 934721, 934723, 934733, 934753, 934763, 934771, 934793, 934799, 934811, 934831, 934837, 934853, 934861, 934883, 934889, 934891, 934897, 934907, 934909, 934919, 934939, 934943, 934951, 934961, 934979, 934981, 935003, 935021, 935023, 935059, 935063, 935071, 935093, 935107, 935113, 935147, 935149, 935167, 935183, 935189, 935197, 935201, 935213, 935243, 935257, 935261, 935303, 935339, 935353, 935359, 935377, 935381, 935393, 935399, 935413, 935423, 935443, 935447, 935461, 935489, 935507, 935513, 935531, 935537, 935581, 935587, 935591, 935593, 935603, 935621, 935639, 935651, 935653, 935677, 935687, 935689, 935699, 935707, 935717, 935719, 935761, 935771, 935777, 935791, 935813, 935819, 935827, 935839, 935843, 935861, 935899, 935903, 935971, 935999, 936007, 936029, 936053, 936097, 936113, 936119, 936127, 936151, 936161, 936179, 936181, 936197, 936203, 936223, 936227, 936233, 936253, 936259, 936281, 936283, 936311, 936319, 936329, 936361, 936379, 936391, 936401, 936407, 936413, 936437, 936451, 936469, 936487, 936493, 936499, 936511, 936521, 936527, 936539, 936557, 936577, 936587, 936599, 936619, 936647, 936659, 936667, 936673, 936679, 936697, 936709, 936713, 936731, 936737, 936739, 936769, 936773, 936779, 936797, 936811, 936827, 936869, 936889, 936907, 936911, 936917, 936919, 936937, 936941, 936953, 936967, 937003, 937007, 937009, 937031, 937033, 937049, 937067, 937121, 937127, 937147, 937151, 937171, 937187, 937207, 937229, 937231, 937241, 937243, 937253, 937331, 937337, 937351, 937373, 937379, 937421, 937429, 937459, 937463, 937477, 937481, 937501, 937511, 937537, 937571, 937577, 937589, 937591, 937613, 937627, 937633, 937637, 937639, 937661, 937663, 937667, 937679, 937681, 937693, 937709, 937721, 937747, 937751, 937777, 937789, 937801, 937813, 937819, 937823, 937841, 937847, 937877, 937883, 937891, 937901, 937903, 937919, 937927, 937943, 937949, 937969, 937991, 938017, 938023, 938027, 938033, 938051, 938053, 938057, 938059, 938071, 938083, 938089, 938099, 938107, 938117, 938129, 938183, 938207, 938219, 938233, 938243, 938251, 938257, 938263, 938279, 938293, 938309, 938323, 938341, 938347, 938351, 938359, 938369, 938387, 938393, 938437, 938447, 938453, 938459, 938491, 938507, 938533, 938537, 938563, 938569, 938573, 938591,

938611, 938617, 938659, 938677, 938681, 938713, 938747, 938761, 938803, 938807, 938827, 938831, 938843, 938857, 938869, 938879, 938881, 938921, 938939, 938947, 938953, 938963, 938969, 938981, 938983, 938989, 939007, 939011, 939019, 939061, 939089, 939091, 939109, 939119, 939121, 939157, 939167, 939179, 939181, 939193, 939203, 939229, 939247, 939287, 939293, 939299, 939317, 939347, 939349, 939359, 939361, 939373, 939377, 939391, 939413, 939431, 939439, 939443, 939451, 939469, 939487, 939511, 939551, 939581, 939599, 939611, 939613, 939623, 939649, 939661, 939677, 939707, 939713, 939737, 939739, 939749, 939767, 939769, 939773, 939791, 939793, 939823, 939839, 939847, 939853, 939871, 939881, 939901, 939923, 939931, 939971, 939973, 939989, 939997, 940001, 940003, 940019, 940031, 940067, 940073, 940087, 940097, 940127, 940157, 940169, 940183, 940189, 940201, 940223, 940229, 940241, 940249, 940259, 940271, 940279, 940297, 940301, 940319, 940327, 940349, 940351, 940361, 940369, 940399, 940403, 940421, 940469, 940477, 940483, 940501, 940523, 940529, 940531, 940543, 940547, 940549, 940553, 940573, 940607, 940619, 940649, 940669, 940691, 940703, 940721, 940727, 940733, 940739, 940759, 940781, 940783, 940787, 940801, 940813, 940817, 940829, 940853, 940871, 940879, 940889, 940903, 940913, 940921, 940931, 940949, 940957, 940981, 940993, 941009, 941011, 941023, 941027, 941041, 941093, 941099, 941117, 941119, 941123, 941131, 941153, 941159, 941167, 941179, 941201, 941207, 941209, 941221, 941249, 941251, 941263, 941267, 941299, 941309, 941323, 941329, 941351, 941359, 941383, 941407, 941429, 941441, 941449, 941453, 941461, 941467, 941471, 941489, 941491, 941503, 941509, 941513, 941519, 941537, 941557, 941561, 941573, 941593, 941599, 941609, 941617, 941641, 941653, 941663, 941669, 941671, 941683, 941701, 941723, 941737, 941741, 941747, 941753, 941771, 941791, 941813, 941839, 941861, 941879, 941903, 941911, 941929, 941933, 941947, 941971, 941981, 941989, 941999, 942013, 942017, 942031, 942037, 942041, 942043, 942049, 942061, 942079, 942091, 942101, 942113, 942143, 942163, 942167, 942169, 942187, 942199, 942217, 942223, 942247, 942257, 942269, 942301, 942311, 942313, 942317, 942341, 942367, 942371, 942401, 942433, 942437, 942439, 942449, 942479, 942509, 942521, 942527, 942541, 942569, 942577, 942583, 942593, 942607, 942637, 942653, 942659, 942661, 942691, 942709, 942719, 942727, 942749, 942763, 942779, 942787, 942811, 942827, 942847, 942853, 942857, 942859, 942869, 942883, 942889, 942899, 942901, 942917, 942943, 942979, 942983, 943003, 943009, 943013, 943031, 943043, 943057, 943073, 943079, 943081, 943091, 943097, 943127, 943139, 943153, 943157, 943183, 943199, 943213, 943219, 943231, 943249, 943273, 943277, 943289, 943301, 943303, 943307, 943321, 943343, 943357, 943363, 943367, 943373, 943387, 943403, 943409, 943421, 943429, 943471, 943477, 943499, 943511, 943541, 943543, 943567, 943571, 943589, 943601, 943603, 943637, 943651, 943693, 943699, 943729, 943741, 943751, 943757, 943763, 943769, 943777, 943781, 943783, 943799, 943801, 943819, 943837, 943841, 943843, 943849, 943871, 943903, 943909, 943913, 943931, 943951, 943967, 944003, 944017, 944029, 944039, 944071, 944077, 944123, 944137, 944143, 944147, 944149, 944161, 944179, 944191, 944233, 944239, 944257, 944261, 944263, 944297, 944309, 944329, 944369, 944387, 944389, 944393, 944399, 944417, 944429, 944431, 944453, 944467, 944473, 944491, 944497, 944519, 944521, 944527, 944533, 944543, 944551, 944561, 944563, 944579, 944591, 944609, 944621, 944651, 944659, 944677, 944687, 944689, 944701, 944711, 944717, 944729, 944731, 944773, 944777, 944803, 944821, 944833, 944857, 944873, 944887, 944893, 944897, 944899, 944929, 944953, 944963, 944969, 944987, 945031, 945037, 945059, 945089, 945103, 945143, 945151, 945179, 945209, 945211, 945227, 945233, 945289, 945293, 945331, 945341, 945349, 945359, 945367, 945377, 945389, 945391, 945397, 945409, 945431, 945457, 945463, 945473, 945479, 945481, 945521, 945547, 945577, 945587, 945589, 945601, 945629, 945631, 945647, 945671, 945673, 945677, 945701, 945731, 945733, 945739, 945767, 945787, 945799, 945809, 945811, 945817, 945823, 945851, 945881, 945883, 945887, 945899, 945907, 945929, 945937, 945941, 945943, 945949, 945961, 945983, 946003, 946021, 946031, 946037, 946079, 946081, 946091, 946093, 946109, 946111, 946123, 946133, 946163, 946177, 946193, 946207, 946223, 946249, 946273, 946291, 946307, 946327, 946331, 946367, 946369, 946391, 946397, 946411, 946417, 946453, 946459, 946469, 946487, 946489, 946507, 946513, 946549, 946573, 946579, 946607, 946661, 946663, 946667, 946669, 946681, 946697, 946717, 946727, 946733, 946741, 946753, 946769, 946783, 946801, 946819, 946823, 946853, 946859, 946861, 946873, 946877, 946901, 946919, 946931, 946943, 946949, 946961, 946969, 946987, 946993, 946997, 947027, 947033, 947083, 947119, 947129, 947137, 947171, 947183, 947197, 947203, 947239, 947263, 947299, 947327, 947341, 947351, 947357, 947369, 947377, 947381, 947383, 947389, 947407, 947411, 947413, 947417, 947423, 947431, 947449, 947483, 947501,

947509, 947539, 947561, 947579, 947603, 947621, 947627, 947641, 947647, 947651, 947659, 947707, 947711,
947719, 947729, 947741, 947743, 947747, 947753, 947773, 947783, 947803, 947819, 947833, 947851, 947857,
947861, 947873, 947893, 947911, 947917, 947927, 947959, 947963, 947987, 948007, 948019, 948029, 948041,
948049, 948053, 948061, 948067, 948089, 948091, 948133, 948139, 948149, 948151, 948169, 948173, 948187,
948247, 948253, 948263, 948281, 948287, 948293, 948317, 948331, 948349, 948377, 948391, 948401, 948403,
948407, 948427, 948439, 948443, 948449, 948457, 948469, 948487, 948517, 948533, 948547, 948551, 948557,
948581, 948593, 948659, 948671, 948707, 948713, 948721, 948749, 948767, 948797, 948799, 948839, 948847,
948853, 948877, 948887, 948901, 948907, 948929, 948943, 948947, 948971, 948973, 948989, 949001, 949019,
949021, 949033, 949037, 949043, 949051, 949111, 949121, 949129, 949147, 949153, 949159, 949171, 949211,
949213, 949241, 949243, 949253, 949261, 949303, 949307, 949381, 949387, 949391, 949409, 949423, 949427,
949439, 949441, 949451, 949453, 949471, 949477, 949513, 949517, 949523, 949567, 949583, 949589, 949607,
949609, 949621, 949631, 949633, 949643, 949649, 949651, 949667, 949673, 949687, 949691, 949699, 949733,
949759, 949771, 949777, 949789, 949811, 949849, 949853, 949889, 949891, 949903, 949931, 949937, 949939,
949951, 949957, 949961, 949967, 949973, 949979, 949987, 949997, 950009, 950023, 950029, 950039, 950041,
950071, 950083, 950099, 950111, 950149, 950161, 950177, 950179, 950207, 950221, 950227, 950231, 950233,
950239, 950251, 950269, 950281, 950329, 950333, 950347, 950357, 950363, 950393, 950401, 950423, 950447,
950459, 950461, 950473, 950479, 950483, 950497, 950501, 950507, 950519, 950527, 950531, 950557, 950569,
950611, 950617, 950633, 950639, 950647, 950671, 950681, 950689, 950693, 950699, 950717, 950723, 950737,
950743, 950753, 950783, 950791, 950809, 950813, 950819, 950837, 950839, 950867, 950869, 950879, 950921,
950927, 950933, 950947, 950953, 950959, 950993, 951001, 951019, 951023, 951029, 951047, 951053, 951059,
951061, 951079, 951089, 951091, 951101, 951107, 951109, 951131, 951151, 951161, 951193, 951221, 951259,
951277, 951281, 951283, 951299, 951331, 951341, 951343, 951361, 951367, 951373, 951389, 951407, 951413,
951427, 951437, 951449, 951469, 951479, 951491, 951497, 951553, 951557, 951571, 951581, 951583, 951589,
951623, 951637, 951641, 951647, 951649, 951659, 951689, 951697, 951749, 951781, 951787, 951791, 951803,
951829, 951851, 951859, 951887, 951893, 951911, 951941, 951943, 951959, 951967, 951997, 952001, 952009,
952037, 952057, 952073, 952087, 952097, 952111, 952117, 952123, 952129, 952141, 952151, 952163, 952169,
952183, 952199, 952207, 952219, 952229, 952247, 952253, 952277, 952279, 952291, 952297, 952313, 952349,
952363, 952379, 952381, 952397, 952423, 952429, 952439, 952481, 952487, 952507, 952513, 952541, 952547,
952559, 952573, 952583, 952597, 952619, 952649, 952657, 952667, 952669, 952681, 952687, 952691, 952709,
952739, 952741, 952753, 952771, 952789, 952811, 952813, 952823, 952829, 952843, 952859, 952873, 952877,
952883, 952921, 952927, 952933, 952937, 952943, 952957, 952967, 952979, 952981, 952997, 953023, 953039,
953041, 953053, 953077, 953081, 953093, 953111, 953131, 953149, 953171, 953179, 953191, 953221, 953237,
953243, 953261, 953273, 953297, 953321, 953333, 953341, 953347, 953399, 953431, 953437, 953443, 953473,
953483, 953497, 953501, 953503, 953507, 953521, 953539, 953543, 953551, 953567, 953593, 953621, 953639,
953647, 953651, 953671, 953681, 953699, 953707, 953731, 953747, 953773, 953789, 953791, 953831, 953851,
953861, 953873, 953881, 953917, 953923, 953929, 953941, 953969, 953977, 953983, 953987, 954001, 954007,
954011, 954043, 954067, 954097, 954103, 954131, 954133, 954139, 954157, 954167, 954181, 954203, 954209,
954221, 954229, 954253, 954257, 954259, 954263, 954269, 954277, 954287, 954307, 954319, 954323, 954367,
954377, 954379, 954391, 954409, 954433, 954451, 954461, 954469, 954491, 954497, 954509, 954517, 954539,
954571, 954599, 954619, 954623, 954641, 954649, 954671, 954677, 954697, 954713, 954719, 954727, 954743,
954757, 954763, 954827, 954829, 954847, 954851, 954853, 954857, 954869, 954871, 954911, 954917, 954923,
954929, 954971, 954973, 954977, 954979, 954991, 955037, 955039, 955051, 955061, 955063, 955091, 955093,
955103, 955127, 955139, 955147, 955153, 955183, 955193, 955211, 955217, 955223, 955243, 955261, 955267,
955271, 955277, 955307, 955309, 955313, 955319, 955333, 955337, 955363, 955379, 955391, 955433, 955439,
955441, 955457, 955469, 955477, 955481, 955483, 955501, 955511, 955541, 955601, 955607, 955613, 955649,
955657, 955693, 955697, 955709, 955711, 955727, 955729, 955769, 955777, 955781, 955793, 955807, 955813,
955819, 955841, 955853, 955879, 955883, 955891, 955901, 955919, 955937, 955939, 955951, 955957, 955963,
955967, 955987, 955991, 955993, 956003, 956051, 956057, 956083, 956107, 956113, 956119, 956143, 956147,
956177, 956231, 956237, 956261, 956269, 956273, 956281, 956303, 956311, 956341, 956353, 956357, 956377,

956383, 956387, 956393, 956399, 956401, 956429, 956477, 956503, 956513, 956521, 956569, 956587, 956617,
956633, 956689, 956699, 956713, 956723, 956749, 956759, 956789, 956801, 956831, 956843, 956849, 956861,
956881, 956903, 956909, 956929, 956941, 956951, 956953, 956987, 956993, 956999, 957031, 957037, 957041,
957043, 957059, 957071, 957091, 957097, 957107, 957109, 957119, 957133, 957139, 957161, 957169, 957181,
957193, 957211, 957221, 957241, 957247, 957263, 957289, 957317, 957331, 957337, 957349, 957361, 957403,
957409, 957413, 957419, 957431, 957433, 957499, 957529, 957547, 957553, 957557, 957563, 957587, 957599,
957601, 957611, 957641, 957643, 957659, 957701, 957703, 957709, 957721, 957731, 957751, 957769, 957773,
957811, 957821, 957823, 957851, 957871, 957877, 957889, 957917, 957937, 957949, 957953, 957959, 957977,
957991, 958007, 958021, 958039, 958043, 958049, 958051, 958057, 958063, 958121, 958123, 958141, 958159,
958163, 958183, 958193, 958213, 958259, 958261, 958289, 958313, 958319, 958327, 958333, 958339, 958343,
958351, 958357, 958361, 958367, 958369, 958381, 958393, 958423, 958439, 958459, 958481, 958487, 958499,
958501, 958519, 958523, 958541, 958543, 958547, 958549, 958553, 958577, 958609, 958627, 958637, 958667,
958669, 958673, 958679, 958687, 958693, 958729, 958739, 958777, 958787, 958807, 958819, 958829, 958843,
958849, 958871, 958877, 958883, 958897, 958901, 958921, 958931, 958933, 958957, 958963, 958967, 958973,
959009, 959083, 959093, 959099, 959131, 959143, 959149, 959159, 959173, 959183, 959207, 959209, 959219,
959227, 959237, 959263, 959267, 959269, 959279, 959323, 959333, 959339, 959351, 959363, 959369, 959377,
959383, 959389, 959449, 959461, 959467, 959471, 959473, 959477, 959479, 959489, 959533, 959561, 959579,
959597, 959603, 959617, 959627, 959659, 959677, 959681, 959689, 959719, 959723, 959737, 959759, 959773,
959779, 959801, 959809, 959831, 959863, 959867, 959869, 959873, 959879, 959887, 959911, 959921, 959927,
959941, 959947, 959953, 959969, 960017, 960019, 960031, 960049, 960053, 960059, 960077, 960119, 960121,
960131, 960137, 960139, 960151, 960173, 960191, 960199, 960217, 960229, 960251, 960259, 960293, 960299,
960329, 960331, 960341, 960353, 960373, 960383, 960389, 960419, 960467, 960493, 960497, 960499, 960521,
960523, 960527, 960569, 960581, 960587, 960593, 960601, 960637, 960643, 960647, 960649, 960667, 960677,
960691, 960703, 960709, 960737, 960763, 960793, 960803, 960809, 960829, 960833, 960863, 960889, 960931,
960937, 960941, 960961, 960977, 960983, 960989, 960991, 961003, 961021, 961033, 961063, 961067, 961069,
961073, 961087, 961091, 961097, 961099, 961109, 961117, 961123, 961133, 961139, 961141, 961151, 961157,
961159, 961183, 961187, 961189, 961201, 961241, 961243, 961273, 961277, 961283, 961313, 961319, 961339,
961393, 961397, 961399, 961427, 961447, 961451, 961453, 961459, 961487, 961507, 961511, 961529, 961531,
961547, 961549, 961567, 961601, 961613, 961619, 961627, 961633, 961637, 961643, 961657, 961661, 961663,
961679, 961687, 961691, 961703, 961729, 961733, 961739, 961747, 961757, 961769, 961777, 961783, 961789,
961811, 961813, 961817, 961841, 961847, 961853, 961861, 961871, 961879, 961927, 961937, 961943, 961957,
961973, 961981, 961991, 961993, 962009, 962011, 962033, 962041, 962051, 962063, 962077, 962099, 962119,
962131, 962161, 962177, 962197, 962233, 962237, 962243, 962257, 962267, 962303, 962309, 962341, 962363,
962413, 962417, 962431, 962441, 962447, 962459, 962461, 962471, 962477, 962497, 962503, 962509, 962537,
962543, 962561, 962569, 962587, 962603, 962609, 962617, 962623, 962627, 962653, 962669, 962671, 962677,
962681, 962683, 962737, 962743, 962747, 962779, 962783, 962789, 962791, 962807, 962837, 962839, 962861,
962867, 962869, 962903, 962909, 962911, 962921, 962959, 962963, 962971, 962993, 963019, 963031, 963043,
963047, 963097, 963103, 963121, 963143, 963163, 963173, 963181, 963187, 963191, 963211, 963223, 963227,
963239, 963241, 963253, 963283, 963299, 963301, 963311, 963323, 963331, 963341, 963343, 963349, 963367,
963379, 963397, 963419, 963427, 963461, 963481, 963491, 963497, 963499, 963559, 963581, 963601, 963607,
963629, 963643, 963653, 963659, 963667, 963689, 963691, 963701, 963707, 963709, 963719, 963731, 963751,
963761, 963763, 963779, 963793, 963799, 963811, 963817, 963839, 963841, 963847, 963863, 963871, 963877,
963899, 963901, 963913, 963943, 963973, 963979, 964009, 964021, 964027, 964039, 964049, 964081, 964097,
964133, 964151, 964153, 964199, 964207, 964213, 964217, 964219, 964253, 964259, 964261, 964267, 964283,
964289, 964297, 964303, 964309, 964333, 964339, 964351, 964357, 964363, 964373, 964417, 964423, 964433,
964463, 964499, 964501, 964507, 964517, 964519, 964531, 964559, 964571, 964577, 964583, 964589, 964609,
964637, 964661, 964679, 964693, 964697, 964703, 964721, 964753, 964757, 964783, 964787, 964793, 964823,
964829, 964861, 964871, 964879, 964883, 964889, 964897, 964913, 964927, 964933, 964939, 964967, 964969,
964973, 964981, 965023, 965047, 965059, 965087, 965089, 965101, 965113, 965117, 965131, 965147, 965161,

965171, 965177, 965179, 965189, 965191, 965197, 965201, 965227, 965233, 965249, 965267, 965291, 965303,
965317, 965329, 965357, 965369, 965399, 965401, 965407, 965411, 965423, 965429, 965443, 965453, 965467,
965483, 965491, 965507, 965519, 965533, 965551, 965567, 965603, 965611, 965621, 965623, 965639, 965647,
965659, 965677, 965711, 965749, 965759, 965773, 965777, 965779, 965791, 965801, 965843, 965851, 965857,
965893, 965927, 965953, 965963, 965969, 965983, 965989, 966011, 966013, 966029, 966041, 966109, 966113,
966139, 966149, 966157, 966191, 966197, 966209, 966211, 966221, 966227, 966233, 966241, 966257, 966271,
966293, 966307, 966313, 966319, 966323, 966337, 966347, 966353, 966373, 966377, 966379, 966389, 966401,
966409, 966419, 966431, 966439, 966463, 966481, 966491, 966499, 966509, 966521, 966527, 966547, 966557,
966583, 966613, 966617, 966619, 966631, 966653, 966659, 966661, 966677, 966727, 966751, 966781, 966803,
966817, 966863, 966869, 966871, 966883, 966893, 966907, 966913, 966919, 966923, 966937, 966961, 966971,
966991, 966997, 967003, 967019, 967049, 967061, 967111, 967129, 967139, 967171, 967201, 967229, 967259,
967261, 967289, 967297, 967319, 967321, 967327, 967333, 967349, 967361, 967363, 967391, 967397, 967427,
967429, 967441, 967451, 967459, 967481, 967493, 967501, 967507, 967511, 967529, 967567, 967583, 967607,
967627, 967663, 967667, 967693, 967699, 967709, 967721, 967739, 967751, 967753, 967763, 967781, 967787,
967819, 967823, 967831, 967843, 967847, 967859, 967873, 967877, 967903, 967919, 967931, 967937, 967951,
967961, 967979, 967999, 968003, 968017, 968021, 968027, 968041, 968063, 968089, 968101, 968111, 968113,
968117, 968137, 968141, 968147, 968159, 968173, 968197, 968213, 968237, 968239, 968251, 968263, 968267,
968273, 968291, 968299, 968311, 968321, 968329, 968333, 968353, 968377, 968381, 968389, 968419, 968423,
968431, 968437, 968459, 968467, 968479, 968501, 968503, 968519, 968521, 968537, 968557, 968567, 968573,
968593, 968641, 968647, 968659, 968663, 968689, 968699, 968713, 968729, 968731, 968761, 968801, 968809,
968819, 968827, 968831, 968857, 968879, 968897, 968909, 968911, 968917, 968939, 968959, 968963, 968971,
969011, 969037, 969041, 969049, 969071, 969083, 969097, 969109, 969113, 969131, 969139, 969167, 969179,
969181, 969233, 969239, 969253, 969257, 969259, 969271, 969301, 969341, 969343, 969347, 969359, 969377,
969403, 969407, 969421, 969431, 969433, 969443, 969457, 969461, 969467, 969481, 969497, 969503, 969509,
969533, 969559, 969569, 969593, 969599, 969637, 969641, 969667, 969671, 969677, 969679, 969713, 969719,
969721, 969743, 969757, 969763, 969767, 969791, 969797, 969809, 969821, 969851, 969863, 969869, 969877,
969889, 969907, 969911, 969919, 969923, 969929, 969977, 969989, 970027, 970031, 970043, 970051, 970061,
970063, 970069, 970087, 970091, 970111, 970133, 970147, 970201, 970213, 970217, 970219, 970231, 970237,
970247, 970259, 970261, 970267, 970279, 970297, 970303, 970313, 970351, 970391, 970421, 970423, 970433,
970441, 970447, 970457, 970469, 970481, 970493, 970537, 970549, 970561, 970573, 970583, 970603, 970633,
970643, 970657, 970667, 970687, 970699, 970721, 970747, 970777, 970787, 970789, 970793, 970799, 970813,
970817, 970829, 970847, 970859, 970861, 970867, 970877, 970883, 970903, 970909, 970927, 970939, 970943,
970961, 970967, 970969, 970987, 970997, 970999, 971021, 971027, 971029, 971039, 971051, 971053, 971063,
971077, 971093, 971099, 971111, 971141, 971143, 971149, 971153, 971171, 971177, 971197, 971207, 971237,
971251, 971263, 971273, 971279, 971281, 971291, 971309, 971339, 971353, 971357, 971371, 971381, 971387,
971389, 971401, 971419, 971429, 971441, 971473, 971479, 971483, 971491, 971501, 971513, 971521, 971549,
971561, 971563, 971569, 971591, 971639, 971651, 971653, 971683, 971693, 971699, 971713, 971723, 971753,
971759, 971767, 971783, 971821, 971833, 971851, 971857, 971863, 971899, 971903, 971917, 971921, 971933,
971939, 971951, 971959, 971977, 971981, 971989, 972001, 972017, 972029, 972031, 972047, 972071, 972079,
972091, 972113, 972119, 972121, 972131, 972133, 972137, 972161, 972163, 972197, 972199, 972221, 972227,
972229, 972259, 972263, 972271, 972277, 972313, 972319, 972329, 972337, 972343, 972347, 972353, 972373,
972403, 972407, 972409, 972427, 972431, 972443, 972469, 972473, 972481, 972493, 972533, 972557, 972577,
972581, 972599, 972611, 972613, 972623, 972637, 972649, 972661, 972679, 972683, 972701, 972721, 972787,
972793, 972799, 972823, 972827, 972833, 972847, 972869, 972887, 972899, 972901, 972941, 972943, 972967,
972977, 972991, 973001, 973003, 973031, 973033, 973051, 973057, 973067, 973069, 973073, 973081, 973099,
973129, 973151, 973169, 973177, 973187, 973213, 973253, 973277, 973279, 973283, 973289, 973321, 973331,
973333, 973367, 973373, 973387, 973397, 973409, 973411, 973421, 973439, 973459, 973487, 973523, 973529,
973537, 973547, 973561, 973591, 973597, 973631, 973657, 973669, 973681, 973691, 973727, 973757, 973759,
973781, 973787, 973789, 973801, 973813, 973823, 973837, 973853, 973891, 973897, 973901, 973919, 973957,

974003, 974009, 974033, 974041, 974053, 974063, 974089, 974107, 974123, 974137, 974143, 974147, 974159,
974161, 974167, 974177, 974179, 974189, 974213, 974249, 974261, 974269, 974273, 974279, 974293, 974317,
974329, 974359, 974383, 974387, 974401, 974411, 974417, 974419, 974431, 974437, 974443, 974459, 974473,
974489, 974497, 974507, 974513, 974531, 974537, 974539, 974551, 974557, 974563, 974581, 974591, 974599,
974651, 974653, 974657, 974707, 974711, 974713, 974737, 974747, 974749, 974761, 974773, 974803, 974819,
974821, 974837, 974849, 974861, 974863, 974867, 974873, 974879, 974887, 974891, 974923, 974927, 974957,
974959, 974969, 974971, 974977, 974983, 974989, 974999, 975011, 975017, 975049, 975053, 975071, 975083,
975089, 975133, 975151, 975157, 975181, 975187, 975193, 975199, 975217, 975257, 975259, 975263, 975277,
975281, 975287, 975313, 975323, 975343, 975367, 975379, 975383, 975389, 975421, 975427, 975433, 975439,
975463, 975493, 975497, 975509, 975521, 975523, 975551, 975553, 975581, 975599, 975619, 975629, 975643,
975649, 975661, 975671, 975691, 975701, 975731, 975739, 975743, 975797, 975803, 975811, 975823, 975827,
975847, 975857, 975869, 975883, 975899, 975901, 975907, 975941, 975943, 975967, 975977, 975991, 976009,
976013, 976033, 976039, 976091, 976093, 976103, 976109, 976117, 976127, 976147, 976177, 976187, 976193,
976211, 976231, 976253, 976271, 976279, 976301, 976303, 976307, 976309, 976351, 976369, 976403, 976411,
976439, 976447, 976453, 976457, 976471, 976477, 976483, 976489, 976501, 976513, 976537, 976553, 976559,
976561, 976571, 976601, 976607, 976621, 976637, 976639, 976643, 976669, 976699, 976709, 976721, 976727,
976777, 976799, 976817, 976823, 976849, 976853, 976883, 976909, 976919, 976933, 976951, 976957, 976991,
977021, 977023, 977047, 977057, 977069, 977087, 977107, 977147, 977149, 977167, 977183, 977191, 977203,
977209, 977233, 977239, 977243, 977257, 977269, 977299, 977323, 977351, 977357, 977359, 977363, 977369,
977407, 977411, 977413, 977437, 977447, 977507, 977513, 977521, 977539, 977567, 977591, 977593, 977609,
977611, 977629, 977671, 977681, 977693, 977719, 977723, 977747, 977761, 977791, 977803, 977813, 977819,
977831, 977849, 977861, 977881, 977897, 977923, 977927, 977971, 978001, 978007, 978011, 978017, 978031,
978037, 978041, 978049, 978053, 978067, 978071, 978073, 978077, 978079, 978091, 978113, 978149, 978151,
978157, 978179, 978181, 978203, 978209, 978217, 978223, 978233, 978239, 978269, 978277, 978283, 978287,
978323, 978337, 978343, 978347, 978349, 978359, 978389, 978403, 978413, 978427, 978449, 978457, 978463,
978473, 978479, 978491, 978511, 978521, 978541, 978569, 978599, 978611, 978617, 978619, 978643, 978647,
978683, 978689, 978697, 978713, 978727, 978743, 978749, 978773, 978797, 978799, 978821, 978839, 978851,
978853, 978863, 978871, 978883, 978907, 978917, 978931, 978947, 978973, 978997, 979001, 979009, 979031,
979037, 979061, 979063, 979093, 979103, 979109, 979117, 979159, 979163, 979171, 979177, 979189, 979201,
979207, 979211, 979219, 979229, 979261, 979273, 979283, 979291, 979313, 979327, 979333, 979337, 979343,
979361, 979369, 979373, 979379, 979403, 979423, 979439, 979457, 979471, 979481, 979519, 979529, 979541,
979543, 979549, 979553, 979567, 979651, 979691, 979709, 979717, 979747, 979757, 979787, 979807, 979819,
979831, 979873, 979883, 979889, 979907, 979919, 979921, 979949, 979969, 979987, 980027, 980047, 980069,
980071, 980081, 980107, 980117, 980131, 980137, 980149, 980159, 980173, 980179, 980197, 980219, 980249,
980261, 980293, 980299, 980321, 980327, 980363, 980377, 980393, 980401, 980417, 980423, 980431, 980449,
980459, 980471, 980489, 980491, 980503, 980549, 980557, 980579, 980587, 980591, 980593, 980599, 980621,
980641, 980677, 980687, 980689, 980711, 980717, 980719, 980729, 980731, 980773, 980801, 980803, 980827,
980831, 980851, 980887, 980893, 980897, 980899, 980909, 980911, 980921, 980957, 980963, 980999, 981011,
981017, 981023, 981037, 981049, 981061, 981067, 981073, 981077, 981091, 981133, 981137, 981139, 981151,
981173, 981187, 981199, 981209, 981221, 981241, 981263, 981271, 981283, 981287, 981289, 981301, 981311,
981319, 981373, 981377, 981391, 981397, 981419, 981437, 981439, 981443, 981451, 981467, 981473, 981481,
981493, 981517, 981523, 981527, 981569, 981577, 981587, 981599, 981601, 981623, 981637, 981653, 981683,
981691, 981697, 981703, 981707, 981713, 981731, 981769, 981797, 981809, 981811, 981817, 981823, 981887,
981889, 981913, 981919, 981941, 981947, 981949, 981961, 981979, 981983, 982021, 982057, 982061, 982063,
982067, 982087, 982097, 982099, 982103, 982117, 982133, 982147, 982151, 982171, 982183, 982187, 982211,
982213, 982217, 982231, 982271, 982273, 982301, 982321, 982337, 982339, 982343, 982351, 982363, 982381,
982393, 982403, 982453, 982489, 982493, 982559, 982571, 982573, 982577, 982589, 982603, 982613, 982621,
982633, 982643, 982687, 982693, 982697, 982703, 982741, 982759, 982769, 982777, 982783, 982789, 982801,
982819, 982829, 982841, 982843, 982847, 982867, 982871, 982903, 982909, 982931, 982939, 982967, 982973,

982981, 983063, 983069, 983083, 983113, 983119, 983123, 983131, 983141, 983149, 983153, 983173, 983179,
983189, 983197, 983209, 983233, 983239, 983243, 983261, 983267, 983299, 983317, 983327, 983329, 983347,
983363, 983371, 983377, 984407, 983429, 983431, 983441, 983443, 983447, 983449, 983461, 983491, 983513,
983519, 983527, 983531, 983533, 983557, 983579, 983581, 983597, 983617, 983659, 983699, 983701, 983737,
983771, 983777, 983783, 983789, 983791, 983803, 983809, 983813, 983819, 983849, 983861, 983863, 983881,
983911, 983923, 983929, 983951, 983987, 983993, 984007, 984017, 984037, 984047, 984059, 984083, 984091,
984119, 984121, 984127, 984149, 984167, 984199, 984211, 984241, 984253, 984299, 984301, 984307, 984323,
984329, 984337, 984341, 984349, 984353, 984359, 984367, 984383, 984391, 984397, 984407, 984413, 984421,
984427, 984437, 984457, 984461, 984481, 984491, 984497, 984539, 984541, 984563, 984583, 984587, 984593,
984611, 984617, 984667, 984689, 984701, 984703, 984707, 984733, 984749, 984757, 984761, 984817, 984847,
984853, 984859, 984877, 984881, 984911, 984913, 984917, 984923, 984931, 984947, 984959, 985003, 985007,
985013, 985027, 985057, 985063, 985079, 985097, 985109, 985121, 985129, 985151, 985177, 985181, 985213,
985219, 985253, 985277, 985279, 985291, 985301, 985307, 985331, 985339, 985351, 985379, 985399, 985403,
985417, 985433, 985447, 985451, 985463, 985471, 985483, 985487, 985493, 985499, 985519, 985529, 985531,
985547, 985571, 985597, 985601, 985613, 985631, 985639, 985657, 985667, 985679, 985703, 985709, 985723,
985729, 985741, 985759, 985781, 985783, 985799, 985807, 985819, 985867, 985871, 985877, 985903, 985921,
985937, 985951, 985969, 985973, 985979, 985981, 985991, 985993, 985997, 986023, 986047, 986053, 986071,
986101, 986113, 986131, 986137, 986143, 986147, 986149, 986177, 986189, 986191, 986197, 986207, 986213,
986239, 986257, 986267, 986281, 986287, 986333, 986339, 986351, 986369, 986411, 986417, 986429, 986437,
986471, 986477, 986497, 986507, 986509, 986519, 986533, 986543, 986563, 986567, 986569, 986581, 986593,
986597, 986599, 986617, 986633, 986641, 986659, 986693, 986707, 986717, 986719, 986729, 986737, 986749,
986759, 986767, 986779, 986801, 986813, 986819, 986837, 986849, 986851, 986857, 986903, 986927, 986929,
986933, 986941, 986959, 986963, 986981, 986983, 986989, 987013, 987023, 987029, 987043, 987053, 987061,
987067, 987079, 987083, 987089, 987097, 987101, 987127, 987143, 987191, 987193, 987199, 987209, 987211,
987227, 987251, 987293, 987299, 987313, 987353, 987361, 987383, 987391, 987433, 987457, 987463, 987473,
987491, 987509, 987523, 987533, 987541, 987551, 987559, 987587, 987593, 987599, 987607, 987631, 987659, 987697,
987713, 987739, 987793, 987797, 987803, 987809, 987821, 987851, 987869, 987911, 987913, 987929, 987941,
987971, 987979, 987983, 987991, 987997, 988007, 988021, 988033, 988051, 988061, 988067, 988069, 988093,
988109, 988111, 988129, 988147, 988157, 988199, 988213, 988217, 988219, 988231, 988237, 988243, 988271,
988279, 988297, 988313, 988319, 988321, 988343, 988357, 988367, 988409, 988417, 988439, 988453, 988459,
988483, 988489, 988501, 988511, 988541, 988549, 988571, 988577, 988579, 988583, 988591, 988607, 988643,
988649, 988651, 988661, 988681, 988693, 988711, 988727, 988733, 988759, 988763, 988783, 988789, 988829,
988837, 988849, 988859, 988861, 988877, 988901, 988909, 988937, 988951, 988963, 988979, 989011, 989029,
989059, 989071, 989081, 989099, 989119, 989123, 989171, 989173, 989231, 989239, 989249, 989251, 989279,
989293, 989309, 989321, 989323, 989327, 989341, 989347, 989353, 989377, 989381, 989411, 989419, 989423,
989441, 989467, 989477, 989479, 989507, 989533, 989557, 989561, 989579, 989581, 989623, 989629, 989641,
989647, 989663, 989671, 989687, 989719, 989743, 989749, 989753, 989761, 989777, 989783, 989797, 989803,
989827, 989831, 989837, 989839, 989869, 989873, 989887, 989909, 989917, 989921, 989929, 989939, 989951,
989959, 989971, 989977, 989981, 989999, 990001, 990013, 990023, 990037, 990043, 990053, 990137, 990151,
990163, 990169, 990179, 990181, 990211, 990239, 990259, 990277, 990281, 990287, 990289, 990293, 990307,
990313, 990323, 990329, 990331, 990349, 990359, 990361, 990371, 990377, 990383, 990389, 990397, 990463,
990469, 990487, 990497, 990503, 990511, 990523, 990529, 990547, 990559, 990589, 990593, 990599, 990631,
990637, 990643, 990673, 990707, 990719, 990733, 990761, 990767, 990797, 990799, 990809, 990841, 990851,
990881, 990887, 990889, 990893, 990917, 990923, 990953, 990961, 990967, 990973, 990989, 991009, 991027,
991031, 991037, 991043, 991057, 991063, 991069, 991073, 991079, 991091, 991127, 991129, 991147, 991171,
991181, 991187, 991201, 991217, 991223, 991229, 991261, 991273, 991313, 991327, 991343, 991357, 991381,
991387, 991409, 991427, 991429, 991447, 991453, 991483, 991493, 991499, 991511, 991531, 991541, 991547,
991567, 991579, 991603, 991607, 991619, 991621, 991633, 991643, 991651, 991663, 991693, 991703, 991717,
991723, 991733, 991741, 991751, 991777, 991811, 991817, 991867, 991871, 991873, 991883, 991889, 991901,

991909, 991927, 991931, 991943, 991951, 991957, 991961, 991973, 991979, 991981, 991987, 991999, 992011, 992021, 992023, 992051, 992087, 992111, 992113, 992129, 992141, 992153, 992179, 992183, 992219, 992231, 992249, 992263, 992267, 992269, 992281, 992309, 992317, 992357, 992359, 992363, 992371, 992393, 992417, 992429, 992437, 992441, 992449, 992461, 992513, 992521, 992539, 992549, 992561, 992591, 992603, 992609, 992623, 992633, 992659, 992689, 992701, 992707, 992723, 992737, 992777, 992801, 992809, 992819, 992843, 992857, 992861, 992863, 992867, 992891, 992903, 992917, 992923, 992941, 992947, 992963, 992983, 993001, 993011, 993037, 993049, 993053, 993079, 993103, 993107, 993121, 993137, 993169, 993197, 993199, 993203, 993211, 993217, 993233, 993241, 993247, 993253, 993269, 993283, 993287, 993319, 993323, 993341, 993367, 993397, 993401, 993407, 993431, 993437, 993451, 993467, 993479, 993481, 993493, 993527, 993541, 993557, 993589, 993611, 993617, 993647, 993679, 993683, 993689, 993703, 993763, 993779, 993781, 993793, 993821, 993823, 993827, 993841, 993851, 993869, 993887, 993893, 993907, 993913, 993919, 993943, 993961, 993977, 993983, 993997, 994013, 994027, 994039, 994051, 994067, 994069, 994073, 994087, 994093, 994141, 994163, 994181, 994183, 994193, 994199, 994229, 994237, 994241, 994247, 994249, 994271, 994297, 994303, 994307, 994309, 994319, 994321, 994337, 994339, 994363, 994369, 994391, 994393, 994417, 994447, 994453, 994457, 994471, 994489, 994501, 994549, 994559, 994561, 994571, 994579, 994583, 994603, 994621, 994657, 994663, 994667, 994691, 994699, 994709, 994711, 994717, 994723, 994751, 994769, 994793, 994811, 994813, 994817, 994831, 994837, 994853, 994867, 994871, 994879, 994901, 994907, 994913, 994927, 994933, 994949, 994963, 994991, 994997, 995009, 995023, 995051, 995053, 995081, 995117, 995119, 995147, 995167, 995173, 995219, 995227, 995237, 995243, 995273, 995303, 995327, 995329, 995339, 995341, 995347, 995363, 995369, 995377, 995381, 995387, 995399, 995431, 995443, 995447, 995461, 995471, 995513, 995531, 995539, 995549, 995551, 995567, 995573, 995587, 995591, 995593, 995611, 995623, 995641, 995651, 995663, 995669, 995677, 995699, 995713, 995719, 995737, 995747, 995783, 995791, 995801, 995833, 995881, 995887, 995903, 995909, 995927, 995941, 995957, 995959, 995983, 995987, 995989, 996001, 996011, 996019, 996049, 996067, 996103, 996109, 996119, 996143, 996157, 996161, 996167, 996169, 996173, 996187, 996197, 996209, 996211, 996253, 996257, 996263, 996271, 996293, 996301, 996311, 996323, 996329, 996361, 996367, 996403, 996407, 996409, 996431, 996461, 996487, 996511, 996529, 996539, 996551, 996563, 996571, 996599, 996601, 996617, 996629, 996631, 996637, 996647, 996649, 996689, 996703, 996739, 996763, 996781, 996803, 996811, 996841, 996847, 996857, 996859, 996871, 996881, 996883, 996887, 996899, 996953, 996967, 996973, 996979, 997001, 997013, 997019, 997021, 997037, 997043, 997057, 997069, 997081, 997091, 997097, 997099, 997103, 997109, 997111, 997121, 997123, 997141, 997147, 997151, 997153, 997163, 997201, 997207, 997219, 997247, 997259, 997267, 997273, 997279, 997307, 997309, 997319, 997327, 997333, 997343, 997357, 997369, 997379, 997391, 997427, 997433, 997439, 997453, 997463, 997511, 997541, 997547, 997553, 997573, 997583, 997589, 997597, 997609, 997627, 997637, 997649, 997651, 997663, 997681, 997693, 997699, 997727, 997739, 997741, 997751, 997769, 997783, 997793, 997807, 997811, 997813, 997877, 997879, 997889, 997891, 997897, 997933, 997949, 997961, 997963, 997973, 997991, 998009, 998017, 998027, 998029, 998069, 998071, 998077, 998083, 998111, 998117, 998147, 998161, 998167, 998197, 998201, 998213, 998219, 998237, 998243, 998273, 998281, 998287, 998311, 998329, 998353, 998377, 998381, 998399, 998411, 998419, 998423, 998429, 998443, 998471, 998497, 998513, 998527, 998537, 998539, 998551, 998561, 998617, 998623, 998629, 998633, 998651, 998653, 998681, 998687, 998689, 998717, 998737, 998743, 998749, 998759, 998779, 998813, 998819, 998831, 998839, 998843, 998857, 998861, 998897, 998909, 998917, 998927, 998941, 998947, 998951, 998957, 998969, 998983, 998989, 999007, 999023, 999029, 999043, 999049, 999067, 999083, 999091, 999101, 999133, 999149, 999169, 999181, 999199, 999217, 999221, 999233, 999239, 999269, 999287, 999307, 999329, 999331, 999359, 999371, 999377, 999389, 999431, 999433, 999437, 999451, 999491, 999499, 999521, 999529, 999541, 999553, 999563, 999599, 999611, 999613, 999623, 999631, 999653, 999667, 999671, 999683, 999721, 999727, 999749, 999763, 999769, 999773, 999809, 999853, 999863, 999883, 999907, 999917, 999931, 999953, 999959, 999961, 999979, 999983

Thank you for reading! I hope this was helpful to you!

# Java Code

Here is the Java code that I created to find these numbers. If you'd like, you can paste the code into an IDE such as jGRASP and find all primes between custom minimum and maximum values. There are ways to optimize the code, and I recommend you do so if you are searching for prime numbers with more than 7 digits. For example, you could edit the code to ignore division by even numbers after performing division by 2 for a value because if a number is not divisible by two, then it's not divisible by any positive even number (and if it is divisible by 2, it's not a prime number). Another modification I recommend would be to search only up to the square root of the number. I hope that this code is helpful to you!

```java
import java.util.Scanner;
public class primeNumbers
{
  public static void main(String[] args)
  {
    Scanner in = new Scanner(System.in);
    System.out.println("This program will list prime numbers between specified minimum and maximum values.");
    System.out.print("Please enter your minimum number to search: ");
    int number = in.nextInt();
    System.out.print("Please enter your maximum number to search: ");
    int maximumNumber = in.nextInt();
    System.out.println();
    System.out.print("The prime numbers in this set are: ");
    int numberOfPrimesInSet = 0;
    for (int counter = number; counter < maximumNumber; counter++)
    {
      if (isPrime(number) == true)
      {
        System.out.print(number + ", ");
        numberOfPrimesInSet++;
      }
      number++;
    }
  }

  public static boolean isPrime(int number)
  {
    for (int i = 2; i < number; i++)
    {
      double divided = (((double) number) / ((double) i));
      if (divided % 1 == 0)
      {
        return false;
      }
    }
    return true;
  }
}
```

# Six Fun Facts about Prime Numbers

1. There are 78,498 prime numbers (in the set of natural numbers) below 1000000.
2. Many mathematicians used to believe that 1 is a prime number, but this view is not commonly held today. Some things, like the sieve of Eratosthenes or the formula for Euler's totient function do not work without redefinition if 1 is assigned primality.
3. The only even prime number is 2. This must be the case, of course, because any larger multiple of 2 would have at least two factors and therefore be composite.
4. Every prime number larger than 6 must be congruent to 1 or 5 modulo 6. This can be shown by process of elimination. A number congruent to 0 modulo 6 is a multiple of 6, and if it is larger than 6, then it would have at least two factors. If a number is congruent to 3 modulo 6, it is a multiple of 3. If a number is congruent to 2 or 4 (or 0, alternatively) modulo 6, it is even.
5. There are no 6-digit prime numbers with repeating digits (e.g. 555555) because 550000+5500+55 = 555555 and all of those numbers are divisible by 11.
6. There are infinitely many primes. Euclid's proof was to show that for every finite list of primes, you could add another prime not in the list.

## Conclusion and About the Author

I made this collection of prime numbers because I love math and want to share it with the world! I hope that this list can be a reference for number theorists and general Math enthusiasts everywhere.

At the time of this book's publishing, I am 16 years old and I study Applied Mathematics at Arizona State University. I am very grateful to be able to take classes at ASU and I hope to continue creating reference books for Mathematicians everywhere.

I am also an author of non-fiction self-help books, and although I prefer to write in Italian, I write in English to reach (and hopefully help) more people. Thank you for your time and I hope that you have a great day!

*A photo of me with my wonderful cousin, Sophia*